T0268946

Astrophysics for Physicists

Designed for teaching astrophysics to physics students at advanced undergraduate or beginning graduate level, this textbook also provides an overview of astrophysics for astrophysics graduate students, before they delve into more specialized volumes.

Assuming background knowledge at the level of a physics major, the textbook develops astrophysics from the basics without requiring any previous study in astronomy or astrophysics. Physical concepts, mathematical derivations and observational data are combined in a balanced way to provide a unified treatment. Topics such as general relativity and plasma physics, which are not usually covered in physics courses but used extensively in astrophysics, are developed from first principles. While the emphasis is on developing the fundamentals thoroughly, recent important discoveries are highlighted at every stage.

ARNAB RAI CHOUDHURI is a Professor of Physics at the Indian Institute of Science. One of the world's leading scientists in the field of solar magnetohydrodynamics, he is author of *The Physics of Fluids and Plasmas* (Cambridge University Press, 1998).

Astrophysics for Physicists

Arnab Rai Choudhuri

Indian Institute of Science

CAMBRIDGE
UNIVERSITY PRESS

University Printing House, Cambridge CB2 8BS, United Kingdom

Cambridge University Press is part of the University of Cambridge.

It furthers the University's mission by disseminating knowledge in the pursuit of education, learning and research at the highest international levels of excellence.

www.cambridge.org
Information on this title: www.cambridge.org/9780521815536

First published 2010
3rd printing with corrections 2011
10th printing 2022

Printed in the United Kingdom by TJ Books Limited, Padstow Cornwall

A catalogue record for this publication is available from the British Library

Library of Congress Cataloguing-in-Publication Data

Choudhuri, Arnab Rai, 1956–
Astrophysics for physicists / Arnab Rai Choudhuri.
p. cm.
ISBN 978-0-521-81553-6 (Hardback)
1. Astrophysics–Textbooks. I. Title.
QB461.C535 2010
523.01–dc22
2009044687

ISBN 978-0-521-81553-6 Hardback

Contents

Preface

Particle physics, condensed matter physics and astrophysics are arguably the three major research frontiers of physics at the present time. It is generally thought that a physics student's training is not complete without an elementary knowledge of particle physics and condensed matter physics. Most physics departments around the world offer one-semester comprehensive courses on particle physics and condensed matter physics (sometimes known by its more traditional name 'solid state physics'). All graduate students of physics and very often advanced undergraduate students also are required to take these courses. Very surprisingly, one-semester comprehensive courses on astrophysics at a similar level are not so frequently offered by many physics departments. If a physics department has general relativists on its faculty, often a one-semester course *General Relativity and Cosmology* would be offered, though this would normally not be a compulsory course for all students. It has thus happened that many students get trained for a professional career in physics without a proper knowledge of astrophysics, one of the most active research areas of modern physics.

Of late, many physics departments are waking up to the fact that this is a very undesirable situation. More and more physics departments around the world are now introducing one-semester comprehensive courses on astrophysics at the advanced undergraduate or beginning graduate level, similar to such courses covering particle physics and solid state physics. The physics department of the Indian Institute of Science, where I have worked for more than two decades by now, has been offering a one-semester course on basic astrophysics for a long time. It is a core course for our Integrated PhD Programme in Physical Sciences as well as our Joint Astronomy and Astrophysics Programme. I must have taught this course to more than half a dozen batches.

Over the years, several excellent textbooks suitable for use in one-semester courses on particle physics and solid state physics have been written. The situation with respect to astrophysics is somewhat peculiar. There are several outstanding elementary textbooks on astrophysics meant for students who do not have much background of physics or mathematics beyond what is taught at the high school level. Then there are well-known specialized textbooks dealing with important sub-areas of astrophysics (such as stars, galaxies, interstellar matter or cosmology). However, there have been few attempts at bridging the gap between these two kinds of textbooks by writing books covering the whole of astrophysics at the level of Kittel's *Solid State Physics* or Perkins's *High Energy Physics* – suitable for a one-semester course meant for students who have already studied mechanics, electromagnetic theory, thermal physics, quantum mechanics and mathematical methods at an advanced level. Whenever I had to teach the course *Fundamentals of Astrophysics* in our department, I found that there was no textbook which was suitable for use in the whole course. The present book has grown out of the material I have taught in this course.

While writing this book, I have kept in mind that most of the students using this book will not aspire to a professional career in astrophysics. So I have tried to stress those aspects of astrophysics which are likely to be of interest to a physicist who is not specializing in astrophysics. Astrophysics is an observational science and an acquaintance with the basic phenomenology is absolutely essential for an appreciation of modern astrophysics. While I have introduced the basic phenomenology throughout the book, I believe that a physics student can appreciate astrophysics without knowing what a T Tauri star or what a BL Lac object is. A student who wishes to be a professional astrophysicist and has to master the terminology of the subject (which is sometimes of the nature of historical baggage) can learn it from other books. Rather than covering the details of too many topics, I have tried to develop the central themes of modern astrophysics fully. The trouble with this approach is that no two astrophysicists will completely agree as to what are central themes and what are details! I have used my judgment to develop what I would consider a balanced account of modern astrophysics. There is no doubt that experts in different areas of astrophysics would feel that I have committed the cardinal sin of not covering something in their area of specialization which they regard vitally important. If I succeed in making experts in all different areas of astrophysics *equally* unhappy, then I would conclude that I have written a balanced book! One other principle I have followed is to give more stress on classical well-established topics rather than topics which are still ill-understood or on which our present views are likely to change drastically in future. To give readers a historical perspective, I have sometimes deliberately chosen figures from original classic papers rather than their more contemporary versions, unless the modern figures supersede the original figures in essential and important ways. I have also intentionally kept away from topics which are too speculative or which do not have close links with observational data at the present time, perhaps reflecting my personal taste.

Virtually all branches of basic physics find applications in some topic of astrophysics or other. I have assumed that the readers of this book would have sufficient knowledge of classical mechanics, electromagnetic theory, optics, special relativity, thermodynamics, statistical mechanics, quantum mechanics, atomic physics and nuclear physics – something that is expected of an advanced student of physics in any good university anywhere. It is a firm belief of the present author that all physics students at this level ought to know some fluid mechanics and plasma physics. However, keeping in mind that this is not the case for physics students in the majority of universities around the world, a background in fluid mechanics or plasma physics has not been assumed and these subjects have been developed from first principles. General relativity is also developed from first principles without assuming any previous knowledge of the subject, though a previous acquaintance with the elementary properties of tensors will help. Some of the other basic physics topics which have been developed in this book without assuming any previous background are the theory of radiative transfer and the kinetic theory of gravitating particles (using the collisionless Boltzmann equation).

I have followed the usual traditional order of first concentrating on stars and then taking up galaxies to end with extragalactic astronomy and cosmology. One issue about which I had to give some thought is the placement of the basic physics topics which I develop in the book. A possible approach would have been to develop all the necessary basic physics topics at the beginning of the book before delving into the world of astrophysics. I personally felt that a more satisfactory approach is to teach these physics topics 'on the way' as we proceed with astrophysics. Since radiative transfer is used so extensively in astrophysics, it comes fairly early in Chapter 2. Two other chapters dealing primarily with basic physics topics are Chapters 7 and 8 devoted respectively to stellar dynamics and plasma astrophysics. These chapters could conceivably be placed somewhere else in the book. I felt that, after learning about our Galaxy and interstellar matter in Chapter 6, students will be in a position to appreciate stellar dynamics and plasma astrophysics particularly well, before they get into extragalactic astronomy where there will be more applications of what they learn in Chapters 7 and 8. However, putting Chapters 7 and 8 where they are has been ultimately my personal choice without very compelling logical reasons behind it.

Now let me comment on the place of general relativity in my book. The course *Fundamentals of Astrophysics* which I have taught in our department on several occasions does not cover general relativity (we have a separate course *General Relativity and Cosmology* in our department). In the *Fundamentals of Astrophysics* course, I basically cover the material of Chapters 1–11, which is more than sufficient for a one-semester course. In Chapters 10–11, I present as much cosmology as can be done without a detailed technical knowledge of general relativity. Initially my plan was to write up only Chapters 1–11. During the course of writing this book, I decided to add the last three chapters – primarily because general relativity is playing an increasingly more important role in many branches of astrophysics. One of the areas of astrophysics which underwent the most explosive growth in the last decade is the study of the Universe at redshifts $z \geq 1$. Issues involved in the study of the high-redshift Universe cannot be appreciated without some technical knowledge of relativistic astrophysics. Another important development in the last decade has been the construction of several large detectors of gravitational radiation – a consequence of general relativity. Because of the increased applications of general relativity to astrophysics and also for the sake of completeness, I finally decided to write Chapters 12–14. After developing general relativity from the first principles in Chapters 12–13, I discuss relativistic cosmology in Chapter 14. So the presentation of cosmology has been somewhat fractured. Topics which can be developed without a technical knowledge of general relativity are presented in Chapters 10–11, while topics requiring general relativity are presented in Chapter 14. Although this arrangement may be intellectually unsatisfactory, I believe that the advantages outweigh the disadvantages. Readers desirous of learning the basics of cosmology without first learning general relativity can go through Chapters 10–11. Instructors wishing to teach a one-semester course of astrophysics to students who do not know general relativity can use the material of Chapters 1–11. On the other hand, a course on general relativity and cosmology can be based on Chapters 10–14 – with some rearrangement of topics and with the inclusion of additional topics like structure formation, which is barely touched upon in this book. Finally, it should be possible to use this as a basic textbook for a two-semester course on astrophysics and relativity – with some additional material thrown in, depending on the choice of the instructor.

This book has been and will probably remain the most ambitious project I have ever undertaken in my life. While writing my previous book *The Physics of Fluids and Plasmas*, I mostly had to deal with topics on which I had some expertise. Now the canvas is much vaster. It is probably not possible today for an individual to have in-depth knowledge of all branches of modern astrophysics. At least, I cannot claim such knowledge. A writer aspiring to cover the whole of astrophysics is, therefore, compelled to write on many subjects on which his/her own knowledge is shaky. Apart from the risk of making actual technical mistakes, one runs the risk of not realizing where the emphasis should be put. I shall be grateful to any reader who brings any mistake to my attention, by sending an e-mail to my address arnab@physics.iisc.ernet.in. I do hope that readers will find that the merits of this book outnumber its flaws.

Acknowledgments

Apart from my gratitude to many authors whose books I consulted when preparing this book (all these authors are mentioned in *Suggestions for Further Reading*), I am grateful to several outstanding teachers I had as a graduate student at the University of Chicago in the early 1980s. I was particularly lucky to have courses on *Plasma Astrophysics* from Eugene Parker, *Relativistic Astrophysics* from James Hartle, *Cosmology* from David Schramm and *Stellar Evolution* from David Arnett. The influence which these teachers left on me provided me guidance when I myself had to teach these subjects to my students and this influence must have eventually percolated into this book.

I asked a few colleagues to read those chapters on which I was feeling particularly unsure. Colleagues who made valuable suggestions on some of these chapters are H.C. Bhatt, Sudip Bhattacharyya, K.S. Dwarakanath, Biman Nath, Tarun Deep Saini and Kandaswamy Subramanian. While these colleagues caught many errors which would have otherwise crept into the book, I am sure that this book still has many errors and mistakes for which I am responsible.

Ramesh Babu, Shashikant Gupta and Bidya Binay Karak prepared many of the figures in this book. I am also grateful to many organizations and individuals who permitted me to reproduce figures under their copyright. The acknowledgments are given in the captions of those figures.

I thank the staff of Cambridge University Press (especially Laura Clark, Vince Higgs, Simon Mitton and Dawn Preston) for their cooperation during the many years I took in preparing this book. Many students who have taken this course from me over the years encouraged me by regularly enquiring about the progress of the book and by giving their feedback on the course. A project of this magnitude would not have been possible without the strong support of my wife Mahua. Two persons who would have been the happiest to see this book left us during the long process of writing this book: my mother and my father. This book is dedicated to their memories.

Arnab Rai Choudhuri

A note on symbols

In discussing astrophysical topics, one often has to combine results from different branches of physics. Historically these branches may have evolved independently and sometimes the same symbol is used for different things in these different branches. In the case of a few symbols, I have added a subscript to make them unambiguous. For example, I use σ_T for the Thomson cross-section (since σ denotes the Stefan–Boltzmann constant), κ_B for the Boltzmann constant (since k denotes the wavenumber in several derivations) and a_B for the blackbody radiation constant (since a denotes the scale factor of the Universe). A look at equation (3.48) of Kolb and Turner (1990) will show the kinds of problems you run into if you use the same symbol to denote different things in a derivation. While I have avoided using the same symbol for different things within one derivation, I sometimes had to use the same symbol for different things in different portions of the book. For example, it has been the custom for many years to use M to denote both mass (of stars or galaxies) and absolute magnitude. Rather than inventing unorthodox symbolism, I have trusted the common sense of readers who should be able to figure out the meaning of the symbol from the context and hopefully will not get confused. I now mention a potential source of confusion. I have used $f(E)$ in §4.2 to denote the probability that particles have energy E and have used $f(p)$ in §5.2 to denote the number density of particles with momentum p (throughout Chapter 7, I use f to denote number density and not probability). While these notations may not be consistent with each other, they happen to be the most convenient notations for the derivations presented in §4.2 and §5.2 (and also the notations used by many previous authors).

Introduction

1.1 Mass, length and time scales in astrophysics

Astrophysics is the science dealing with stars, galaxies and the entire Universe. The aim of this book is to present astrophysics as a serious science based on quantitative measurements and rigorous theoretical reasoning.

The standard units of mass, length and time that we use (cgs or SI units) are appropriate for our everyday life. For expressing results of astrophysical measurements, however, they are not the most convenient units. Let us begin with a discussion of the basic units we use in astrophysics and the scales of various astrophysical objects we encounter.

Unit of mass

The mass of the Sun is denoted by the symbol $M_\odot$ and is often used as the unit of mass in astrophysics. Its value is

$$M_\odot = 1.99 \times 10^{30} \text{ kg}. \tag{1.1}$$

Although intrinsic brightnesses and sizes of stars vary over several orders of magnitude, the masses of most stars lie within a relatively narrow range from $0.1 M_\odot$ to $20 M_\odot$. The reason behind this will be discussed in §3.6.1. Hence the solar mass happens to be a very convenient unit in stellar astrophysics. Sometimes, however, we have to deal with objects much more massive than stars. The mass of a typical galaxy can be $10^{11} M_\odot$. Globular clusters, which are dense clusters of stars having nearly spherical shapes, typically have masses around $10^5 M_\odot$.

Unit of length

The average distance of the Earth from the Sun is called the *Astronomical Unit* (abbrev. AU). Its value is

$$AU = 1.50 \times 10^{11} \text{ m}. \tag{1.2}$$

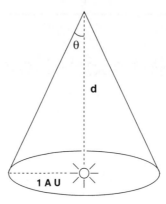

Fig. 1.1 Definition of parsec.

It is a very useful unit for measuring distances within the solar system. But it is too small a unit to express the distances to stars and galaxies.

As the Earth goes around the Sun, the nearby stars seem to change their positions very slightly with respect to the faraway stars. This phenomenon is known as *parallax*. Let us consider a star on the polar axis of the Earth's orbit at a distance d away, as shown in Figure 1.1. The angle θ is half of the angle by which this star appears to shift with the annual motion of the Earth and is defined to be the parallax. It is obviously given by

$$\theta = \frac{1\,\text{AU}}{d}. \tag{1.3}$$

The *parsec* (abbrev. pc) is the distance where the star has to be so that its parallax turns out to be $1''$. Keeping in mind that $1''$ is equal to $\pi/(180 \times 60 \times 60)$ radians, it is easily found from (1.3) that

$$\text{pc} = 3.09 \times 10^{16}\,\text{m}. \tag{1.4}$$

It may be noted that 1 pc is equal to 3.26 light years – a unit very popular with popular science writers, but rarely used in serious technical literature. For even larger distances, the standard units are kiloparsec (10^3 pc, abbrev. kpc), megaparsec (10^6 pc, abbrev. Mpc) and gigaparsec (10^9 pc, abbrev. Gpc).

The star nearest to us, Proxima Centauri, is at about a distance of 1.31 pc. Our Galaxy and many other galaxies like ours are shaped like disks with thickness of order 100 pc and radius of order 10 kpc. The geometric mean between these two distances, which is 1 kpc, may be taken as a measure of the galactic size. The Andromeda Galaxy, one of the nearby bright galaxies, is at a distance of about 0.74 Mpc. The distances to very faraway galaxies are of order Gpc. It should be kept in mind that light from very distant galaxies started when the Universe was much younger and the concept of distance to such galaxies is not a very straightforward concept, as we shall see in §14.4.1. It is useful to keep

Table 1.1 Approximate conversion factors to be memorized.

$M_\odot$	$\approx$	2×10^{30} kg
pc	$\approx$	3×10^{16} m
yr	$\approx$	3×10^{7} s

the following rule of thumb in mind: pc is a measure of interstellar distances, kpc is a measure of galactic sizes, Mpc is a measure of intergalactic distances and Gpc is a measure of the visible Universe.

Unit of time

Astrophysicists have to deal with very different time scales. On the one hand, the age of the Universe is of the order of a few billion years. On the other hand, there are pulsars which emit pulses periodically after intervals of fractions of a second. There is no special unit of time. Astrophysicists use years for large time scales and seconds for small time scales, the conversion factor being

$$\text{yr} = 3.16 \times 10^{7} \text{ s}. \tag{1.5}$$

The stars typically live for millions to billions of years. Occasionally, one uses the unit gigayear (10^9 yr, abbrev. Gyr). The age of the Sun is believed to be about 4.5 Gyr.

The importance of order of magnitude estimates

We can often have good guesses of the values of various quantities around us even without making accurate measurements. By looking at a table, I may make a rough estimate that its side is about 1 m long. By lifting a sack of potatoes, I may make a rough estimate that it weighs about 5 kg. Careful measurements usually show that such guesses are not very much off the mark. We never have the suspicion that a measurement of the length of a table would yield values like either 10^{-2} cm or 100 km. For astrophysical quantities, we usually do not have any such direct feeling. If somebody tells us that the mass of the Sun is either 10^{20} kg or 10^{40} kg, there would be nothing in our everyday experiences on the basis of which we could say that these values are unreasonable. Hence, in astrophysics, it is often very useful first to make order of magnitude estimates of various quantities before embarking on a more detailed calculation. Throughout this book, we shall be making various order of magnitude estimates. For such purposes, it is useful to remember the conversion factors given in Table 1.1. The accurate values of these conversion factors are given in (1.1), (1.4) and (1.5).

Although the emphasis in this book will be on understanding things and not memorizing things, we would urge the readers to commit the conversion factors of Table 1.1 to memory. They are used too often in making various order of magnitude estimates!

1.2 The emergence of modern astrophysics

From the dawn of civilization, human beings have wondered about the starry sky. Astronomy is one of the most ancient sciences. Perhaps mathematics and medicine are the only other sciences which can claim as ancient a tradition as astronomy. But modern astrophysics, which arose out of a union between astronomy and physics, is a fairly recent science; it can be said to have been born in the middle of the nineteenth century.

Let us say a few words about ancient astronomy. Early humans noticed that most stars did not seem to change their positions with respect to each other. The seven stars of the Great Bear occupy the same relative positions night after night. But a handful of starlike objects – the planets – kept on changing their positions with respect to the background stars. It was noticed that there was a certain regularity in the movements of the planets. Building a model of the planetary motions was the outstanding problem of ancient astronomy, which reached its culmination in the geocentric theory of Hipparchus (second century BC) and Ptolemy (second century AD). Ptolemy's *Almagest*, which luckily survived the ravages of time, has come down to us as one of the greatest classics of science and provides the definitive account of the geocentric model. The scientific Renaissance of Europe began with Copernicus (1543) showing that a heliocentric model provided a simpler explanation of the planetary motions than the geocentric model. The new physics developed by Galileo and Newton finally provided a dynamical theory which could be used to calculate the orbits of planets around the Sun.

Only very rarely a branch of science reaches a phase when the practitioners of that science feel that all the problems which that branch of science had set out to solve had been adequately solved. With the development of Newtonian mechanics, planetary astronomy reached a kind of finality. Even the complicated techniques of calculating perturbations to planetary orbits due to the larger planets got perfected by the nineteenth century. Astronomers then turned their attention beyond the solar system. Telescopes also became sufficiently large by the middle of the nineteenth century to reveal some of the secrets of the stellar world to us. It may be mentioned that, with the heralding of the Space Age in the middle of the twentieth century, research in planetary science has blossomed again. However, modern planetary science has become a scientific discipline quite distinct from astrophysics and we shall not discuss about planets in this book.

If stars are distributed in a three-dimensional space and the Earth is going round the Sun, then nearby stars should appear to change their positions with the movement of the Earth, i.e. they should display parallax. Now we know that even the nearest stars have too little parallax to be detected by the naked eye. Certainly no parallax observations were available at the time of Copernicus. While proposing that the Earth moves around the Sun, Copernicus (1543) himself was bothered by the question why stars showed no parallax and correctly guessed that the stars may just be too far away. Ever since the invention of the telescope, astronomers have been on the lookout for parallax. Finally, in the fateful year 1838, three astronomers working in three different countries almost simultaneously reported the first parallax measurements (Bessel in Germany, Struve in Russia and Henderson in South Africa). This forever demolished the Aristotelian belief that stars are studded on the two-dimensional inner surface of a crystal sphere. Suddenly the sky ceased to be a two-dimensional globe and opened into an apparently limitless three-dimensional space! The stars are not static objects in space. The component of velocity perpendicular to the line of sight would lead to the change of position of a star in the sky. Such motions in the globe of the sky are called *proper motions*. Even Barnard's star, which has the largest proper motion of about $10''$ per yr, would take 360 yr to move through $1°$ in the sky. Most stars have much smaller proper motions and it is no wonder the appearance of the sky has not changed that much in the last 2000 yr. Some of the first measurements of proper motions were also made in the middle of the nineteenth century and it became clear that stars are luminous objects wandering around in the vast, dark three-dimensional space.

Another momentous event took place in the middle of the nineteenth century. Bunsen and Kirchhoff (1861) provided the first correct explanation of the dark lines observed by Fraunhofer (1817) in the solar spectrum and realized that the presence of various chemical elements in the Sun can be inferred from those dark lines. As soon as astronomers started looking carefully at the stellar spectra, it became clear that the Sun and the stars are made up of the same chemical elements which are found on the Earth. This discovery provided a death blow to the other Aristotelian doctrine that heavenly bodies are made up of the element ether which is different from terrestrial elements and obeyed different laws of physics. Newton had shown that planets obeyed the same laws of physics as falling objects at the Earth's surface. It now became clear that stars are made up of the same stuff as the Earth and the laws of physics discovered in the terrestrial laboratories should hold for them.

With the realization that the laws of physics can be applied to understand the behaviour of stars, the modern science of astrophysics was born. Nowadays the words 'astronomy' and 'astrophysics' are used almost interchangeably. Although modern astrophysicists study problems completely different from the problems studied by ancient astronomers, two very useful concepts introduced by ancient astronomers are still universally used. One is the concept of celestial coordinates, and the other is the magnitude scale for describing the brightness of a celestial object. We now turn to these two topics.

1.3 Celestial coordinates

The sky appears as a spherical surface above our heads. We call it the *celestial sphere*. Just as the position of a place on the Earth's surface can be specified with the latitude and longitude, the position of an astronomical object on the celestial sphere can be specified with two similar coordinates. These coordinates are defined in such a way that faraway stars which appear immovable with respect to each other have fixed coordinates. Objects like planets which move with respect to them will have their coordinates changing with time.

The coordinate corresponding to latitude is called the *declination*. The points where the Earth's rotation axis would pierce the celestial sphere are called *celestial poles*. The north celestial pole is at present close to the pole star. The great circle on the celestial sphere vertically above the Earth's equator is called the *celestial equator*. The declination is essentially the latitude on the celestial sphere defined with respect to the celestial poles and equator. Something lying on the celestial equator has declination zero, whereas the north pole has declination $+\pi/2$.

The coordinate corresponding to longitude is called the *right ascension* (R.A. in brief). Just as the zero of longitude is fixed by taking the longitude of Greenwich as zero, we need to fix the zero of R.A. for defining it. This is done with the help of a great circle called the *ecliptic*. Since the Earth goes around the Sun in a year, the Sun's position with respect to the distant stars, as seen by us, keeps changing and traces out a great circle in the sky. The ecliptic is this great circle. Twelve famous constellations (known as the *signs of the zodiac*) appear on the ecliptic. It was noted from almost prehistoric times that the Sun happens to be in different constellations in different times of the year. We cannot, of course, directly see a constellation when the Sun lies in it. But, by looking at the stars just after sunset and just before sunrise, ancient astronomers could infer the position of the Sun in the celestial sphere. The celestial equator and the ecliptic are inclined at an angle of about $23\frac{1}{2}^{\circ}$ and intersect at two points, as shown in Figure 1.2. One of these points, lying in the constellation Aries, is taken as the zero of R.A. When the Sun is at this point, we have the vernal equinox. It is a standard convention to express the R.A. in hours rather than in degrees. The celestial sphere rotates around the polar axis by 15° in one hour. Hence one hour of R.A. corresponds to 15°.

The declination and R.A. are basically defined with respect to the rotation axis of the Earth, which fixes the celestial poles and equator. One problematic aspect of introducing coordinates in this way is that the Earth's rotation axis

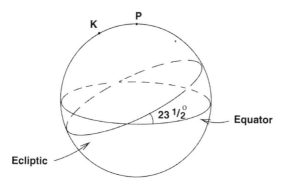

Fig. 1.2 The celestial sphere with the equator and the ecliptic indicated on it. The celestial pole is denoted by P, whereas K is the pole of the ecliptic.

is not fixed, but precesses around an axis perpendicular to the plane of the Earth's orbit around the Sun. This means that the point P in Figure 1.2 traces out an approximate circle in the celestial sphere slowly in about 25,800 years, around the pole K of the ecliptic. This phenomenon is called *precession* and was discovered by Hipparchus (second century BC) by comparing his observations with the observations made by earlier astronomers about 150 years previously. The precession is caused by the gravitational torque due to the Sun acting on the Earth and can be explained from the dynamics of rigid bodies (see, for example, Goldstein, 1980, §5–8). Due to precession, the positions of the celestial poles and the celestial equator keep changing slowly with respect to fixed stars. Hence, if the declination and the R.A. of an astronomical object at a time are defined with respect to the poles and the equator at that time, then certainly the values of these coordinates will keep changing with time. The current convention is to use the coordinates defined with respect to the positions of the poles and the equator in the year 2000.

Many ground-based optical telescopes have been traditionally designed to have *equatorial mounting*, which means that the main axis of the telescope is parallel to the rotation axis of the Earth. The telescope is designed such that it can have two kinds of motion. Firstly, it can be rotated towards or away from the axis of mounting (which is the Earth's rotation axis). Secondly, the telescope can be moved to generate a conical surface with this axis as the central axis. Suppose we want to turn the telescope towards an object of which the declination and R.A. are known. The first kind of motion enables us to set the telescope at the correct declination. The second kind of motion allows us to turn it to various values of R.A. at that declination.

The main advantage of using the declination and R.A. is that an equatorially mounted telescope can easily be turned to an object of which we know the declination and R.A. However, there is another coordinate system, called *galactic coordinates*, widely used in galactic studies. In this system, the plane of our

Galaxy is taken as the equator and the direction of the galactic centre as seen by us (in the constellation Sagittarius) is used to define the zero of longitude.

1.4 Magnitude scale

Suppose we have two series of lamps – the first series with lamps having intensities $I_0, 2I_0, 3I_0, 4I_0 \ldots$, whereas the lamps in the second series have intensities $I_0, 2I_0, 4I_0, 8I_0 \ldots$. When we look at the two series of lamps, it is the second series which will appear to have lamps of steadily increasing intensity. In other words, the human eye is more sensitive to a geometric progression of intensity rather than an arithmetic progression. The magnitude scale for describing apparent brightnesses of celestial objects is based on this fact.

On the basis of naked eye observations, the Greek astronomer Hipparchus (second century BC) classified all the stars into six classes according to their apparent brightnesses. We can now of course easily measure the apparent brightness quantitatively. It appears that stars in any two successive classes, on the average, differ in apparent brightness by the same common factor. A quantitative basis of the magnitude scale was given by Pogson (1856) by noting that the faintest stars visible to the naked eye are about 100 times fainter compared to the brightest stars. Since the brightest and faintest stars differ by five magnitude classes, stars in two successive classes should differ in apparent brightness by a factor $(100)^{1/5}$. Suppose two stars have apparent brightnesses l_1 and l_2, whereas their magnitude classes are m_1 and m_2. It is clear that

$$\frac{l_2}{l_1} = (100)^{\frac{1}{5}(m_1 - m_2)}. \tag{1.6}$$

Note that the magnitude scale is defined in such a fashion that a fainter object has a higher value of magnitude. On taking the logarithm of (1.6), we find

$$m_1 - m_2 = 2.5 \log_{10} \frac{l_2}{l_1}. \tag{1.7}$$

This can be taken as the definition of *apparent magnitude* denoted by m, which is a measure of the apparent brightness of an object in the sky.

Since a star emits electromagnetic radiation in different wavelengths, one important question is: what is the wavelength range over which we consider the electromagnetic radiation emitted by a star to measure its apparent brightness quantitatively? If we use apparent brightnesses based on the radiation in all wavelengths, then the magnitude defined from it is called the *bolometric magnitude*. Since any device for measuring intensity of light does not respond to all wavelengths in the same way, finding the bolometric magnitude from measurements with a particular device is not straightforward. A much more convenient system, called the *Ultraviolet–Blue–Visual system* or the *UBV system*, was introduced by Johnson and Morgan (1953) and is now universally used by astronomers. In this system, the light from a star is made to pass through filters which allow only light in narrow wavelength bands around the three wavelengths: 3650 Å, 4400 Å and 5500 Å. From the measurements of the intensity of light that has passed through these filters, we get magnitudes in ultraviolet, blue and visual, usually denoted by U, B and V. Typical examples of V magnitudes are: the Sun, $V = -26.74$; Sirius, the brightest star, $V = -1.45$; faintest stars measured, $V \approx 27$.

Suppose we consider a reddish star. It will have less brightness in B band compared to V band. Hence its B magnitude should have a larger numerical value than its V magnitude. So we can use $(B - V)$ as an indication of a star's colour. The more reddish a star, the larger will be the value of $(B - V)$.

The *absolute magnitude* of a celestial object is defined as the magnitude it would have if it were placed at a distance of 10 pc. The relation between relative magnitude m and absolute magnitude M can easily be found from (1.7). If the object is at a distance d pc, then $(10/d)^2$ is the ratio of its apparent brightness and the brightness it would have if it were at a distance of 10 pc. Hence

$$m - M = 2.5 \log_{10} \frac{d^2}{10^2},$$

from which

$$m - M = 5 \log_{10} \frac{d}{10}. \tag{1.8}$$

The absolute magnitude in the V band, denoted by M_V, is often used as a convenient quantity to indicate the intrinsic brightness of an object.

1.5 Application of physics to astrophysics. Relevance of general relativity

Astrophysics is a supreme example of applied physics. To be a competent astrophysicist, first and foremost one has to be a competent physicist. Virtually all branches of physics are needed in the study of astrophysics. Classical mechanics, electromagnetic theory, optics, thermodynamics, statistical mechanics, fluid dynamics, plasma physics, quantum mechanics, atomic physics, nuclear physics, particle physics, special and general relativity – there is no branch of physics which does not find application in some astrophysical problem or other. We shall use results from all these branches of physics in this book. In the astrophysical setting, however, the laws of physics are often applied to extremes of various physical conditions like density, pressure, temperature, velocity, angular velocity, gravitational field, magnetic field, etc. – well beyond the limits for which the laws have been tested in the laboratory. For example, the vacuousness of the intergalactic space is much more than the best vacuums we can create at the present time, whereas the interiors of neutron stars may have the almost inconceivable density of 10^{17} kg m^{-3}. Only in one case, human beings may have been able to surpass Nature. There are good reasons to suspect that temperatures lower than 2.73 K never existed anywhere in the Universe until scientists succeeded in creating such temperatures about a century ago.

At the first sight, it may seem that the astrophysicists are concerned with the macro-world of very large systems like stars and galaxies, which is far removed from the micro-world of atoms, nuclei and elementary particles. However, it turns out very often that we need the physics of the micro-world to make sense of the macro-world of astrophysics. One example is the famous Chandrasekhar mass limit of white dwarf stars, which will be derived in §5.3. It was found by Chandrasekhar (1931) that the maximum mass which white dwarfs (which are compact dead stars in which no more energy generation takes place) can have is given by

$$M_{\text{Ch}} = 2.018 \frac{\sqrt{6}}{8\pi} \left(\frac{hc}{G} \right)^{3/2} \frac{1}{m_{\text{H}}^2 \mu_e^2}, \tag{1.9}$$

where h is Planck's constant, m_{H} is the mass of hydrogen atom and μ_e is something called the mean molecular weight of electrons (to be introduced in §5.2) having a value close to 2. On putting numerical values of various quantities, M_{Ch} turns out to be about $1.4 M_{\odot}$. Thus the constants of the atomic world like h and m_{H} determine the mass limit of a vast object like a white dwarf star. It is this interplay between the physics of the micro-world and the physics of the macro-world which makes modern astrophysics such a fascinating scientific discipline. Very often major breakthroughs in micro-physics have a big impact in astrophysics, and occasionally discoveries in astrophysics have provided new insights in micro-physics.

We shall assume the readers of this book to have a working knowledge of mechanics, electromagnetic theory, thermal physics and quantum physics at an advanced undergraduate or beginning graduate level. General relativity happens to be a branch of physics which is often not included in a regular physics curriculum, but which is applied in some areas of astrophysics. Till Chapter 11, we proceed without assuming any background of general relativity. Then, only in the last three chapters of this book, we give an introduction to general relativity and consider its applications to astrophysical problems. Readers unwilling to learn general relativity can still get a reasonably rounded background of modern astrophysics from this book by studying till Chapter 11. We now make a few comments on the circumstances in which general relativity is expected to be important and what a reader misses if he or she is ignorant of general relativity.

Even readers without any technical knowledge of general relativity would have heard of black holes, which are objects with gravitational fields so strong that even light cannot escape. Let us try to find out when this happens.

Newtonian theory does not tell us how to calculate the effect of gravity of light. So let us figure out when a particle moving with speed c will get trapped, according to Newtonian theory. Suppose we have a spherical mass M of radius r and a particle of mass m is ejected from its surface with speed c. The gravitational potential energy of the particle is

$$-\frac{GMm}{r}.$$

If we use the non-relativistic expression for kinetic energy for a crude estimate (we should actually use special relativity for a particle moving with c!), then the total energy of the particle is

$$E = \frac{1}{2}mc^2 - \frac{GMm}{r}.$$

Newtonian theory tells us that the particle will escape from the gravitational field if $E > 0$ and will get trapped if $E < 0$. In other words, the condition of trapping is

$$\frac{2GM}{c^2 r} > 1. \tag{1.10}$$

It turns out that more accurate calculations using general relativity gives exactly the same condition (1.10) for light trapping, which was first obtained by Laplace (1795) by the arguments which we have given. General relativity is needed when this factor

$$f = \frac{2GM}{c^2 r} \tag{1.11}$$

is of order unity. On the other hand, Newtonian theory is quite adequate if this factor is much smaller than 1. For the Sun with mass 1.99×10^{30} kg and radius 6.96×10^8 m, this factor f turns out to be only 4.24×10^{-6}. Hence Newtonian theory is almost adequate for all phenomena in the solar system. Only if we want to calculate very accurate orbits of planets close to the Sun (such as Mercury), we have to bother about general relativity.

   Are there situations in astrophysics where general relativity is essential? We can use (1.11) to calculate the radius to which the solar mass has to be shrunk such that light emitted at its surface gets trapped. This radius turns out to be 2.95 km. As we shall discuss in more detail in Chapters 4–5, when the energy source of a star is exhausted, the star can collapse to very compact configurations like neutron stars or black holes. General relativity is needed to study such objects. If matter is distributed uniformly with density ρ inside radius r, then we can write

$$M = \frac{4}{3}\pi r^3 \rho$$

and (1.11) becomes

$$f = \frac{8\pi}{3} \frac{Gr^2\rho}{c^2}. \tag{1.12}$$

We note that f is large when either ρ is large or r is large (for given ρ). The density ρ is very high inside objects like neutron stars. Can there be situations where general relativity is important due to large r? We know of one object with very large size – our Universe itself. The distance to the farthest galaxies is of order 1 Gpc. It is difficult to estimate the average density of the universe accurately. Probably it is of order 10^{-26} kg m^{-3}, as we shall discuss in §10.5. Substituting these values in (1.12), we get

$$f \approx 0.06.$$

This tells us that we should use general relativity to study the dynamics of the whole Universe, which comes under cosmology. Thus, in astrophysics, we have two clear situations in which general relativity is important – the study of collapsed stars and the study of the whole Universe (or cosmology). In most other circumstances, we can get good results by applying Newtonian theory of gravity.

Even though general relativity is needed to study the structure of a collapsed star, we do not require general relativity to study some of the physical phenomena in the surrounding space or to figure out the conditions under which the collapse takes place. Again, we shall see in Chapter 10 that Newtonian mechanics allows us a formulation of the dynamics of the Universe, which is conceptually incomplete, but the crucial equation surprisingly turns out to be identical with the equation derived from general relativity (see §10.4). We are thus able to do quite a bit of astrophysics without general relativity. However, general relativity becomes essential when we want to make a conceptually satisfactory investigation of the properties of the Universe as revealed by very faraway galaxies. This subject will be taken up in Chapter 14 for those readers who are willing to learn general relativity in Chapters 12–13.

1.6 Sources of astronomical information

In most branches of science, controlled experiments play a very important role. Astrophysics is a peculiar science in which astronomical observations take the place of controlled experiments. An astronomer can only observe an astronomical object with the help of the signals reaching us from the object. We list below four kinds of possible sources of astronomical information.

1. Electromagnetic radiation
To this day, the electromagnetic radiation reaching us from celestial objects gives us the most extensive information about these objects. Until the time of

World War II, all astronomical observations were primarily based on visible light. However, in the last few decades, virtually all the bands of electromagnetic radiation have become available for the astronomer. Instruments and methods for detection of electromagnetic radiation (or photons) are discussed in §1.7.

2. Neutrinos

Nuclear reactions inside stars produce neutrinos, as we shall discuss in detail in Chapter 4. Since neutrinos take part in weak interactions alone (and not in strong or electromagnetic interactions), the cross-section of any neutrino process is very small. Hence most of the neutrinos created at the centre of a star can come out without interacting with the stellar matter. Unlike photons which come from the outer layers of a star and cannot tell us anything directly about the stellar core, neutrinos come out of the core unmodified. However, the very small cross-section of interaction between matter and neutrinos also makes it difficult to detect neutrinos. Only when a neutrino has interacted with the detector, can we be sure of its presence. Because of this difficulty of detecting neutrinos, we expect to detect neutrinos only either from very nearby sources or from sources which emit exceptionally large fluxes of neutrinos (like a supernova explosion) if the source is not too nearby.

For detecting neutrinos, we need a huge amount of some substance with atoms having nuclei with which neutrinos interact. In the 1960s Davis started a famous experiment to detect neutrinos from the Sun by using a huge underground tank of cleaning liquid C_2Cl_4 as the detector. Initially Davis detected fewer neutrinos than what is expected theoretically. The puzzling solar neutrino problem and its subsequent resolution is described in §4.4.2. In the late 1980s and the early 1990s, other neutrino detection experiments started, one of the most important being Kamiokande in Japan. Apart from the Sun, the only other astronomical source from which it has so far been possible to detect neutrinos is the Supernova 1987A, as discussed in §4.7. Only about 20 neutrinos detected in two terrestrial experiments could be ascribed to this supernova! Neutrino astronomy is, therefore, very much in its infancy.

3. Gravitational radiation

According to general relativity, a disturbance in a gravitational field can propagate in the form of a wave with speed c (to be shown in §13.4). Indirect evidence for the existence of gravitational radiation has come from the binary pulsar discovered by Hulse and Taylor (1975), as discussed in §5.5.1. The binary pulsar is a system in which two neutron stars are orbiting around each other with an orbital period of about 8 hours. This system continuously emits gravitation radiation and keeps on losing energy, thereby causing the two neutron stars to come closer together. This results in a decrease in the orbital period, which has been measured and is found to be in good agreement with the theoretical prediction from general relativity. This is, however, an indirect confirmation of the theory of gravitational radiation. One would like to directly measure the gravitational radiation reaching the Earth from astronomical sources.

As we shall see in §13.5, gravitational radiation impinging on an object causes a deformation of the object. Even a supernova explosion in our Galaxy is expected to produce a size deformation which may be at most of order only 10^{-18} part of the size of the object. Even if the detector has a size of the order of a km, the deformation will be of the order of 10^{-15} m only. One needs very sensitive interferometric techniques to measure such tiny deformations. As discussed in §13.5, several gravitational radiation detectors are now being constructed around the world, but there is yet no unambiguous detection of gravitational radiation from any astronomical source. In contrast to neutrino astronomy which is in its infancy, gravitational wave astronomy is still waiting to be born.

4. Cosmic rays

These are highly energetic charged particles (electrons, protons and heavier nuclei) continuously bombarding the Earth from all directions. As we shall discuss in §8.10, we believe that these charged particles are accelerated primarily in the shock waves produced in supernova explosions. Afterwards, however, they spiral around the magnetic field of the Galaxy and, by the time they reach us, they appear to be coming from directions totally different from the direction of their original source. In the case of electromagnetic radiation reaching us from outer space, usually the astronomical source can be identified without too much ambiguity. In contrast, we cannot identify the astronomical source from which a cosmic ray particle has come. Cosmic rays, therefore, have limited applications as a source of astronomical information.

1.7 Astronomy in different bands of electromagnetic radiation

We now consider astronomy with electromagnetic radiation, which is so far our main source of astronomical information. The Earth's atmosphere is an annoying inconvenience for the astronomer. The atmosphere is transparent to only small bands of electromagnetic radiation. Even though visible light passes through the atmosphere, the light rays are affected by the disturbances in the atmosphere, leading to a degradation of the astronomical image. Figure 1.3 indicates the heights above the sea level which we have to climb before we can receive radiation of a particular wavelength from the outer space. Apart from visible light, radio waves in a certain wavelength band can reach the Earth's surface. However, radio waves with wavelengths larger than about 10 m cannot reach us from astronomical sources, since this wavelength corresponds to the plasma frequency of the ionosphere such that the ionosphere reflects radio waves with wavelengths larger than about 10 m (see §8.13.2). In fact, this is the reason why faraway regions on the Earth's surface can communicate with

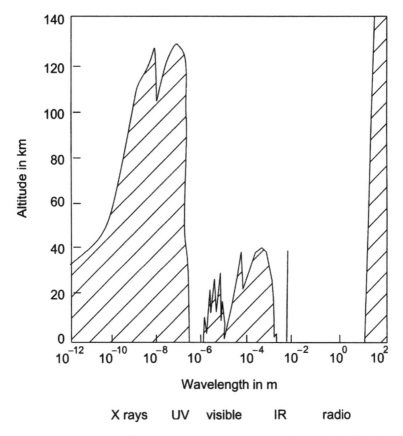

Fig. 1.3 The penetrating ability of electromagnetic wave through the Earth's atmosphere. The altitudes against different wavelengths indicate the heights above the sea level we have to climb to receive radiation of that wavelength from astronomical sources. Adapted from Shu (1982, p. 17).

long-wavelength radio waves in spite of the curvature of the surface. We, therefore, need to use shorter wavelengths for doing radio astronomy and longer wavelengths for communicating with distant regions on the Earth's surface. Near infrared radiation is absorbed mainly by water vapour, which remains confined in the lower layers of the atmosphere. Hence it is possible to do astronomy in the near infrared by going to the top of a mountain in a dry region. However, we need to go above the Earth's atmosphere to do ultraviolet or X-ray astronomy, since radiation in these wavelengths is absorbed by the upper atmosphere. We now say a few words about the instruments for doing astronomy in different wavelength bands.

1.7.1 Optical astronomy

This is astronomy in visible light. Although human beings have been observing the starry sky from prehistoric times, modern optical astronomy can be said to have been born when Galileo turned his telescope to the night sky in 1609. While Galileo's telescope was of the refracting type, Newton developed the reflecting telescope around 1668. The crucial optical component in a refracting telescope is a lens and in a reflecting telescope is a parabolic mirror. Most of the large telescopes constructed in the last one century are of the reflecting type. It is not difficult to understand the reasons. Firstly, a mirror is free from chromatic aberration, which affects a lens. Secondly, making a large lens of high quality is much more difficult than making a large mirror, since a mirror requires only a defect-free surface, whereas a lens involves a volume of glass that has to be perfectly uniform and defect-free. Finally, a mirror can be supported from the whole of its back-side, unlike a lens which is principally supported only along its outer circumference. For a large optical component which can bend under its own weight, a proper mechanical support is crucial.

The size of a telescope is indicated by the diameter of its main optical component (the lens or the mirror). The great refractor of Yerkes Observatory near Chicago, which was built in 1897 and has a diameter of 1 m, still remains the world's largest refracting telescope. From the beginning of the twentieth century, reflecting telescopes started becoming large enough for accurate extra-galactic studies. The 2.5 m reflector at Mount Wilson Observatory in California, commissioned in 1917, was probably one of the most important telescopes in the history of astronomy. It was used by astronomers like Hubble to make several path-breaking discoveries. The 5 m reflector of the nearby Mount Palomar Observatory, completed in 1948, remained the world's largest telescope for several years. Only in the last few years, it has been possible to build much larger telescopes by using new technology. The largest telescope at present is the Keck Telescope in Hawaii, which started operating from 1993. Instead of a single mirror, it has 36 hexagonal adjustable segments which together make up a large parabolic mirror of 10 m diameter.

Why do we try to build bigger and bigger telescopes? There are basically two reasons – to achieve higher resolution and to collect more light. Let us look at these two issues.

The resolving power of a mirror or a lens of diameter D is given by

$$\theta = 1.22 \frac{\lambda}{D}, \tag{1.13}$$

where λ is the wavelength of the light used (see, for example, Born and Wolf, 1980, §8.6.2). For a 1 m telescope, the resolving power at a wavelength of 5000 Å should be of order $0.12''$. Telescopes which are of this size and larger, however, produce images much less sharp than what is theoretically expected. This is because the air through which the light rays pass before reaching the telescope is always in turbulent motion. As a result, the paths of light rays become slightly deflected, giving rise to blurred images. Astronomers use the term *seeing* to indicate the quality of image under a given atmospheric

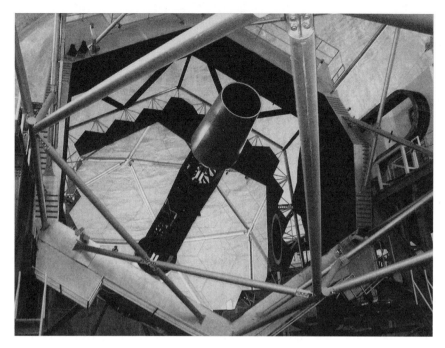

Fig. 1.4 A view of the Keck Telescope in Hawaii showing its mirror made up of 36 segments. Courtesy: W. M. Keck Observatory.

condition. Seeing is rarely good enough to allow images which are sharp enough to resolve more than 0.5″. Only if we can place a telescope above the Earth's atmosphere, is it possible to achieve the theoretical resolution given by (1.13). That is why the Hubble Space Telescope (HST), which was placed into orbit in 1990 and had an initial problem of image formation rectified in 1993, produced much sharper and crisper images than any ground-based telescope, even though its mirror has a diameter of only 2.4 m.

It may be mentioned that during the last couple of decades astronomers have come up with ingenious techniques for producing images even with ground-based telescopes that are sharper than what they would be if we were limited by seeing. In *speckle imaging*, which is possible only for fairly bright sources, very short exposure images are first produced. Since air above the telescope would not move much during the short exposure, the image would be sharp but dim. Combining many such images, a proper sharp image is constructed. The other technique is *adaptive optics*, which involves putting a deformable mirror in the light path within the telescope. A computer which gets information from a sensor about the deflection of light paths caused by turbulence keeps adjusting the mirror to correct for the effect of seeing.

It is clear that a bigger ground-based telescope cannot achieve higher resolutions beyond a certain limit. However, the light-gathering ability of a telescope – which obviously increases with the area of the mirror and therefore goes as D^2 – turns out to be crucial when we want to produce images of very faint objects like faraway galaxies. Anybody who has been fascinated by beautiful photographs of galaxies in books usually becomes very disappointed when he or she looks at a galaxy through a telescope for the first time in life. Beautiful galaxy pictures are usually produced only after long exposures. One needs a large telescope to produce photographs or spectra of very faint galaxies.

1.7.2 Radio astronomy

Radio astronomy – the first of the new astronomies – began when Jansky (1933) discovered radio signals coming from the direction of the constellation Sagittarius, where the galactic centre is located. Reber (1940) later built a primitive radio telescope in his backyard and found that radio signals were coming from the Sun and also from some other directions in the sky. The development of radar technology during World War II provided a major boost for the blossoming of radio astronomy after the war.

The main component of a radio telescope is an antenna in the form of a dish, which focuses the radio waves at a focal point, where receiving instruments can be kept. The early radio telescopes consisted of single dishes. The famous radio telescope of Jodrell Bank near Manchester, constructed in 1957, has a fully steerable single dish of 76 m diameter. Since radio waves are not affected by the atmospheric turbulence (though radio waves at wavelengths longer than 20 cm are affected by the plasma irregularities in the ionosphere and the solar wind), the resolving power of a radio telescope is not limited by atmospheric seeing and can achieve the theoretical value given by (1.13). With the development of interferometric techniques by Ryle and others, it became possible to combine signals received by different dishes and to produce images of which the resolution was determined by the maximum separation amongst the dishes. At a wavelength of 10 cm, antennas spread over an area of 1 km give a resolution of order $2.4''$. Perhaps the world's most important radio telescope in the last few years has been the Very Large Array (VLA) in New Mexico, which became operational around 1980 and consists of 27 radio antennas in a Y-shaped configuration spread over a few km. To achieve even higher resolution, one can combine signals from different radio telescopes around the Earth operating together in a mode called the Very Long Baseline Interferometry (VLBI). Then essentially the diameter of the Earth becomes the D that you put in (1.13). VLBI can achieve much higher resolution than what is possible in optical astronomy.

Let us say a few words about the kinds of astronomical sources from which one expects radio waves. Surfaces of stars have temperatures of order a few thousand degrees and emit primarily in visible wavelengths. The visible radiation received by optical telescopes from hot bodies (like stars) is emitted by them because of their temperature – the type of radiation usually called *thermal*

Fig. 1.5 The Very Large Array (VLA) radio telescope in New Mexico, made up of several dish antennas. Courtesy: NRAO/AUI/NSF.

radiation in astronomy. There are, however, many *non-thermal* processes due to which an object may emit radiation. Some of the most intriguing objects discovered by radio telescopes – pulsars (§5.5) and quasars (§9.4) – emit not because they have temperatures appropriate for the emission of radio waves, but because of non-thermal processes. One very important example of non-thermal radiation in astronomy is synchrotron radiation, which is emitted by relativistic electrons spiralling around magnetic field lines (§8.11). All signals received by radio telescopes, however, are not non-thermal. One of the most famous discoveries in the history of radio astronomy is that of the thermal radiation with 2.73 K temperature filling the whole Universe (§10.5).

1.7.3   X-ray astronomy

Since X-rays are absorbed by the Earth's ionosphere, it is necessary to send an X-ray telescope completely above the Earth's atmosphere in order to receive X-rays from astronomical objects. The first extraterrestrial X-ray signals were received by Geiger counters flown in a rocket (Giacconi *et al.*, 1962). X-ray astronomy really came of age when the satellite Uhuru, completely devoted to X-ray astronomy, was launched in 1970. The Chandra X-ray Observatory (named after S. Chandrasekhar), which was lifted into orbit in 1999, is capable of producing much sharper X-ray images than any of the previous X-ray telescopes.

X-rays are reflected from metal surfaces only when they are incident at grazing angles (otherwise, they pass through metals). Hence X-ray telescopes

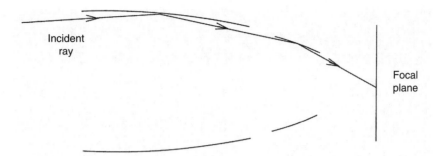

Incident
ray

Focal
plane

Fig. 1.6 A schematic representation of the optics of an X-ray telescope, in which X-rays are focused by two successive reflections at grazing incidence.

are designed very differently from optical telescopes. Figure 1.6 shows a sketch of an X-ray telescope in which X-rays are brought to a focus after two reflections at grazing angles. Also, mirrors in X-ray telescopes have to be much smoother than mirrors in optical telescopes because of the small wavelength of X-rays. Hence building powerful X-ray telescopes has been a formidable technological challenge.

X-rays are mainly emitted by very hot gases in astronomical systems. As we shall see in §5.6, one of the most important sources of astronomical X-rays is the type of binary star system in which one is a compact star gravitationally pulling off gas from its inflated binary companion.

1.7.4   Other new astronomies

After this brief discussion of the three bands of electromagnetic radiation which have yielded the maximum amount of astronomical information (optical, radio and X-ray), let us make a few remarks about the other bands. By now, virtually all wavelengths of electromagnetic radiation have been explored by astronomers.

Since star-forming regions are much less hot than the surfaces of stars, they are expected to emit infrared radiation. Therefore infrared astronomy is very important in understanding the star formation process, amongst other things. As we already mentioned, near infrared astronomy can be done from telescopes located at sufficiently high altitudes. One difficulty with infrared astronomy is that all objects around in the observatory emit infrared radiation and one has to pick up the signals from astronomical sources out of all these. It is like doing optical astronomy with lights around. There is no doubt that space is a better place for infrared astronomy. The Infrared Astronomy Satellite (IRAS) was launched in 1983. It has been followed by the Space Infrared Telescope Facility (SIRTF) launched in 2003.

Other important satellite missions devoted to studying other bands of electromagnetic radiation are the International Ultraviolet Explorer (IUE), launched in 1978 to explore the Universe in the ultraviolet, and the Compton Gamma Ray Observatory, launched in 1991 to detect gamma rays from outer space.

1.8 Astronomical nomenclature

Somebody embarking on a first study of astronomy may get confused by the names of various astronomical objects. Only a few of the brightest stars were given names in various ancient civilizations. Some of these names are still in use. For stars which do not have names and for all other astronomical objects, astronomers had to invent schemes by which an astronomical object can be identified unambiguously. There are several famous catalogues of astronomical objects. Very often an astronomical object is identified by the entry number in a well-known catalogue.

Stars down to about ninth magnitude were listed in the famous *Henry Draper Catalogue*, published during 1918–1924. It gives the celestial coordinates and spectroscopic classification (to be discussed in §3.5.1) of about 225,000 stars. A star listed in this catalogue is indicated by 'HD' followed by its listing number. For example, Sirius, the brightest star in the sky, can also be referred to as HD 48915, since it is listed as the object number 48915 in the *Henry Draper Catalogue*.

As will be clear from this book, modern astrophysicists are very much interested in objects other than stars visible in the sky. During 1774–1781 the French astronomer Charles Messier compiled a famous list of more than 100 non-stellar objects visible through a small telescope. This list includes some of the most widely studied galaxies, star clusters, supernova remnants and nebulae of various types. These objects are indicated by 'M' followed by the number in the Messier catalogue. The Andromeda Galaxy is M31, whereas the Crab Nebula, the remnant of a supernova seen from the Earth in 1054, is M1. A much bigger catalogue for non-stellar objects with nearly 8000 entries was compiled by Dreyer (1888) based primarily on observations of Hershel. This is known as the *New General Catalogue*, abbreviated as NGC. Galaxies not listed by Messier but listed in NGC are usually indicated by 'NGC' followed by the number in this catalogue.

After the development of radio and X-ray astronomies, astronomers had to devise schemes for identifying objects discovered in the radio and X-ray wavelengths. Initially when only a few objects emitting radio or X-rays were known, they were often named after the constellation in which they were found. The strongest radio source and the strongest X-ray source in the constellation Cygnus, for example, are known respectively as Cygnus A and Cygnus X-1. A very useful catalogue of radio sources is the *Third Cambridge Catalogue of Radio Sources*, known as 3C (Edge *et al.*, 1959). Radio sources listed in this catalogue are often indicated by '3C' followed by the number in the catalogue. The object 3C 273 is the brightest quasar (to be discussed in §9.4).

Lastly, some astronomical objects are named by their celestial coordinates. For example, PSR 1913 + 16 is the name of the pulsar (to be discussed in §5.5) having the right ascension (R.A.) 19 hours 13 minutes and the declination +16°.

Exercises

1.1 The Sun is at a distance of about 8 kpc from the galactic centre and moves around the galactic centre in a circular path with a velocity of about $220 \, \mathrm{km \, s^{-1}}$. Make a rough estimate of the mass of the Galaxy.

1.2 A star at a distance of 4 pc has an apparent magnitude 2. What is its absolute magnitude? Given the fact that the Sun has a luminosity 3.9×10^{26} W and has an absolute magnitude of about 5, find the luminosity of the star.

1.3 The Giant Metrewave Radio Telescope (GMRT) near Pune has several antennas spread over a region of size about 10 km. Make an estimate of the resolution (in arcseconds) which this telescope is expected to have. How large will an optical telescope have to be to achieve similar resolution in visible light?

2

Interaction of radiation
with matter

2.1 Introduction

As we pointed out in §1.6, most of our knowledge about the astrophysical
Universe is based on the electromagnetic radiation that reaches us from the sky.
By analysing this radiation, we infer various characteristics of the astrophysical
systems from which the radiation was emitted or through which the radiation
passed. Hence an understanding of how radiation interacts with matter is very
vital in the study of astrophysics. Such an interaction between matter and
radiation can be studied at two levels: macroscopic and microscopic. At the
macroscopic level, we introduce suitably defined emission and absorption coefficients, and then try to solve our basic equations assuming these coefficients to
be given. This subject is known as *radiative transfer*. At the microscopic level,
on the other hand, we try to calculate the emission and absorption coefficients
from the fundamental physics of the atom. Much of this chapter is devoted
to the macroscopic theory of radiative transfer. Only in §2.6, do we discuss how
the absorption coefficient of matter can be calculated from microscopic physics.
The emission coefficient directly follows from the absorption coefficient if the
matter is in thermodynamic equilibrium, as we shall see in §2.2.4.

2.2 Theory of radiative transfer

2.2.1 Radiation field

Let us first consider how we can provide the mathematical description of
radiation at a given point in space. It is particularly easy to give a mathematical
description of blackbody radiation, which is homogeneous and isotropic inside
a container. We shall assume the reader to be familiar with the basic physics of

23

Fig. 2.1 Illustration of specific intensity.

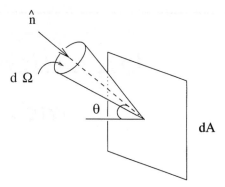

blackbody radiation, which is discussed in many excellent textbooks on thermal physics (see, for example, Saha and Srivastava, 1965, Ch. XV; Reif, 1965, pp. 373–388). One of the most famous results in the theory of blackbody radiation is Planck's law (Planck, 1900), which specifies the energy density U_ν in the frequency range $\nu, \nu + d\nu$:

$$U_\nu d\nu = \frac{8\pi h}{c^3} \frac{\nu^3 d\nu}{\exp\left(\frac{h\nu}{\kappa_B T}\right) - 1}. \tag{2.1}$$

This law more or less provides us with complete information about blackbody radiation at a given temperature T. Since blackbody radiation is isotropic, we do not have to provide any directional information. In general, however, the radiation in an arbitrary situation is not isotropic. When we have sunlight streaming into a room, we obviously have a non-isotropic situation involving the flow of radiation from a preferred direction. We require a more complicated prescription to describe such radiation mathematically.

We consider a small area dA at a point in space as shown in Figure 2.1. Let us consider the amount of radiation $dE_\nu d\nu$ passing through this area in time dt from the solid angle $d\Omega$ and lying in the frequency range $\nu, \nu + d\nu$. It is obvious that $dE_\nu d\nu$ should be proportional to the projected area $dA\cos\theta$, as well as proportional to dt, $d\Omega$ and $d\nu$. Hence we can write

$$dE_\nu d\nu = I_\nu(\mathbf{r}, t, \hat{\mathbf{n}})\cos\theta\, dA\, dt\, d\Omega\, d\nu, \tag{2.2}$$

where $\hat{\mathbf{n}}$ is the unit vector indicating the direction from which the radiation is coming. The quantity $I_\nu(\mathbf{r}, t, \hat{\mathbf{n}})$, which can be a function of position $\mathbf{r}$, time t and direction $\hat{\mathbf{n}}$, is called the *specific intensity*. If $I_\nu(\mathbf{r}, t, \hat{\mathbf{n}})$ is specified for all directions at every point of a region at a time, then we have a complete prescription of the *radiation field* in that region at that time. In this elementary treatment, we shall restrict ourselves only to radiation fields which are independent of time.

It is possible to calculate various quantities like flux, energy density and pressure of radiation if we know the radiation field at a point in space. For example, radiation flux is simply the total energy of radiation coming from all directions at a point per unit area per unit time. Hence we simply have to divide (2.2) by $dA\,dt$ and integrate over all solid angles. It is easy to see that the radiation flux associated with frequency ν is

$$F_\nu = \int I_\nu \cos\theta\,d\Omega, \tag{2.3}$$

whereas the total radiation flux is

$$F = \int F_\nu\,d\nu. \tag{2.4}$$

Energy density of radiation

Let us consider energy dE_ν of radiation associated with frequency ν as given by (2.2). This energy passes through area dA in time dt in the direction $\hat{\mathbf{n}}$. Since the radiation traverses a distance $c\,dt$ in time dt, we expect this radiation dE_ν to fill up a cylinder with base dA and axis of length $c\,dt$ in the direction $\hat{\mathbf{n}}$. The volume of such a cylinder being $\cos\theta\,dA\,c\,dt$, the energy density of this radiation

$$\frac{dE_\nu}{\cos\theta\,dA\,c\,dt} = \frac{I_\nu}{c}d\Omega$$

follows from (2.2). To get the total energy density of radiation at a point associated with frequency ν, we have to integrate over radiation coming from different directions so that

$$U_\nu = \int \frac{I_\nu}{c}d\Omega. \tag{2.5}$$

We now apply (2.5) to blackbody radiation to find its specific intensity. Since blackbody radiation is isotropic, the specific intensity of blackbody radiation, usually denoted by $B_\nu(T)$, should be independent of direction. Hence, on applying (2.5) to blackbody radiation, we get

$$U_\nu = \frac{4\pi}{c}B_\nu(T),$$

where 4π comes from the integration over Ω. Making use of the expression (2.1), we now conclude that the specific intensity of blackbody radiation is given by

$$B_\nu(T) = \frac{2h\nu^3}{c^2}\frac{1}{\exp\left(\frac{h\nu}{\kappa_B T}\right) - 1}. \tag{2.6}$$

Pressure due to radiation

The pressure of the radiation field over a surface is given by the flux of momentum perpendicular to that surface. The momentum associated with energy dE_ν

is dE_ν/c and its component normal to the surface dA is $dE_\nu \cos\theta/c$. By dividing this by $dA\,dt$, we get the momentum flux associated with dE_ν, which is

$$\frac{dE_\nu \cos\theta}{c}\,\frac{1}{dA\,dt} = \frac{I_\nu}{c}\cos^2\theta\,d\Omega$$

on making use of (2.2). The pressure P_ν is obtained by integrating this over all directions, i.e.

$$P_\nu = \frac{1}{c}\int I_\nu \cos^2\theta\,d\Omega. \tag{2.7}$$

If the radiation field is isotropic, then we get

$$P_\nu = \frac{I_\nu}{c}\int \cos^2\theta\,d\Omega = \frac{4\pi}{3}\frac{I_\nu}{c}. \tag{2.8}$$

It follows from (2.5) that

$$U_\nu = 4\pi\frac{I_\nu}{c}$$

for isotropic radiation. Combining this with (2.8), we have

$$P_\nu = \frac{1}{3}U_\nu \tag{2.9}$$

for isotropic radiation.

2.2.2 Radiative transfer equation

If matter is present, then in general the specific intensity keeps changing as we move along a ray path. Before we consider the effect of matter, first let us find out what happens to the specific intensity in empty space as we move along a ray path.

Let dA_1 and dA_2 be two area elements separated by a distance R and placed perpendicularly to a ray path, as shown in Figure 2.2. Let $I_{\nu 1}$ and $I_{\nu 2}$ be the specific intensity of radiation in the direction of the ray path at dA_1 and dA_2. We want to find out the amount of radiation passing through both dA_1 and dA_2 in time dt in the frequency range $\nu, \nu + d\nu$. If $d\Omega_2$ is the solid angle subtended

Fig. 2.2 Two area elements
perpendicular to a ray path.

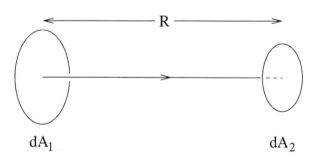

dA$_1$ dA$_2$

by dA_1 at dA_2, then according to (2.2) the radiation falling on dA_2 in time dt after passing through dA_1 is

$$I_{\nu 2} \, dA_2 \, dt \, d\Omega_2 \, d\nu.$$

From considerations of symmetry, this should also be equal to

$$I_{\nu 1} \, dA_1 \, dt \, d\Omega_1 \, d\nu,$$

where $d\Omega_1$ is the solid angle subtended by dA_2 at dA_1. Equating these two expressions and noting that

$$d\Omega_1 = \frac{dA_2}{R^2}, \quad d\Omega_2 = \frac{dA_1}{R^2},$$

we get

$$I_{\nu 1} = I_{\nu 2}. \tag{2.10}$$

In other words, in empty space the specific intensity along a ray path does not change as we move along the ray path. If s is the distance measured along the ray path, then we can write

$$\frac{dI_\nu}{ds} = 0 \tag{2.11}$$

in empty space.

At first sight, this may appear like a surprising result. We know that the intensity falls off as we move further and further away from a source of radiation. Can the specific intensity remain constant? The mystery is cleared up when we keep in mind that the specific intensity due to a source is essentially its intensity divided by the solid angle it subtends, which means that specific intensity is a measure of the surface brightness. As we move further away from a source of radiation, both its intensity and angular size fall as $(\text{distance})^2$. Hence the surface brightness, which is the ratio of these two, does not change. Suppose you are standing on a street in a dark night and are looking at the street lights. The lights further away would appear smaller in size, but their surfaces would appear as bright as the surfaces of nearby lights. This result has an important astronomical implication. If we neglect intergalactic extinction, then the surface brightness of a galaxy which is resolved by a telescope is independent of distance. Whether the galaxy is nearby or far away, its surface would appear equally bright to us. We may expect a similar consideration to hold for stars also. Then why do distant stars look dimmer? Since the theory of radiative transfer is based on the concept of ray path, we are tacitly assuming geometrical optics in all our derivations. So our results hold as long as geometrical optics is valid. If the star is very far away, then its disk is not resolved and geometrical optics no longer holds. The angular size of the star may be caused by the diffraction of light or the seeing (§1.7.1). As the star is moved further away, its intensity diminishes, but the angular size due to diffraction does not change much. Hence a decreasing amount of radiation gets spread over an image of the same angular size, making the star appear dimmer. It may be noted that very faraway galaxies also look dimmer due to general relativistic effects to be discussed in §14.4.1.

Let us now consider what happens if matter is present along the ray path. If the matter emits, we expect that it will add to the specific intensity. This can be taken care of by adding an emission coefficient j_ν on the right-hand side of (2.11). On the other hand, absorption by matter would lead to a diminution of specific intensity and the diminution rate must be proportional to the specific intensity itself. In other words, the stronger the beam, the more energy there is for absorption. Hence the absorption term on the right-hand side of (2.11) should be negative and proportional to I_ν. Thus, in the presence of matter, (2.11) gets modified to the following form

$$\frac{dI_\nu}{ds} = j_\nu - \alpha_\nu I_\nu, \tag{2.12}$$

where α_ν is the absorption coefficient. This is the celebrated radiative transfer equation and provides the basis for our understanding of interaction between radiation and matter.

In the early years of spectral research, many astronomers held the view that the Sun was surrounded by a cool layer of gas which only absorbed radiation at certain frequencies to produce the dark lines. Schuster (1905) recognized the importance of treating emission and absorption simultaneously by the same layer of gas. A primitive version of radiative transfer theory was formulated by Schuster (1905) by considering only two beams of radiation – one moving upward and one moving downward. Schwarzschild (1914) was the first to formulate a proper radiative transfer theory by considering the specific intensity of radiation.

It is fairly trivial to solve the radiative transfer equation (2.12) if either the emission coefficient or the absorption coefficient is zero. Let us consider the case of $j_\nu = 0$, i.e. matter is assumed to absorb only but not to emit. Then (2.12) becomes

$$\frac{dI_\nu}{ds} = -\alpha_\nu I_\nu. \tag{2.13}$$

On integrating this equation over the ray path from s_0 to s, we get

$$I_\nu(s) = I_\nu(s_0) \exp\left[-\int_{s_0}^{s} \alpha_\nu(s')\, ds'\right]. \tag{2.14}$$

More general solutions of the radiative transfer equation will be discussed now.

2.2.3 Optical depth. Solution of radiative transfer equation

We define *optical depth* τ_ν through the relation

$$d\tau_\nu = \alpha_\nu\, ds \tag{2.15}$$

such that the optical depth along the ray path between s_0 and s becomes

$$\tau_\nu = \int_{s_0}^{s} \alpha_\nu(s')\, ds'.$$ (2.16)

From (2.14) and (2.16), it follows that the specific intensity along the ray path falls as

$$I_\nu(\tau_\nu) = I_\nu(0) e^{-\tau_\nu}$$ (2.17)

if matter does not emit.

If the optical depth $\tau_\nu \gg 1$ along a ray path through an object, then the object is known as *optically thick*. On the other hand, an object is known as *optically thin* if $\tau_\nu \ll 1$ for a ray path through it. It follows from (2.17) that an optically thick object extinguishes the light of a source behind it, whereas an optically thin object does not decrease the light much. Hence the terms optically thick and optically thin roughly mean opaque and transparent at the frequency of electromagnetic radiation we are considering.

We now define the *source function*

$$S_\nu = \frac{j_\nu}{\alpha_\nu}.$$ (2.18)

Dividing the radiative transfer equation (2.12) by α_ν, we get

$$\frac{dI_\nu}{d\tau_\nu} = -I_\nu + S_\nu$$ (2.19)

on making use of (2.15) and (2.18). Multiplying this equation by e^{τ_ν}, we obtain

$$\frac{d}{d\tau_\nu}(I_\nu e^{\tau_\nu}) = S_\nu e^{\tau_\nu}.$$

Integrating this equation from optical path 0 to τ_ν (i.e. from s_0 to s along the ray path), we get

$$I_\nu(\tau_\nu) = I_\nu(0)\, e^{-\tau_\nu} + \int_0^{\tau_\nu} e^{-(\tau_\nu - \tau_\nu')} S_\nu(\tau_\nu')\, d\tau_\nu'.$$ (2.20)

This is the general solution of the radiative transfer equation.

If matter through which the radiation is passing has constant properties, then we can take S_ν constant and work out the integral in (2.20). This gives

$$I_\nu(\tau_\nu) = I_\nu(0)\, e^{-\tau_\nu} + S_\nu\, (1 - e^{-\tau_\nu}).$$

We are now interested in studying the emission and absorption properties of an object itself without a source behind it. Then we take $I_\nu(0) = 0$ and write

$$I_\nu(\tau_\nu) = S_\nu\, (1 - e^{-\tau_\nu}).$$ (2.21)

Let us consider the cases of optically thin and thick objects. If the object is optically thin (i.e. $\tau_\nu \ll 1$), then we write $1 - \tau_\nu$ for $e^{-\tau_\nu}$ such that

$$I_\nu(\tau_\nu) = S_\nu \tau_\nu.$$

For matter with constant properties, we take $\tau_\nu = \alpha_\nu L$, where L is the total length of the ray path. Making use of (2.18), we get

$$\text{Optically thin: } I_\nu = j_\nu L. \tag{2.22}$$

On the other hand, if the object is optically thick, then we neglect $e^{-\tau_\nu}$ compared to 1 in (2.21). Then

$$\text{Optically thick: } I_\nu = S_\nu. \tag{2.23}$$

We have derived two tremendously important results (2.22) and (2.23). To understand their physical significance, we have to look at some thermodynamic considerations.

2.2.4 Kirchhoff's law

Suppose we have a box kept in thermodynamic equilibrium. If we make a small hole on its side, we know that the radiation coming out of the hole will be blackbody radiation. Hence the specific intensity of radiation coming out of the hole is simply

$$I_\nu = B_\nu(T), \tag{2.24}$$

where $B_\nu(T)$ is given by (2.6). We now keep an optically thick object behind the hole as shown in Figure 2.3. If this object is in thermodynamic equilibrium with the surroundings, then it will not disturb the environment and the radiation coming out of the hole will still be blackbody radiation, with specific intensity given by (2.24). On the other hand, we have seen in (2.23) that the radiation coming out of an optically thick object has the specific intensity equal to the source function. From (2.23) and (2.24), we conclude

$$S_\nu = B_\nu(T) \tag{2.25}$$

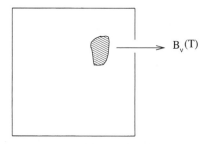

Fig. 2.3 Blackbody radiation coming out of a hole in a box with an optically thick obstacle placed behind the hole.

when matter is in thermodynamic equilibrium. On using (2.18), we finally have

$$j_\nu = \alpha_\nu B_\nu(T). \tag{2.26}$$

This famous result is known as *Kirchhoff's law* (Kirchhoff, 1860). The relevance of this law in the radiative transfer theory was recognized by the pioneers Schuster (1905) and Schwarzschild (1914).

Let us now stop and try to understand what we have derived. Very often matter tends to emit and absorb more at specific frequencies corresponding to spectral lines. Hence both j_ν and α_ν are expected to have peaks at spectral lines. But, according to (2.26), the ratio of these coefficients should be the smooth blackbody function $B_\nu(T)$. We now look at the results (2.22) and (2.23). The radiation coming out of an optically thin source is essentially determined by its emission coefficient. Since the emission coefficient is expected to have peaks at spectral lines, we find the emission from an optically thin system like a hot transparent gas to be mainly in spectral lines. On the other hand, the specific intensity of radiation coming out of an optically thick source is its source function, which has been shown to be equal to the blackbody function $B_\nu(T)$. Hence we expect an optically thick object like a hot piece of iron to emit roughly like a blackbody. The theory of radiative transfer is important not only in astrophysics. If we want to understand rigorously and quantitatively many everyday phenomena such as why hot transparent gases emit in spectral lines whereas hot pieces of iron emit like blackbodies, then we need to invoke the theory of radiative transfer. The nature of radiation from an astrophysical source crucially depends on whether the source is optically thin or optically thick. Emission from a tenuous nebula is usually in spectral lines. On the other hand, a star emits very much like a blackbody. Why is the radiation from a star not exactly blackbody radiation? Why do we see absorption lines? We derived (2.23) by assuming the source to have constant properties. This is certainly not true for a star. As we go down from the star's surface, temperature keeps increasing. Hence (2.23) should be only approximately true. It is the temperature gradient near the star's surface which gives rise to the absorption lines. This will be shown in §2.4.3.

2.3 Thermodynamic equilibrium revisited

By assuming thermodynamic equilibrium, we have derived the tremendously important result (2.25) that the source function should be equal to the blackbody function $B_\nu(T)$. In a realistic situation, we rarely have strict thermodynamic equilibrium. The temperature inside a star is not constant, but varies with its radius. In such a situation, will (2.25) hold? Before answering this question, let us look at some basic characteristics of thermodynamic equilibrium.

2.3.1 Basic characteristics of thermodynamic equilibrium

If a system is in thermodynamic equilibrium, then certain important principles of physics can be applied to that system. Let us recapitulate some of these important principles. We assume the reader to be familiar with them and do not present derivations or detailed discussion.

Maxwellian velocity distribution

Different particles in a gas move around with different velocities. If the gas is in thermodynamic equilibrium with temperature T, then the number of particles having speeds between v and $v + dv$ is given by

$$dn_v = 4\pi n \left(\frac{m}{2\pi \kappa_B T}\right)^{3/2} v^2 \exp\left(-\frac{mv^2}{2\kappa_B T}\right) dv, \qquad (2.27)$$

where n is the number of particles per unit volume and m is the mass of each particle. This is the celebrated law of the Maxwellian velocity distribution (Maxwell, 1860).

Boltzmann and Saha equations

We know that a hydrogen atom has several different energy levels. It is also possible to break the hydrogen atom into a proton and an electron. This process of removing an electron from the atom is called ionization. If a gas of hydrogen atoms is kept in thermodynamic equilibrium, then we shall find that a certain fraction of the atoms will occupy a particular energy state and also a certain fraction will be ionized. The same considerations hold for other gases besides hydrogen.

If n_0 is the number density of atoms in the ground state, then the number density n_e of atoms in an excited state with energy E above the ground state is given by

$$\frac{n_e}{n_0} = \exp\left(-\frac{E}{\kappa_B T}\right). \qquad (2.28)$$

This is the Boltzmann distribution law.

Saha (1920) derived the equation which tells us what fraction of a gas will be ionized at a certain temperature T and pressure P. Derivation of this equation can be found in books such as Mihalas (1978, §5-1) or Rybicki and Lightman (1979, §9.5). If χ is the ionization potential (i.e. the amount of energy to be supplied to an atom to ionize it), then the fraction x of atoms which are ionized is given by

$$\frac{x^2}{1-x} = \frac{(2\pi m_e)^{3/2}}{h^3} \frac{(\kappa_B T)^{5/2}}{P} \exp\left(-\frac{\chi}{\kappa_B T}\right), \qquad (2.29)$$

where h is Planck's constant and m_e is the mass of electron.

Planck's law of blackbody spectrum

When radiation is in thermodynamic equilibrium with matter, it is called black-body radiation. The spectral distribution of energy in blackbody radiation is given by the famous law derived by Planck (1900). We have already written down the law in (2.1).

2.3.2 Concept of local thermodynamic equilibrium

Now we come to the all-important question: when can we expect a system to be in thermodynamic equilibrium and when can we expect the above principles (Maxwellian velocity distribution, Boltzmann equation, Saha equation, Planck's law) to hold? If a box filled with gas and radiation is kept isolated from the surroundings, then we know that thermodynamic equilibrium will get established inside, and all the above principles will hold. However, a realistic system is always more complicated. Inside a star, the temperature keeps decreasing as we go from the central region to the outside. Can the above principles be applied in such a situation? To answer that question, let us try to understand how thermodynamic equilibrium gets established.

We again consider a box filled with gas and radiations. Even if the gas particles initially do not obey the Maxwellian distribution, they will relax to it after undergoing a few collisions. In other words, collisions – or rather interactions amongst the constituents of the system – are vital in establishing thermodynamic equilibrium. We assume the reader to be familiar with the concept of mean free path. When collisions are frequent, the mean free path turns out to be small. Hence, the smallness of mean free path is a measure of how important collisions are. If the mean free path is small, then particles in a gas will interact with each other more effectively and we expect that principles like the Maxwellian velocity distribution, the Boltzmann equation or the Saha equation will hold. But how small will the mean free path have to be? Suppose the temperature is varying inside a gas and we consider a point X with the left side hotter and right side colder. Then gas particles coming to X from the left side will be more energetic than the gas particles coming from the right side. This will make the velocity distribution at X different from the Maxwellian, provided particles are able to come directly to X from regions where temperatures are significantly different from the temperature at X. However, if the mean free path is small and the temperature does not vary much over that distance, then these considerations will be unimportant and we shall have the Maxwellian velocity distribution. Hence the condition of validity of the Maxwellian velocity distribution (as well as the Boltzmann equation and the Saha equation) is that the mean free path has to be small enough such that the temperature does not vary much over the mean free path. For Planck's law to be established for radiation, the radiation has to be in equilibrium with matter. This is possible only when radiation interacts efficiently with matter. The absorption coefficient α_ν in the radiative transfer equation (2.12) is a measure of the interaction between radiation and matter. We note from (2.12) that α_ν has the dimension of inverse length. Its inverse α_ν^{-1} gives the distance over which a significant part of a beam of radiation would get absorbed by matter. Often this distance α_ν^{-1} is referred to as the mean free path of photons, since this is the typical distance a photon is expected to traverse freely before interacting with an atom. The smaller the value of α_ν^{-1}, the more efficient is the interaction between matter and radiation. If α_ν^{-1} is sufficiently small such that the temperature can be taken as constant over such distances, then we expect Planck's law of blackbody radiation to hold.

Let us consider a simple example from everyday life – sunlight streaming into a room through a window. Is this a system in thermodynamic equilibrium? The mean free path of air molecules is only of the order of 10^{-7} m. Hence we expect the Maxwellian velocity distribution and the Boltzmann equation to hold for air molecules. The ionization level in air at room temperature is negligible, which is completely consistent with the Saha equation. However, if the system were in complete thermodynamic equilibrium, then radiation in the room should obey Planck's law at the room temperature. This is definitely not the case. Since air is virtually transparent to visible light, the photons do not interact with air molecules at all. The photons in the beam of sunlight have come directly from the surface of the Sun and have not interacted with matter at all after they left the solar surface. If we analyse the spectrum of sunlight, then we find that it is not like a blackbody spectrum at room temperature, but the shape of the spectrum is rather like a blackbody spectrum at a temperature of 6000 K (the surface temperature of the Sun), although the energy density in sunlight is obviously much less than the energy density in blackbody radiation at 6000 K. Although thermodynamic equilibrium is a very useful concept, this example would make one realize that we usually do not have full thermodynamic equilibrium around us.

If the temperature is varying within a system, then it is not in full thermodynamic equilibrium. However, we can have a situation where both α_ν^{-1} and the mean free path of particles are small compared to the length over which the temperature varies appreciably. In a such situation, all the important laws of thermodynamic equilibrium are expected to hold within a local region, provided we use the local temperature T in the expressions (2.1), (2.27), (2.28) and (2.29). Such a situation is known as *local thermodynamic equilibrium*, abbreviated as LTE. Sunlight streaming into the room is obviously not a case of LTE because the air is almost transparent to radiation and hence α_ν^{-1} must be very large. Inside a star, however, we expect LTE to be a very good approximation and we can assume (2.25) to hold when we solve the radiative transfer equation inside the star. In the outermost atmosphere of a star, LTE may fail and it often becomes necessary to consider departures from LTE when studying the transfer of radiation there. In our elementary treatment, we shall consider radiative transfer only in situations of LTE.

2.4 Radiative transfer through stellar atmospheres

Several treatises have been written on radiative transfer theory, one of the most famous being by Chandrasekhar (1950). Now, we have written down the general solution of the radiative transfer equation in (2.20). If it is so easy to write down the general solution, then what is the necessity of writing treatises on this subject? Well, we get a complete solution of the problem from (2.20) only if we know in advance the source function S_ν everywhere. This is almost never the case! Even in LTE when (2.25) holds, we need to know the temperatures at different points to find out the source functions there. We usually have the problem of finding out the radiation field and the temperature simultaneously. If the radiation field is strong in a region, then we expect the temperature to be high there. Hence the radiation field determines the temperature. On the other hand, the temperature determines the source function and thereby the radiation field through (2.20). It is the simultaneous solution of temperature and radiation field which is a tremendously challenging problem. To give an idea how this problem can be solved, we discuss the application of radiative transfer theory to stellar atmospheres.

Traditionally, the study of stellar astronomy has been divided into two branches: *stellar interior* and *stellar atmosphere*. One may wonder, at what depth inside a star does the stellar atmosphere end and the stellar interior begin. As it happens, the terms stellar interior and stellar atmosphere do not correspond to physically distinct regions of a star, but to two different scientific subjects which address two different sets of questions. We shall see in the next chapter that there exist certain relationships amongst such quantities as the total mass of a star and its luminosity (more massive stars tend to be more luminous). Since we need to look at the physical processes in the interior of the star to understand such global relationships, studying and understanding such global characteristics of a star constitutes the subject of stellar interior. On the other hand, to explain and analyse the spectrum of a star, we need to consider the passage of radiation through the outer layers of a star. This is the subject of stellar atmospheres. We now give a very brief introduction to this subject.

2.4.1 Plane parallel atmosphere

When we focus our attention on the local region of a stellar atmosphere, we can neglect the curvature and assume the various thermodynamic quantities like the temperature T to be constant over horizontal planes. Let us take the z axis in the vertical direction, with z increasing above. Any thermodynamic variable of the atmosphere can be a function of z alone. Let us consider an element of a ray path ds as shown in Figure 2.4. If dz is the change in z corresponding to ds, then we have

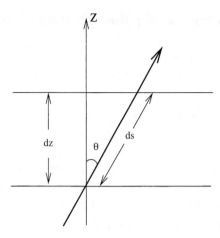

Fig. 2.4 A ray path through a plane parallel atmosphere.

$$ds = \frac{dz}{\cos \theta} = \frac{dz}{\mu}, \qquad (2.30)$$

where

$$\theta = \cos^{-1} \mu$$

is the angle subtended by the ray path with the vertical direction.

We have seen in §2.2.1 that the specific intensity $I_\nu(\mathbf{r}, t, \hat{\mathbf{n}})$ can in general be a function of position, time and direction. We are considering a static situation. In the plane parallel stellar atmosphere, nothing varies in the horizontal directions and all direction vectors lying on a cone around the vertical axis are symmetrical. Hence we expect the specific intensity $I_\nu(z, \mu)$ to be a function of z and $\mu = \cos \theta$ only. On using (2.30), the radiative transfer equation (2.12) gives us the equation

$$\mu \frac{\partial I_\nu(z, \mu)}{\partial z} = j_\nu - \alpha_\nu I_\nu \qquad (2.31)$$

for the plane parallel atmosphere problem.

We now define the optical depth of a plane parallel atmosphere slightly differently from the way it was defined in (2.15). We define

$$d\tau_\nu = -\alpha_\nu \, dz. \qquad (2.32)$$

In other words, the optical depth is now defined as a function of the vertical distance z rather than the distance along ray path s. The negative sign in (2.32) implies that the optical depth increases as we go deeper down, which corresponds to the usual notion of depth. The normal convention is to take $\tau_\nu = 0$ at the top of the stellar atmosphere.

On dividing (2.31) by α_ν and using the definition (2.18) of the source function, we get

$$\mu \frac{\partial I_\nu(\tau_\nu, \mu)}{\partial \tau_\nu} = I_\nu - S_\nu. \tag{2.33}$$

Multiplying (2.33) by $e^{-(\tau_\nu/\mu)}$ and rewriting it slightly, we have

$$\mu \frac{d}{d\tau_\nu}\left(I_\nu e^{-\frac{\tau_\nu}{\mu}}\right) = -S_\nu e^{-\frac{\tau_\nu}{\mu}}.$$

Integrating this equation from some reference optical depth $\tau_{\nu,0}$, we obtain

$$I_\nu e^{-\frac{\tau_\nu}{\mu}}\big|_{\tau_{\nu,0}}^{\tau_\nu} = -\int_{\tau_{\nu,0}}^{\tau_\nu} \frac{S_\nu}{\mu} e^{-\frac{\tau_\nu}{\mu}} dt_\nu. \tag{2.34}$$

We now consider two cases separately: (I) $0 \le \mu \le 1$, which corresponds to a ray path proceeding in the outward direction in the stellar atmosphere; and (II) $-1 \le \mu \le 0$, which corresponds to a ray path going inward in the stellar atmosphere. In case (I) the ray path can be assumed to begin from a great depth inside the star and we can take $\tau_{\nu,0} = \infty$. On the other hand, the ray path in case (II) starts receiving contributions beginning with the top of the stellar atmosphere and we can take $\tau_{\nu,0} = 0$. In these two cases, (2.34) reduces to

$$0 \le \mu \le 1: \quad I_\nu(\tau_\nu, \mu) = \int_{\tau_\nu}^{\infty} S_\nu e^{-(t_\nu - \tau_\nu)/\mu} \frac{dt_\nu}{\mu}, \tag{2.35}$$

$$-1 \le \mu \le 0: \quad I_\nu(\tau_\nu, \mu) = \int_0^{\tau_\nu} S_\nu e^{-(\tau_\nu - t_\nu)/(-\mu)} \frac{dt_\nu}{(-\mu)}. \tag{2.36}$$

It may be noted that (2.36) was obtained by using the boundary condition $I_\nu(0, \mu) = 0$ for negative μ, which implies that the specific intensity for a downward ray path is zero at the top of the stellar atmosphere.

So far we have not used any thermodynamics. To proceed further, we have to use some suitable expression for the source function S_ν in (2.35) and (2.36). Let us assume LTE throughout the stellar atmosphere so that the source function everywhere is equal to the Planck function there, in accordance with Kirchhoff's law (2.25). Suppose we want to find out the radiation field at some optical depth τ_ν. The source function there is given by $B_\nu(T(\tau_\nu))$, which we write as $B_\nu(\tau_\nu)$ for simplification. The source function at a nearby optical depth t_ν can be written in the form of a Taylor expansion around the optical depth τ_ν, i.e.

$$S_\nu(t_\nu) = B_\nu(\tau_\nu) + (t_\nu - \tau_\nu)\frac{dB_\nu}{d\tau_\nu} + \ldots \tag{2.37}$$

Truncating (2.37) after the linear term and substituting it in both (2.35) and (2.36), we get for both positive and negative μ the very important equation

$$I_\nu(\tau_\nu, \mu) = B_\nu(\tau_\nu) + \mu\frac{dB_\nu}{d\tau_\nu}, \tag{2.38}$$

provided the point considered is sufficiently inside the atmosphere such that $\tau_\nu \gg 1$ and we can take $e^{-\tau_\nu}$ to be zero. It is the second term on the right-hand side of (2.38) which depends on μ and makes the radiation field anisotropic. If there was no variation of temperature within the stellar atmosphere, then $dB_\nu/d\tau_\nu$ would vanish and the radiation field would become an isotropic black-body radiation. It is the presence of a temperature gradient in the atmosphere which makes the radiation field depart from the Planckian distribution, making it anisotropic. An estimate of the anisotropy will be given below.

We have seen in (2.3), (2.5) and (2.7) that the radiation flux, the energy density and the pressure of a radiation field can be calculated from the specific intensity. In the case of a plane parallel atmosphere, the integration over the solid angle becomes simplified due to symmetry. If $A(\cos\theta)$ is any function of angle in a plane parallel atmosphere, then

$$\int A(\cos\theta)\, d\Omega = \int_{\theta=0}^{\pi} \int_{\phi=0}^{2\pi} A(\cos\theta)\, \sin\theta\, d\theta\, d\phi$$

$$= 2\pi \int_{+1}^{-1} A(\mu)\, d(-\mu) = 2\pi \int_{-1}^{+1} A(\mu)\, d\mu.$$

On making use of this, (2.3), (2.5) and (2.7) give

$$U_\nu = \frac{2\pi}{c} \int_{-1}^{1} I_\nu\, d\mu, \tag{2.39}$$

$$F_\nu = 2\pi \int_{-1}^{1} I_\nu\, \mu\, d\mu, \tag{2.40}$$

$$P_\nu = \frac{2\pi}{c} \int_{-1}^{1} I_\nu\, \mu^2\, d\mu. \tag{2.41}$$

If we substitute (2.38) in the above three equations, then we get

$$U_\nu = \frac{4\pi}{c} B_\nu(\tau_\nu), \tag{2.42}$$

$$F_\nu = \frac{4\pi}{3} \frac{dB_\nu}{d\tau_\nu}, \tag{2.43}$$

$$P_\nu = \frac{4\pi}{3c} B_\nu(\tau_\nu). \tag{2.44}$$

It is clear from (2.38) that the ratio of the anisotropic part in the radiation field to the isotropic part is of order

$$\frac{dB_\nu/d\tau_\nu}{B_\nu} \approx \frac{3F_\nu}{cU_\nu}$$

on making use of (2.42) and (2.43). Approximating F_ν/U_ν by F/U (where F and U are respectively the total flux and the total energy density integrated over all wavelengths), we have

$$\frac{\text{Anisotropic term}}{\text{Isotropic term}} \approx \frac{3F}{c\,U}. \tag{2.45}$$

We now use the standard results from thermal physics that the total energy density of blackbody radiation is given by

$$U = a_{\text{B}} T^4, \tag{2.46}$$

whereas the total flux is the flux which eventually emerges out of the surface and is given by the Stefan–Boltzmann law (Stefan, 1879; Boltzmann, 1884):

$$F = \sigma T_{\text{eff}}^4, \tag{2.47}$$

where T_{eff} is the effective temperature of the surface and

$$\sigma = \frac{c a_{\text{B}}}{4} \tag{2.48}$$

(see, for example, Saha and Srivastava, 1965, §15.21; also Exercise 2.1). On making use of (2.46), (2.47) and (2.48), it follows from (2.45) that

$$\frac{\text{Anisotropic term}}{\text{Isotropic term}} \approx \frac{3}{4}\left(\frac{T_{\text{eff}}}{T}\right)^4. \tag{2.49}$$

As we go deeper in a stellar atmosphere, T becomes much larger than T_{eff}, making the anisotropic term negligible compared to the isotropic term. In other words, the radiation field is nearly isotropic in sufficiently deep layers of a stellar atmosphere where the temperature is considerably higher than the surface temperature.

The expression (2.38) would give us the radiation field inside a stellar atmosphere, if we knew how temperature varied with depth and could calculate the Planck function $B_\nu(\tau_\nu)$ at different depths. As we already pointed out, this is not known a priori in general and the real challenge of studying stellar atmospheres is to solve the radiation field and the temperature structure of the stellar atmosphere simultaneously. We now show how this can be done in a simplified idealized model known as the *grey atmosphere*.

2.4.2   The grey atmosphere problem

If the absorption coefficient α_ν is constant for all frequencies, then the atmosphere is called a grey atmosphere. There is no real stellar atmosphere which has this property. The grey atmosphere is an idealized mathematical model which is much simpler to treat than a more realistic stellar atmosphere and gives us some insight into the nature of the problem, as we shall see below.

If α_ν is independent of frequency, then it follows from (2.32) that the value of optical depth at some physical depth will be the same for all frequencies. In such a situation, denoting the optical depth by τ, (2.33) can be integrated over v to give

$$\mu \frac{\partial I(\tau, \mu)}{\partial \tau} = I - S, \tag{2.50}$$

where

$$I = \int I_v \, dv \tag{2.51}$$

and

$$S = \int S_v \, dv \tag{2.52}$$

are the total specific intensity and total source function integrated over all frequencies. Just similar to (2.39), (2.40) and (2.41), we write down the total energy density, total radiation flux and total radiation pressure integrated over all frequencies:

$$U = \frac{2\pi}{c} \int_{-1}^{1} I \, d\mu, \tag{2.53}$$

$$F = 2\pi \int_{-1}^{1} I \, \mu \, d\mu, \tag{2.54}$$

$$P = \frac{2\pi}{c} \int_{-1}^{1} I \, \mu^2 \, d\mu. \tag{2.55}$$

We also define the average specific intensity J averaged over all angles

$$J = \frac{1}{2} \int_{-1}^{1} I \, d\mu. \tag{2.56}$$

It follows from (2.53) and (2.56) that

$$J = \frac{c}{4\pi} U. \tag{2.57}$$

We now obtain two important moment equations of (2.50). Multiplying (2.50) by 1/2 and integrating over μ, we get

$$\frac{1}{4\pi} \frac{dF}{d\tau} = J - S \tag{2.58}$$

on making use of (2.54) and (2.56). On the other hand, multiplying (2.50) by $2\pi \mu/c$ and integrating over μ, we get

$$\frac{dP}{d\tau} = \frac{F}{c}. \tag{2.59}$$

Although I is a function of both τ and μ, it may be noted that F and P are functions of τ alone. Hence we have used ordinary derivatives in (2.58) and (2.59) instead of partial derivatives as in (2.50). Very often, within a stellar atmosphere, we do not have a source or sink of energy. The energy generated in the stellar interior passes out in the form of a constant energy flux through the outer layers of the stellar atmosphere. In other words, F has to be independent of depth. It follows from (2.58) that this is possible only if

$$J = S, \tag{2.60}$$

i.e. the average specific intensity has to be equal to the source function. This is called the condition of *radiative equilibrium*. Using (2.56) and (2.60), we write (2.50) in the form

$$\mu \frac{\partial I(\tau, \mu)}{\partial \tau} = I - \frac{1}{2} \int_{-1}^{1} I \, d\mu \tag{2.61}$$

valid under radiative equilibrium. This is an integro-differential equation for $I(\tau, \mu)$. There are techniques for solving (2.61) exactly and obtaining I for all τ and μ. Readers are referred to Chandrasekhar (1950, Ch. III) and Mihalas (1978, pp. 64–74) for a discussion of the exact solution of the grey atmosphere problem. Since the method of exact solution is somewhat complicated and beyond the scope of this elementary treatment, we now discuss an approximate method of solving the grey atmosphere problem.

Since F is constant under the condition of radiative equilibrium, we can easily integrate (2.59) to obtain

$$P = \frac{F}{c}(\tau + q), \tag{2.62}$$

where q is the constant of integration. It follows from (2.9) that the total pressure and total energy density of an isotropic radiation field are related by

$$P = \frac{1}{3}U. \tag{2.63}$$

We note from (2.49) that the radiation field becomes nearly isotropic as we go somewhat below the surface and (2.63) holds. Just underneath the surface, however, we do not expect (2.63) to hold. If we assume (2.63) to hold *everywhere*, then finding a full solution to our problem becomes straightforward. This is known as the *Eddington approximation* (Eddington, 1926, §226). Under this approximation, we can combine (2.57), (2.60), (2.62) and (2.63) to obtain

$$S = \frac{3F}{4\pi}(\tau + q). \tag{2.64}$$

We have seen from equations like (2.35) and (2.36) that the specific intensity can easily be written down if the source function is given. The main challenge is to obtain the solution for the source function (which depends on the temperature) at different depths along with the specific intensity. If we can evaluate the constant of integration q, then (2.64) will finally provide us with the solution for the source function for the grey atmosphere problem (albeit under the Eddington approximation). We now show below that q can be evaluated by calculating the flux from (2.64) and setting it equal to F.

Just as (2.33) can be solved to obtain (2.35) for the specific intensity in the upward direction, we can similarly write down the solution of (2.50):

$$I(\tau, \mu \geq 0) = \int_{\tau}^{\infty} S e^{-(t-\tau)/\mu} \frac{dt}{\mu}. \tag{2.65}$$

The specific intensity of radiation coming out of the stellar surface is obtained by setting $\tau = 0$ in this equation, i.e.

$$I(0, \mu) = \int_{0}^{\infty} S e^{-t/\mu} \frac{dt}{\mu}.$$

Substituting from (2.64), we get

$$I(0, \mu) = \frac{3F}{4\pi} \int_{0}^{\infty} (t + q) e^{-t/\mu} \frac{dt}{\mu} = \frac{3F}{4\pi} (\mu + q). \tag{2.66}$$

From (2.54), the flux coming out of the upper surface of the stellar atmosphere is

$$F = 2\pi \int_{0}^{1} I \mu \, d\mu.$$

On substituting for I from (2.66), we get

$$F = \frac{3F}{2} \left(\frac{1}{3} + \frac{q}{2} \right),$$

which gives the value of the constant of integration to be

$$q = \frac{2}{3}. \tag{2.67}$$

On putting this value of q in (2.64), the source function as a function of depth inside the stellar atmosphere is finally given by

$$S = \frac{3F}{4\pi} \left(\tau + \frac{2}{3} \right). \tag{2.68}$$

On making use of (2.57) and (2.60), we have from (2.68)

$$cU = 3F \left(\tau + \frac{2}{3} \right).$$

Using (2.46), (2.47) and (2.48), we get

$$T^4 = \frac{3}{4} T_{\text{eff}}^4 \left(\tau + \frac{2}{3} \right). \tag{2.69}$$

This equation tells us how temperature varies inside a grey atmosphere.

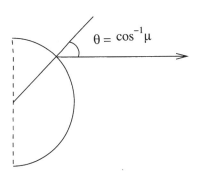

Fig. 2.5 A ray coming to an observer from the solar disk.

$$\theta = \cos^{-1}\mu$$

Finally we derive an important result for radiation coming out from the stellar surface. Substituting from (2.67) into (2.66), we have

$$I(0, \mu) = \frac{3F}{4\pi}\left(\mu + \frac{2}{3}\right),$$

which implies

$$\frac{I(0, \mu)}{I(0, 1)} = \frac{3}{5}\left(\mu + \frac{2}{3}\right). \tag{2.70}$$

This equation has a very important physical significance. Suppose we consider the intensity of radiation coming from different points on the disk of the Sun as seen by us. The ray coming from the central point of the solar disk emerges out of the solar surface in the vertical direction and the specific intensity for this ray will be $I(0, 1)$. On the other hand, the ray coming from an off-centre point must emerge from the solar surface at an angle $\theta = \cos^{-1}\mu$ with the vertical, as seen in Figure 2.5, and the corresponding specific intensity will be $I(0, \mu)$.

Hence (2.70) gives the variation of intensity on the solar disk as we move from the centre to the edge. In astronomical jargon, the region near the edge of the solar disk is referred to as the limb of the Sun. Therefore, a law giving the variation of intensity over the solar disk is called the *limb-darkening law*. The theoretical limb-darkening law predicts that the intensity at the edge of the solar disk will be about 40% of the intensity at the centre. In Figure 2.6, we show the observationally determined limb-darkening along with the plot of (2.70) obtained by the Eddington approximation as well as the theoretical limb-darkening law derived by an exact solution of the grey atmosphere problem (i.e. derived by solving the integro-differential equation (2.61) exactly). Although theory matches the observational data reasonably well, the discrepancy between the two is due to the fact that the solar atmosphere is not grey.

2.4.3 Formation of spectral lines

The grey atmosphere problem provides us with an example of how the radiative transfer equation can be solved consistently to give the source function along with the radiation field. As we have pointed out, one of the key problems of

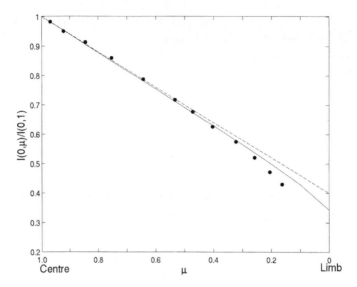

Fig. 2.6 The observed limb-darkening of the solar disk (indicated by dots) along with theoretical limb-darkening laws obtained by the Eddington approximation (dashed line) and by exact solution of the grey atmosphere problem (solid line). The observational data (indicated by dots) are for wavelength $\lambda = 5485$ Å as given by Pierce *et al.* (1950).

stellar atmospheres is a quantitative understanding of spectral line formation. The grey atmosphere problem does not throw any light directly on the problem of line formation. It is the constancy of the total radiation flux F which allowed us to integrate equations like (2.59), thus paving the way for a complete solution of the grey atmosphere problem. In the case of the general non-grey stellar atmosphere, we have frequency-dependent equations exactly analogous to (2.58) and (2.59). Those equations, however, cannot be solved in the same way as the equations of the grey atmosphere problem, since the flux F_ν associated with frequency ν is not in general a constant, even when the total flux F is a constant. If there is no source or sink in the stellar atmosphere, then a constant energy passes through the layers of stellar atmosphere, but the energy continuously gets redistributed amongst different frequencies. For example, in the interior of the Sun where the temperature is of order 10^7 K, the radiation field is mainly made up of X-ray photons. By the time the energy flux reaches the solar surface, it mainly consists of visible light. In a rigorous treatment of stellar atmospheres, it is also necessary to split the absorption coefficient α_ν into two parts: scattering and true absorption. In radiative equilibrium, true absorption is followed by complete re-emission of the absorbed radiation. One important difference between scattering and true absorption is that in scattering the frequency of radiation does not change, whereas in true absorption followed by re-emission the frequency changes. Also, Kirchhoff's law is applicable only in the case of true absorption. We refer the reader to Mihalas (1978) for a treatment of the radiative transfer problem by treating scattering and true absorption separately.

Although a proper treatment of the non-grey atmosphere is beyond the scope of our elementary presentation, we give some idea about the line formation problem. It follows from (2.38) that

$$I_\nu(0, 1) = B_\nu(\tau_\nu = 0) + \frac{dB_\nu}{d\tau_\nu}.$$

If we expand $B_\nu(\tau_\nu = 1)$ in a Taylor series around $\tau_\nu = 0$ and keep only the linear term, then it becomes equal to the right-hand side of the above equation. Hence

$$I_\nu(0, 1) \approx B_\nu(\tau_\nu = 1). \qquad (2.71)$$

This very important equation tells us that the specific intensity of radiation at a frequency ν coming out of a stellar atmosphere is approximately equal to the Planck function at a depth of the atmosphere where the optical depth for that frequency ν equals unity. We now show how (2.71) can be used to explain the formation of spectral lines.

Let us consider an idealized situation that the absorption coefficient in the outer layers of a stellar atmosphere is equal to α_C at all frequencies except a narrow frequency range around ν_L where it has a larger value α_L. This is sketched in Figure 2.7(a). We now use (2.71) to find the spectrum of the radiation emerging out of this atmosphere. For frequencies in the continuum outside the spectral line, the optical depth becomes unity at a depth α_C^{-1}. If the temperature there is T_C, then the spectrum in the continuum region will be like the blackbody spectrum $B_\nu(T_C)$. For frequencies within the spectral line, the optical depth becomes unity at a shallower depth α_L^{-1}, where the temperature must have a lower value T_L. Figure 2.7(b) shows both the functions $B_\nu(T_C)$ and $B_\nu(T_L)$. Since the specific intensity in the continuum is given by $B_\nu(T_C)$ and the specific intensity in the spectral line by $B_\nu(T_L)$, the full spectrum looks as indicated by the dark line in Figure 2.7(b). We saw in §2.2.4 that the spectrum of radiation coming out of an object with strictly uniform temperature inside is pure blackbody radiation. Since any object radiating from the surface is expected to have a temperature gradient in the layers underneath the surface, we conclude that the existence of spectral lines should be a very common occurrence.

One of the aims of stellar atmosphere studies is to estimate the abundances of various elements in an atmosphere from a spectroscopic analysis. Suppose an element has a spectral line at frequency ν_L. Then, the larger the number of atoms of that element per unit volume, the higher will be the value of the absorption coefficient α_L at that frequency and consequently the spectral line will be stronger. Thus, from the strengths of spectral lines, it is possible to calculate the abundances of elements. Although it is beyond the scope of this book to discuss the quantitative details of this subject, we shall give some basic ideas of spectral line analysis in §2.7.

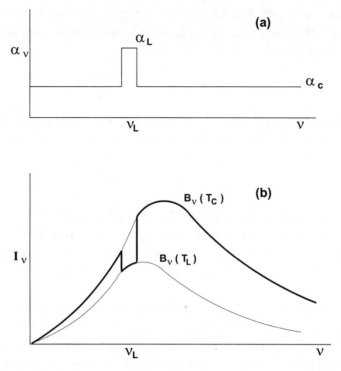

Fig. 2.7 For an absorption coefficient indicated in (a), the emergent spectrum is indicated in (b) by the dark line. The two blackbody spectra shown in (b) correspond to the temperatures T_C (upper curve) and T_L (lower curve) explained in the text.

2.5 Radiative energy transport in the stellar interior

In §2.4, we have discussed radiative transfer in the outer layers of a star. Astrophysicists studying stellar interiors have to consider radiative transfer in the stellar interior as well. In a typical star, energy is usually produced by nuclear reactions in the innermost core of the star. This energy in the form of radiation is then transported outward. We shall see later in §3.2.4 that sometimes convection can transport energy in a stellar interior. For the time being, let us consider a stellar interior in which energy is transported outward by radiative transfer. In the study of stellar atmospheres, one has to bother about the distribution of energy in different wavelengths, since the ultimate goal is to understand the spectrum of radiation coming out of the stellar atmosphere. While studying stellar interiors, however, one is primarily interested in finding out how an energy flux is driven outward by the gradient of the radiation field. While discussing the grey atmosphere problem, we had derived (2.59) relating the flux of radiation to the gradient of radiation pressure. Can this equation be applied to stellar interiors even in situations when α_ν varies with ν and the grey atmosphere assumption does not hold?

We now want to show that (2.59) holds even for a non-grey atmosphere if an average of α_ν over ν is taken in a suitable way. In a non-grey situation, we shall have the frequency-dependent version of (2.59):

$$\frac{dP_\nu}{d\tau_\nu} = \frac{F_\nu}{c},\tag{2.72}$$

which can be obtained from (2.33) in exactly the same way (2.59) was obtained from (2.50). From (2.72), it follows that

$$F_\nu = -\frac{c}{\alpha_\nu}\frac{dP_\nu}{dz}$$

on making use of (2.32). Integrating over all frequencies, the total radiation flux is

$$F = \int F_\nu d\nu = -c \int \frac{1}{\alpha_\nu}\frac{dP_\nu}{dz}d\nu.\tag{2.73}$$

We now want F to satisfy an equation of the form (2.59), i.e.

$$F = -c\frac{1}{\alpha_R}\frac{dP}{dz},\tag{2.74}$$

where α_R is a suitable average of α_ν. To figure out how this averaging has to be done, we need to equate (2.73) and (2.74), which gives

$$\frac{1}{\alpha_R} = \frac{\int \frac{1}{\alpha_\nu}\frac{dP_\nu}{dz}d\nu}{\int \frac{dP_\nu}{dz}d\nu}.\tag{2.75}$$

Now P_ν is proportional to the Planck function B_ν as seen from (2.44). So we can write

$$\frac{dP_\nu}{dz} = \frac{4\pi}{3c}\frac{\partial B_\nu}{\partial T}\frac{dT}{dz}.$$

We substitute this both in the numerator and denominator of (2.75), and cancel out dT/dz. This finally gives

$$\frac{1}{\alpha_R} = \frac{\int \frac{1}{\alpha_\nu}\frac{\partial B_\nu}{\partial T}d\nu}{\int \frac{\partial B_\nu}{\partial T}d\nu}.\tag{2.76}$$

The mean absorption coefficient α_R defined in this way is known as the *Rosseland mean* (Rosseland, 1924). When α_R is defined in this way, the flux of radiant energy is given by (2.74). We often write

$$\alpha_R = \rho\chi,\tag{2.77}$$

where χ is known as the *opacity* of the stellar matter. On using (2.46), (2.63) and (2.77), we can put (2.74) in the form

$$F = -\frac{c}{\chi\rho}\frac{d}{dz}\left(\frac{a_B}{3}T^4\right).\tag{2.78}$$

As we shall see in the next chapter, (2.78) is one of the fundamental equations for studying stellar interiors, which was first derived by Eddington (1916).

2.6 Calculation of opacity

To build a model of the stellar interior, it is necessary to solve a slightly modified version of (2.78) as discussed in the §3.2.3. To solve this equation, we need to know the value of opacity χ. The gas in the interior of a star exists under such conditions of temperature and pressure which cannot be easily reproduced in the laboratory. Hence we cannot experimentally find out χ for conditions appropriate for the stellar interior. The opacity χ, therefore, has to be calculated theoretically. This is a fairly complicated calculation. With improvements in stellar models, more and more accurate computations of opacity are demanded. This has become a highly specialized and technical subject, with very few groups in the world who have the right expertise for calculating opacity accurately. Other scientists who need values of opacity for their research almost never try to calculate the opacity themselves, but use the values computed by the groups who specialize in these computations. For several decades, the so-called Los Alamos opacity tables (Cox and Stewart, 1970) remained the last word on this subject. There is no point in discussing details of opacity calculation methods here. We summarize below only some of the main ideas. For a clear discussion of the quantum mechanical principles involved in opacity calculations, the readers are referred to Clayton (1983, §3–3).

Suppose we have a gas of a certain composition kept at a given density and temperature. We want to calculate its opacity theoretically. We can apply the Boltzmann law (2.28) and the Saha equation (2.29) to find out the numbers of atoms and electrons in various energy levels and in various stages of ionization. When electromagnetic radiation of frequency ν impinges on the system, atoms can absorb this radiation if electrons associated with the atoms are pushed to levels which have energies higher by an amount $h\nu$ compared to previous levels. We know from quantum mechanics that atomic energy levels can be either bound (discrete levels) or free (continuum). Hence the absorption of radiation by an atom can be due to three kinds of upward electronic transitions: (i) bound-bound, (ii) bound-free and (iii) free-free. One can apply Fermi's golden rule of quantum mechanics with a semiclassical treatment of radiation to calculate the absorption cross-sections for these processes (see, for example, Mihalas, 1978, §4–2; Clayton, 1983, §3–3). Finally one adds up the cross-sections for all atoms and electrons at different excitation and ionization levels present in a unit volume. After including the effect of stimulated emission (which is discussed in §6.6), one gets the absorption coefficient α_ν, from which the opacity is obtained by applying (2.76) and (2.77).

If certain approximations are made, then it can be shown that both bound-free and free-free transitions (which are the dominant processes for the opacity) lead to an opacity which varies with density ρ and temperature T in the following way:

$$\chi \propto \frac{\rho}{T^{3.5}}.$$ (2.79)

This is called Kramers's law, after Kramers (1923) who arrived at this law while studying the absorption of X-rays by matter. This approximate law certainly could not be true for all temperatures. For example, when the temperature is sufficiently low, most of the atoms will be in their lowest energy levels. In such a situation, it will be possible for radiation to be absorbed only if there are sufficiently energetic photons to knock off electrons from these lowest energy levels. Since the radiation falling on the system is very close to blackbody radiation and since blackbody radiation at a low temperature will not have many energetic photons to knock off the atomic electrons from the lowest levels, we conclude that not much radiation will be absorbed. Hence opacity is expected to drop at low temperatures and to depart from Kramers's law.

Figure 2.8 gives the opacity of material of solar composition, based on detailed calculations. Each curve is for a definite density and shows how the

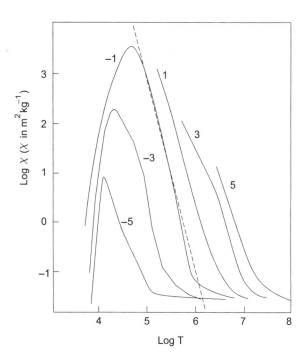

Fig. 2.8 Opacity of material of solar composition as a function of temperature. Different curves correspond to different densities, with the values of $\log \rho$ (ρ in kg m^{-3}) indicated next to the curves. The dashed line indicates the slope that would result if opacity varied as $T^{-3.5}$ for a fixed density. Adapted from Tayler (1994, p. 101).

opacity varies with temperature for that density. As we expect, the opacity is negligible at low temperatures. When the temperature is about a few thousand degrees, the opacity is maximum. The curves for higher densities lie higher, as expected from Kramers's law (2.79). On the right side of the peak, we find that the opacity falls sharply with temperature and a $T^{-3.5}$ dependence in accordance with Kramers's law (indicated by the dashed line) is not a bad fit for this. However, Kramers's law would suggest that opacity should keep on going down with temperature and should be very small at high temperatures. But that does not seem to be happening. At high temperature, the opacity seems to become independent of density and reaches an asymptotic value. We now turn to an explanation for this.

2.6.1 Thomson scattering

At sufficiently high temperatures, many atoms in a gas become ionized and there is a supply of free electrons. It is well known that a free electron can scatter radiation by a process called *Thomson scattering* (Thomson, 1906), which turns out to be extremely important in many astrophysical processes. Since many advanced textbooks on electrodynamics present a derivation of the Thomson scattering cross-section (see, for example, Panofsky and Phillips, 1962, §22-2–22-4; Rybicki and Lightman, §3.4–3.6), we merely quote the result without reproducing the derivation here.

Suppose an electromagnetic wave of frequency ω falls on an electron bound to an atom by spring constant $m_e\omega_0^2$, where m_e is the mass of the electron. The equation of motion of the electron subject to an electric field $\mathbf{E}$ is

$$m_e\left(\frac{d^2\mathbf{x}}{dt^2} + \gamma\frac{d\mathbf{x}}{dt} + \omega_0^2\mathbf{x}\right) = -e\mathbf{E},$$

where γ is the damping constant. The electric field of the electromagnetic wave will force the electron to undergo an oscillatory motion. We know that a charge in an oscillatory motion emits electromagnetic waves. The energy of this emitted wave must come from the energy of the incident electromagnetic wave. In other words, some energy of the incident wave gets scattered in other directions. A completely classical treatment shows that the scattering cross-section is given by

$$\sigma = \frac{8\pi}{3}\left(\frac{e^2}{4\pi\epsilon_0 m_e c^2}\right)^2 \frac{\omega^4}{(\omega^2 - \omega_0^2)^2 + \gamma^2\omega^2}. \tag{2.80}$$

When the frequency of the incident electromagnetic wave is very high ($\omega \gg \omega_0, \gamma$), the electron is forced to move like a free electron and the cross-section reduces to the Thomson cross-section for free electrons, which is

$$\sigma_T = \frac{8\pi}{3} \left(\frac{e^2}{4\pi \epsilon_0 m_e c^2} \right)^2. \tag{2.81}$$

Before discussing the contribution of Thomson scattering to the opacity, let us briefly consider the other limit of the electron being tightly bound to the atom ($\omega_0 \gg \omega, \gamma$). In that limit, (2.80) reduces to

$$\sigma_R = \sigma_T \left(\frac{\omega}{\omega_0} \right)^4. \tag{2.82}$$

This is the celebrated *Rayleigh scattering*, in which the cross-section goes as ω^4 or as λ^{-4}, where λ is the wavelength of the incident electromagnetic wave. Rayleigh scattering provides explanations for many natural as well as astronomical phenomena. In the visible spectrum, blue light is scattered more than red light because the wavelength of blue light is shorter. This explains why the setting Sun looks reddish. The rays of the setting Sun have to traverse through a larger distance of the atmosphere, where blue light is selectively scattered away, leaving more red light in the beam compared to the blue light. On the other hand, the daytime sky looks blue because the dust particles in the sky scatter more blue colour from the sunlight into our eyes. When starlight passes through interstellar dust, it also becomes redder due to the selective scattering of blue light by the dust particles. However, as we shall discuss in §6.1.3, the interstellar extinction of starlight seems to go as λ^{-1} rather than λ^{-4}.

On substituting the values of different fundamental quantities in (2.81), the numerical value of the Thomson cross-section is found to be

$$\sigma_T = 6.65 \times 10^{-29} \text{ m}^2. \tag{2.83}$$

If there are n_e free electrons per unit volume, then the 'absorption coefficient' due to Thomson scattering is $n_e \sigma_T$ (remember that scattering is not true absorption as we briefly point out in §2.4.3 and there is no corresponding emission coefficient satisfying (2.26)). Hence, by (2.77), the opacity χ_T due to Thomson scattering is given by

$$\chi_T = \frac{n_e}{\rho} \sigma_T. \tag{2.84}$$

At sufficiently high temperatures when a gas is fully ionized, n_e is proportional to the density and n_e/ρ depends on the composition alone. For example, for fully ionized hydrogen, n_e/ρ is equal to $1/m_H$ (where m_H is the mass of hydrogen atom) so that from (2.83) and (2.84) we conclude that the opacity for fully ionized hydrogen is

$$\chi_T = 3.98 \times 10^{-2} \text{ m}^2 \text{ kg}^{-1} \tag{2.85}$$

if the temperature is sufficiently high to make other sources of opacity unimportant.

While computating opacity, the contribution due to Thomson scattering is added to the contributions from bound-free and free-free transitions (keeping in mind that there is no stimulated emission associated with Thomson scattering). However, when Thomson scattering is present in an atmosphere, some special care has to be taken while solving the radiative transfer equation, since Kirchhoff's law (2.26) will not hold for Thomson scattering. It should be clear from Figure 2.8 that at the typical temperatures of stellar surfaces Thomson scattering should contribute very little to the opacity. Hence, while studying radiative transfer through stellar atmospheres, one can usually neglect Thomson scattering and take (2.26) to be fully valid.

Readers are urged to work out Exercise 2.7 to get a feeling about the role of Thomson scattering in making a gas opaque. The air around us is transparent only because all the electrons are locked inside atoms. If all the atomic electrons were to come out of atoms, then air would be opaque in a few metres. Apart from stellar interiors, Thomson scattering plays a very important role in the early Universe. When all matter in the early Universe was ionized (due to the high temperature), matter was sufficiently opaque to keep the matter and radiation coupled together. Once the temperature fell with the expansion of the Universe and atoms formed, locking up the free electrons inside them, the Universe suddenly became transparent. We shall discuss the consequences of this in §11.7.

2.6.2 Negative hydrogen ion

The temperature of the solar surface is about 6000 K. It appears that the solar surface is sufficiently opaque and we cannot see anything underneath it. One important question is what makes the solar gases so opaque at a temperature of 6000 K? At that temperature, hydrogen and helium atoms (which are the most abundant atoms) are not ionized and mostly occupy the lowest energy levels. To force transitions to higher energy levels, one needs photons having energy of the order of a few eV. Blackbody radiation at 6000 K does not have enough photons with such energy. So, at first sight, it seems that matter at 6000 K should not be able to absorb radiation and should be transparent. It mystified astrophysicists for some time as to what causes the opacity of the solar surface, until a clever idea was suggested by Wildt (1939) and later confirmed by Chandrasekhar and Breen (1946) through detailed calculations. The electron inside a hydrogen atom is not able to screen the electrostatic force of the nucleus completely. So it is possible for the hydrogen atom to attract an additional electron and form a loosely bound negative ion H^-. The binding energy of the negative hydrogen ion is only about 0.75 eV – much smaller than the ionization energy of 13.6 eV. So blackbody radiation at 6000 K has enough photons to knock off this loosely bound electron and can get absorbed in this process. It is estimated that there are enough negative hydrogen ions at the solar surface and that they are providing the opacity there.

2.7 Analysis of spectral lines

In §2.4.3 we have given a qualitative idea of how spectral lines form. Astronomers, however, require a quantitative theory of spectral lines in order to analyse them to determine the composition of the source. A quantitative theory of spectral lines in a stellar atmosphere involves certain difficulties because we need to consider both absorption and emission at spectral lines in the outer layers of the star. A simpler problem is to consider the passage of radiation through a medium which absorbs only at a spectral line and does not emit. We shall now present an analysis of this simpler problem to give an idea of this subject. Even this simpler problem is often of considerable practical relevance. For example, we may consider the passage of visible light from a star through the interstellar medium. Since parts of the interstellar medium are made up of gas having fairly low temperatures like $100\,\mathrm{K}$ (see §6.6), there is negligible emission of visible light from this gas which may absorb starlight at particular frequencies. Since this gas is cold, it produces spectral lines which are much narrower than typical spectral lines produced in the stellar atmosphere. The extreme narrowness of a line in a stellar spectrum is indicative that it is produced during the passage of light through the interstellar medium rather than in the stellar atmosphere.

Let n be the number density of atoms of a certain kind in the absorbing medium having energy levels differing by $h\nu_0$. We expect these atoms to absorb at the frequency ν_0 and produce a spectral line. It is customary to write the absorption cross-section of the atom as

$$\sigma = \frac{e^2}{4\epsilon_0 m_e c} f,\qquad(2.86)$$

where f is called the *oscillator strength*. Each spectral line will be characterized by an oscillator strength f. The larger the value of f, the stronger the spectral line is expected to be. We also expect the absorption coefficient to have a normalized profile $\phi(\Delta\nu)$ where $\Delta\nu$ is the departure of the frequency from the line centre at ν_0 and $\int \phi(\Delta\nu)\,d\nu = 1$. Then the absorption coefficient is given by

$$\alpha_\nu = n\sigma\phi(\Delta\nu) = \frac{e^2}{4\epsilon_0 m_e c} n f \phi(\Delta\nu)$$

so that the optical depth through the absorbing medium, as given by (2.16), is

$$\tau_\nu = \frac{e^2}{4\epsilon_0 m_e c} N f \phi(\Delta\nu),\qquad(2.87)$$

where $N = \int n\,ds$ is the column density of the atoms along the line of sight through the absorbing medium. As we shall see in §6.6, one has to subtract the effect of induced emission in a full calculation of the absorption coefficient. For visible light passing through a gas at temperature of order 100 K, the induced emission is negligible (because of the very low population of the upper level) and we do not consider it here. If we assume that there is no emission in the medium, then the intensity is given by (2.17). The intensity I_c of the continuum just outside the spectral line will be equal to $I_\nu(0)$ appearing in (2.17). Hence we can write (2.17) as

$$I_\nu = I_c e^{-\tau_\nu} \tag{2.88}$$

with τ_ν given by (2.87).

It is clear that $(I_c - I_\nu)/I_c$ is the fractional dip in intensity at some frequency ν inside the spectral line. We can get an estimate of the strength of the spectral line by integrating this fractional dip over the spectral line. This is called the *equivalent width* of the spectral line, defined as

$$W_\lambda = \int \frac{I_c - I_\nu}{I_c}\, d\lambda. \tag{2.89}$$

On using (2.88) and changing the integration variable from λ to ν, we get

$$W_\lambda = \frac{\lambda^2}{c} \int [1 - e^{-\tau_\nu}]\, d\nu, \tag{2.90}$$

where λ is the wavelength of the spectral line which is taken outside the integral because it does not vary much over the spectral line.

Certain simplifications are possible if the spectral line is weak, when we can take $e^{-\tau_\nu} \approx 1 - \tau_\nu$ so that (2.90) becomes

$$W_\lambda = \frac{\lambda^2}{c} \int \tau_\nu\, d\nu.$$

Substituting from (2.87) and remembering that $\phi(\Delta\nu)$ is normalized, we get

$$\frac{W_\lambda}{\lambda} = \frac{e^2}{4\epsilon_0 m_e c^2} N f \lambda. \tag{2.91}$$

For a weak spectral line of which we know the oscillator strength f, we can use (2.91) to determine the column density N of the atoms producing the spectral line when we have a measurement of the equivalent width W_λ.

Suppose the absorbing medium has certain atoms producing several spectral lines with different oscillator strengths f. It follows from (2.91) that W_λ/λ will be proportional to $N f \lambda$ for weak spectral lines. Even if the spectral lines are not weak, we can plot W_λ/λ as a function of $N f \lambda$ for all the spectral lines. Such a plot is shown in Figure 2.9. The curve passing through the data points is called the *curve of growth*. The left side of the curve of growth shows a linearly increasing regime corresponding to weak spectral lines for which we have the

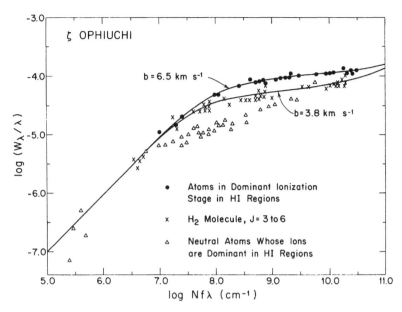

Fig. 2.9 Equivalent widths of various spectral lines produced in the spectrum of the star ζ Ophiuchi by absorption in the interstellar medium, plotted against $Nf\lambda$. The curves of growth for hydrogen atoms and hydrogen molecules are shown. From Spitzer and Jenkins (1975). (©Annual Reviews Inc. Reproduced with permission from *Annual Reviews of Astronomy and Astrophysics*.)

proportionality to $Nf\lambda$. For stronger spectral lines, the curve of growth saturates to a horizontal plateau. The reason behind this saturation is that the fractional dip $(I_c - I_\nu)/I_c$ appearing in (2.89) can never be more than 1, no matter how strong the spectral line is. Eventually, for very strong spectral lines, the curve of growth again shows a tendency of rising because very strong spectral lines have some absorption in the wings (i.e. two sides of the core of a spectral line) which are not saturated.

2.8   Photon diffusion inside the Sun

We close our discussion on the interaction of radiation with matter by working out a curious example. Suppose the energy generation rate at the centre of the Sun were to increase or decrease suddenly due to some reason. We expect that eventually the surface of the Sun will become brighter or dimmer as a consequence of this. How much time will it take before the effect of this sudden change at the centre becomes visible at the surface?

The photons created at the centre of the Sun interact with the neighbouring atoms. Atoms which have absorbed photons will de-excite by giving out photons. In this process, photons diffuse from the centre of the Sun towards

the surface. The absorption and re-emission of photons by atoms can be far
from simple. The atom may spend some time in the excited state and when
it de-excites, it may not come back to exactly the same state in which it was
originally in. As a result, the emitted photon may have a frequency different
from what was the frequency of the absorbed photon. This is necessary because
the initial photons at the centre at temperature of order 10^7 would typically be
X-ray photons, whereas the photons which reach the outer surface are more
likely to be photons of visible light. We now make a rough estimate of the
diffusion time by making a simplifying assumption that photons merely do a
random walk through stellar matter where an encounter with an atom simply
changes the direction of flight of the photon.

Let us first make an estimate of the mean free path of photons between
encounters with atoms. As pointed out in §2.3.2, the inverse of the absorption
coefficient gives this mean free path. So we can take $\alpha_R^{-1} = (\rho \chi)^{-1}$ to be the
mean free path. While this mean free path would be a function of radius, we
simplify our life further by using an approximate average value. It is seen from
Figure 2.8 that $10^{-1}\,\mathrm{m}^2\,\mathrm{kg}^{-1}$ would be an appropriate value for χ to use in the
solar interior. Taking an average density of order $10^3\,\mathrm{kg\,m}^{-3}$, we get a mean
free path of about 1 cm.

Suppose an average photon has to take N steps to diffuse from the centre to
the surface. If $\mathbf{l}_1, \mathbf{l}_2, \ldots, \mathbf{l}_N$ are the displacements in these steps, then the total
displacement is

$$\mathbf{L} = \mathbf{l}_1 + \mathbf{l}_2 + \cdots + \mathbf{l}_N.$$

On squaring and averaging both sides, we would have

$$\langle L^2 \rangle = \langle l_1^2 \rangle + \langle l_2^2 \rangle + \cdots + \langle l_N^2 \rangle, \tag{2.92}$$

since it is obvious that the cross-terms will give zero on averaging over different
photons. As we are making the simplifying assumption that all the steps are
equal, (2.92) becomes

$$\langle L^2 \rangle = N \langle l^2 \rangle.$$

Taking $l = 1\,\mathrm{cm}$ and L to be equal to the solar radius, N turns out to be of
order 10^{22}. With a step size of 1 cm, an average photon would travel over a
distance 10^{20} m in order to escape from the centre to the surface. Dividing this
by the speed of light, we get a diffusion time of order 10^4 years. A more careful
calculation shows that the photon diffusion time inside the Sun is actually a few
times larger than this.

If the energy generation rate at the centre of the Sun were to change
suddenly, that information will take tens of thousands years to reach the surface.
The sunlight that we receive today was created by nuclear reactions at the centre
of the Sun at a time when our ancestors were fighting woolly mammoths and
sabre-toothed tigers.

Exercises

2.1 Assuming that the spectrum of blackbody radiation is given by Planck's law (2.1), prove the following.

(a) Show that the total energy density of blackbody at temperature T is given by

$$U = a_B T^4,$$

where

$$a_B = \frac{8\pi\kappa_B^4}{c^3 h^3} \int_0^\infty \frac{x^3 dx}{e^x - 1}.$$

(Note: this integral can be evaluated exactly and can be shown to be equal to $\pi^4/15$.)

(b) Show that the total energy radiated in unit time from unit area on the surface of a blackbody is given by σT^4, where

$$\sigma = \frac{c a_B}{4}.$$

(c) Show that the frequency ν_{max} at which the energy density U_ν is maximum is given by

$$\frac{\nu_{max}}{T} = 5.88 \times 10^{10} \text{ Hz K}^{-1}.$$

2.2 Consider a 'pinhole camera' having a small circular hole of diameter d in its front and having a 'film plane' at a distance L behind it (see Figure 2.10). Show that the flux F_ν at the film plane is related to the incident intensity $I_\nu(\theta, \phi)$ in the following way

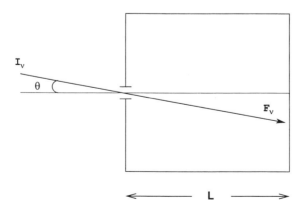

Fig. 2.10 See Exercise 2.2.

$$F_v = \frac{\pi \cos^4 \theta}{4 f^2} I_v(\theta, \phi),$$

where $f = L/d$ is the 'focal ratio'.

2.3 Consider hydrogen gas having the same density as the density of air under normal temperature and pressure ($\rho = 1.29 \text{ kg m}^{-3}$). Given the fact that the ionization potential χ of hydrogen is 13.6 eV, use the Saha equation (2.29) to calculate the fraction of ionization x at different temperatures T and make a plot of x versus T.

2.4 Find out the specific intensity $I(\tau, \mu > 0)$ at an arbitrary optical depth τ inside a plane-parallel grey atmosphere obeying the Eddington approximation. You may assume a constant energy flux F passing through the atmosphere.

2.5 Consider a spherical cloud of gas with a radius R and a constant inside temperature T far away from the observer. (a) Assuming the cloud to be optically thin, find out how the brightness seen by the observer would vary as a function of distance b from the cloud centre. (b) What is the overall effective temperature of the cloud surface? (c) How will the answers to (a)–(b) be modified if the cloud were optically thick?

2.6 How will you calculate the spectrum of radiation emerging from a grey atmosphere assuming Eddington approximation? Those of you who are comfortable with numerical computations may like to write a small computer program to compute $I_v(0, 1)$ as a function of v and plot it.

Suppose G shows the spectrum from a grey atmosphere and R the spectrum from a real atmosphere (see Figure 2.11). What can you say about the variation of α_v with frequency?

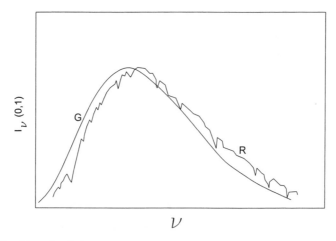

Fig. 2.11 See Exercise 2.6.

2.7 Consider an atmosphere of completely ionized hydrogen having the same density as the density of the Earth's atmosphere. Using the fact that a beam of light passing through this atmosphere will be attenuated due to Thomson scattering by free electrons, calculate the path length which this beam has to traverse before its intensity is reduced to half its original strength. (This problem should give you an idea of why the matter-radiation decoupling to be discussed in §11.7 took place after the number of free electrons was reduced due to the formation of atoms.)

3

Stellar astrophysics I: Basic theoretical ideas and observational data

3.1 Introduction

At the beginning of §2.4, we pointed out the scope of the subject *stellar interior*. It appears from observational data (to be discussed in detail later) that various quantities pertaining to stars have some relations amongst each other. For example, a more massive star usually has a higher luminosity and also a higher surface temperature. To explain such observed relations theoretically, we have to figure out the equations which should hold inside a star and then solve them to construct models of stellar structure.

The years $\approx$1920–1940 constituted the golden period of research in this field, when theoretical developments led to elegant explanations of a vast mass of observational data pertaining to stars. Ever since that time, the subject of stellar interior or stellar structure has remained a cornerstone of modern astrophysics and improved computational powers have led to more detailed models. This is a subject in which theory and observations are intimately combined together to build up an imposing edifice. While presenting a subject like this, the first question that a teacher or a writer has to face is this: from a purely pedagogical point of view, is it better to start with a discussion of observational data or with a discussion of basic theoretical ideas?

It follows from simple theoretical considerations that there must be objects like stars, provided energy can be generated by some mechanism in the central regions. We need not know the details of the energy generation mechanism to make this prediction. Eddington, who played the leading role in establishing the theoretical discipline of stellar structure, has imagined a physicist on a cloud-bound planet, who has never seen stars but makes theoretical predictions of stars on the basis of his calculations (Eddington, 1926, p. 16). Then one day the veil of cloud is removed and the physicist is able to look at the stars he has predicted. Although important trends in observational data were discovered before their theoretical explanation and, in fact, provided a motivation for developing the theory, here we shall proceed somewhat like Eddington's physicist on a cloud-bound planet. First we shall discuss some of the basic theoretical ideas. Then we shall present the observations and discuss whether our theoretical results are confirmed by observations. Although some of the patterns in the observational data can be explained by very simple theoretical considerations, we shall see that it is necessary to delve deeper into theory to have a more complete picture. After familiarizing ourselves with observational data in the later parts of this chapter, we shall return to some of the deeper theoretical issues in the next chapter.

3.2   Basic equations of stellar structure

We now establish the basic equations of stellar structure by assuming the star to be spherically symmetric. If the star is rotating sufficiently rapidly, then there will be some flattening in the direction of the rotation axis. Again, if the star has strong magnetic fields, that can be another cause of departure from spherical symmetry. Such complications are neglected in the first treatment of stellar structure. When we look at our own Sun, we find spherical symmetry to be a fairly good approximation. The rotational flattening of the Sun is negligible. Although the solar corona is highly non-spherical due to the presence of magnetic fields, the magnetic fields are not strong enough to cause departures from spherical symmetry below the surface of the Sun.

3.2.1   Hydrostatic equilibrium in stars

Let M_r be the mass inside the radius r of a star. Then the mass inside radius $r + dr$ should be $M_r + dM_r$, which means that dM_r is the mass of the spherical shell between radii r and $r + dr$. If ρ is the density at radius r, then the mass of this shell is $\rho \times 4\pi r^2 dr$, i.e.

$$dM_r = 4\pi r^2 \rho \, dr,$$

from which

$$\frac{dM_r}{dr} = 4\pi r^2 \rho. \tag{3.1}$$

This is the first of our stellar structure equations.

Let us now consider a small portion of the shell between r and $r + dr$. If dA is the transverse area of this small element, the forces exerted by pressure acting on its inward and outward surfaces are $P\, dA$ and $-(P + dP)\, dA$, where P and $P + dP$ are respectively the pressures at radii r and $r + dr$. So the net force arising out of pressure is $-dP\, dA$, which should be balanced by gravity under equilibrium conditions. The gravitational field at r is caused by the mass M_r inside r and is equal to $-GM_r/r^2$. Since the mass of the small element under consideration is $\rho\, dr\, dA$, the force balance condition for it is

$$-dP\, dA - \frac{GM_r}{r^2}\rho\, dr\, dA = 0,$$

from which

$$\frac{dP}{dr} = -\frac{GM_r}{r^2}\rho. \tag{3.2}$$

This is the second of the stellar structure equations.

A look at (3.1) and (3.2) will show that they involve three variable functions of the radial coordinate r: M_r, ρ and P. Certainly two equations are not enough to solve for three variable functions. We shall see in Chapter 5 that there are special kinds of dense stars like white dwarfs and neutron stars inside which pressure becomes a function of density alone. In such cases, the number of independent variables becomes two rather than three, and the above equations (3.1) and (3.2) can be solved to find the stellar structure. In normal stars, however, the stellar material behaves very much like a perfect gas, and pressure is a function of both density and temperature, having the form $P \propto \rho T$. In such a situation, we need additional equations for temperature and energy generation to solve the stellar structure. These additional equations will be derived in §3.2.3.

Central pressure and temperature of the Sun

Although the hydrostatic equilibrium equation (3.2) does not tell us the whole story about the stellar interior, we now show that it can nevertheless provide us with valuable clues about the interior conditions of stars. In the astrophysical Universe, we often have to deal with quantities for whose magnitudes we have no a priori feeling. For example, what are the temperature T_c and pressure P_c at the centre of the Sun? Nothing from our everyday life gives even a clue for the values of these quantities. So even an order of magnitude estimate correct within a factor of 10 is an important first step. We now show that (3.2) allows us to make an approximate estimate of P_c and T_c. Throughout this book, we shall again and again make such order of magnitude estimates of various astrophysical quantities without solving the equations exactly. For various order of magnitude estimates involving stars, we shall use the following approximate values of solar luminosity $L_\odot$ and solar radius $R_\odot$:

$$L_\odot \approx 4 \times 10^{26} \text{ W}, \tag{3.3}$$

$$R_\odot \approx 7 \times 10^{8} \text{ m}. \tag{3.4}$$

Their accurate values are given in Appendix A. Some of the other quantities needed in order of magnitude estimates are listed in Table 1.1.

For the purpose of an order of magnitude estimate, the derivative dP/dr can be replaced by $-P_c/R_\odot$. The various quantities on the right-hand side of (3.2) have to be replaced by their appropriate averages. Taking $M_\odot/2$ and $R_\odot/2$ to be the averages of M_r and r, (3.2) reduces to the approximate equation

$$\frac{P_c}{R_\odot} \approx \frac{G(M_\odot/2)}{(R_\odot/2)^2}\left(\frac{M_\odot}{\frac{4}{3}\pi R_\odot^3}\right).$$

On substituting the values of $M_\odot$ and $R_\odot$, we find

$$P_c \approx 6 \times 10^{14} \text{ N m}^{-2}. \tag{3.5}$$

Since the gas inside the Sun behaves very much like a perfect gas, we can use $P = n\kappa_B T$, where n is the number density of gas particles. Assuming the gas to contain hydrogen predominantly, the number of atoms per unit volume is ρ/m_H. Since hydrogen is completely ionized in the deep solar interior and each hydrogen atom contributes two particles (a proton and an electron), we have $n = 2\rho/m_H$ so that

$$P = \frac{2\kappa_B}{m_H}\rho T.$$

If we take the central density to be about twice the mean density, then at the centre of the Sun

$$P_c = \frac{4\kappa_B}{m_H}\left(\frac{M_\odot}{\frac{4}{3}\pi R_\odot^3}\right)T_c.$$

On taking the value of P_c from (3.5) and substituting the values of other quantities, we obtain

$$T_c \approx 10^7 \text{ K}. \tag{3.6}$$

Thus we have estimated the values of central pressure and temperature of the Sun in a relatively painless way. These values compare quite favourably with the values which one obtains from a detailed solution of all stellar structure equations. This example should illustrate the power of an order of magnitude estimate, which is such a favourite tool of the working astrophysicist!

3.2.2 Virial theorem for stars

In ordinary stars like the Sun, the inward gravitational pull is balanced by the excess pressure of the hot interior. In other words, it is the thermal energy of the interior which balances gravity. We, therefore, expect that the total thermal energy should be of the same order as the total gravitational energy. This can be rigorously established from the hydrostatic equilibrium equation (3.2). We multiply both sides of (3.2) by $4\pi r^3$ and then integrate from the centre of the star to its outer radius R. This gives

$$\int_0^R \frac{dP}{dr} 4\pi r^3 dr = \int_0^R \left(-\frac{GM_r}{r^2}\rho\right) 4\pi r^3 dr.$$

The left-hand side can be easily integrated by parts, leading to

$$-\int_0^R 3P \times 4\pi r^2 dr = \int_0^R \left(-\frac{GM_r}{r}\right) 4\pi r^2 \rho \, dr. \tag{3.7}$$

The right-hand side is clearly the total gravitational energy E_G of the star, i.e.

$$E_G = \int_0^R \left(-\frac{GM_r}{r}\right) 4\pi \rho \, r^2 dr. \tag{3.8}$$

Since $(3/2)\kappa_B T$ is the mean energy of thermal motion per particle in a region of temperature T and hence $(3/2)n\kappa_B T$ is the thermal energy per unit volume, the total thermal energy of the star is given by

$$E_T = \int_0^R \frac{3}{2} n\kappa_B T \times 4\pi r^2 dr = \int_0^R \frac{3}{2} P \times 4\pi r^2 dr. \tag{3.9}$$

Using (3.8) and (3.9), we can write (3.7) in the form

$$2E_T + E_G = 0. \tag{3.10}$$

This elegant and famous result is known as the *virial theorem*.

From (3.10), we get

$$E_T = -\frac{1}{2} E_G = \frac{1}{2}|E_G|, \tag{3.11}$$

since the total gravitational energy E_G, as given by (3.8), is clearly a negative quantity. The sum of thermal and gravitational energies

$$E = E_G + E_T = \frac{1}{2} E_G = -\frac{1}{2}|E_G| \tag{3.12}$$

is also negative. It is not difficult to understand why E should be negative. Suppose that a star formed by slow gravitational contraction of material which was initially spread over a much larger volume. As the star contracts, it must become hotter and radiates away some energy so that the energy of the star has to become negative.

We now know that a normal star radiates energy which is produced by nuclear reactions in the interior (to be discussed in the next chapter). So, apart from thermal and gravitational energies, a star has another store of energy, i.e. nuclear energy. In the early years of stellar research, however, this additional source of energy was not known and E as given by (3.12) was regarded as the total energy. When Helmholtz (1854) and Kelvin (1861) first addressed the question of the source of stellar energy, it was believed that the thermal and gravitational energies were all that one needed to consider. In such a scenario, a star could gradually contract and a part of the gravitational potential energy released in the process could radiate away. We expect the star to be in approximate hydrostatic equilibrium as it collapses slowly and hence the virial theorem (3.10) should always hold approximately. As the star contracts, it becomes more gravitationally bound making $|E_G|$ larger and it follows from (3.11) that E_T also becomes larger, implying that the star becomes hotter. Now, the gravitational potential energy lost during the contraction of the star has to be transformed into other forms of energy. It is clear from (3.11) that exactly half of the gravitational energy released is transformed into thermal energy. The other half must leave the system so that the total energy E can be given by (3.12). We thus arrive at a very beautiful result. If there was no such thing as nuclear energy, then all stars had to contract slowly. Half of the gravitational potential energy released in the process has to be converted to thermal energy, whereas the other half should leave the system, presumably in the form of radiation. Helmholtz (1854) and Kelvin (1861) suggested that this is how stars shine.

If we estimate the lifetime of a star on the basis of this theory, then we can at once see that this theory could not possibly be correct. According to this theory, the Sun has so far radiated away an amount of energy equal to $(1/2)|E_G|$. Assuming that the Sun always radiated energy at the present rate $L_\odot$, we conclude that the age of the Sun should be

$$\tau_{KH} \approx \frac{\frac{1}{2}|E_G|}{L_\odot}. \tag{3.13}$$

An approximate value of $|E_G|$ can be easily calculated from (3.8). Replacing M_r and r by their average values, we have

$$|E_G| \approx \frac{G(M_\odot/2)}{(R_\odot/2)} \times M_\odot \approx 4 \times 10^{41} \text{ J}. \tag{3.14}$$

Putting this in (3.13), we find the Kelvin–Helmholtz time scale to be

$$\tau_{KH} \approx 10^7 \text{ yr}.$$

Even in the days of Kelvin and Helmholtz, there was enough geological evidence that the Earth was much older than this. So the age of the Sun definitely could not be so short!

3.2.3 Energy transport inside stars

The energy generated by nuclear reactions in the central region of a star is transported outward. We now have to derive equations which describe this process. Let L_r be the total amount of energy flux per unit time which flows outward across a spherical surface of radius r inside the star (the spherical

surface being centred at the centre of the star). We expect L_r to be equal to
the luminosity L of the star at the outer radius $r = R$ of the star. If $L_r + dL_r$ is
the outward energy flux at radius $r + dr$, then dL_r is obviously the additional
input to the energy flux made by the spherical shell between r and $r + dr$. If
ε is the rate of energy generation per unit mass per unit time (presumably by
nuclear reactions), then we should have

$$dL_r = 4\pi r^2 dr \times \rho \varepsilon,$$

from which

$$\frac{dL_r}{dr} = 4\pi r^2 \rho \varepsilon. \tag{3.15}$$

After (3.1) and (3.2), this is the third of the important stellar structure equations.

The energy flux is driven by the temperature gradient inside the star. We
need an equation for that as well. We know that there are three important modes
of heat transfer in nature: conduction, convection and radiation. Although conduction is important in compact stars like white dwarfs, it turns out to be totally
unimportant in the interiors of normal stars. In the next subsection, we shall
discuss the possibility of convection. Right now, let us consider a region in the
interior of a star where heat is transported outward only by radiative transfer. We
have already derived an expression for the energy flux per unit area by radiative
transfer in (2.78). Replacing z by r, the energy flux L_r across the spherical
surface of radius r is given by

$$L_r = 4\pi r^2 F = -4\pi r^2 \frac{c}{\chi\rho} \frac{d}{dr} \left(\frac{a_B}{3} T^4 \right),$$

from which

$$\frac{dT}{dr} = -\frac{3}{4a_Bc} \frac{\chi\rho}{T^3} \frac{L_r}{4\pi r^2}. \tag{3.16}$$

This is the fourth equation of stellar structure if the heat flux is carried outward
by radiative transfer. We need to replace (3.16) by a different equation if the
heat flux is carried by convection. We shall derive this alternative equation in
the next subsection.

It may be noted that the first three equations of stellar structure – (3.1), (3.2)
and (3.15) – follow from fairly straightforward considerations. Only (3.16),
which was obtained by Eddington (1916), is somewhat non-trivial. It may be
useful for the reader to look at an instructive alternative derivation of (3.16)
given by Eddington (1926, §71).

3.2.4 Convection inside stars

In radiative transfer, energy is transported without any material motion. Convection, on the other hand, involves up and down motions of the gas. Hot blobs of

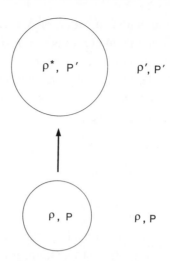

Fig. 3.1 Vertical displacement of a blob of gas in a stratified atmosphere.

gas move upward and cold blobs of gas move downward, thereby transporting heat. Let us now find out under what circumstances this is likely to happen.

Suppose we have a perfect gas in hydrostatic equilibrium inside a star. We now consider a blob of gas which has been displaced vertically upward as shown in Figure 3.1. Initially the blob of gas had the same density ρ and the same pressure P as the surroundings. The external gas density and pressure at the new position of the blob are ρ' and P'. We know that pressure imbalances in a gas are rather quickly removed by acoustic waves, but heat exchange between different parts of the gas takes more time. Hence it is not unreasonable to consider the blob to have been displaced *adiabatically* and to have the same pressure P' as the surroundings in its new position. Let ρ^* be its density in the new position. If $\rho^* < \rho'$, then the displaced blob will be buoyant and will continue to move upward further away from its initial position, making the system unstable and giving rise to convection. On the other hand, if $\rho^* > \rho'$, then the blob will try to return to its original position so that the system will be stable and there will be no convection. So convection is of the nature of an instability in the system. To find the condition for convective instability, we have to determine whether ρ^* is greater than or less than the surrounding density ρ'.

From the assumption that the blob has been displaced adiabatically, it follows that

$$\rho^* = \rho \left(\frac{P'}{P} \right)^{1/\gamma}. \tag{3.17}$$

If dP/dr is the pressure gradient in the atmosphere, we can substitute

$$P' = P + \frac{dP}{dr} \Delta r$$

and make a binomial expansion keeping terms to the linear order in Δr. This gives

$$\rho^* = \rho + \frac{\rho}{\gamma P}\frac{dP}{dr}\Delta r. \tag{3.18}$$

On the other hand,

$$\rho' = \rho + \frac{d\rho}{dr}\Delta r.$$

Using $\rho = P/RT$, we get

$$\rho' = \rho + \frac{\rho}{P}\frac{dP}{dr}\Delta r - \frac{\rho}{T}\frac{dT}{dr}\Delta r. \tag{3.19}$$

Here $d\rho/dr$ and dT/dr are the density and temperature gradients in the atmosphere. From (3.18) and (3.19),

$$\rho^* - \rho' = \left[-\left(1 - \frac{1}{\gamma}\right)\frac{\rho}{P}\frac{dP}{dr} + \frac{\rho}{T}\frac{dT}{dr}\right]\Delta r. \tag{3.20}$$

Keeping in mind that dT/dr and dP/dr are both negative, the atmosphere is stable if

$$\left|\frac{dT}{dr}\right| < \left(1 - \frac{1}{\gamma}\right)\frac{T}{P}\left|\frac{dP}{dr}\right|. \tag{3.21}$$

This is the famous *Schwarzschild stability condition* (Schwarzschild, 1906). If the temperature gradient of the atmosphere is steeper than the critical value $(1 - 1/\gamma)(T/P)|dP/dr|$, then the atmosphere is unstable to convection.

Convection is an extremely efficient mechanism for transporting energy. The temperature gradient has to be only slightly steeper than the critical gradient to drive the typical stellar energy flux. We would not be very far off the mark if we take

$$\frac{dT}{dr} = \left(1 - \frac{1}{\gamma}\right)\frac{T}{P}\frac{dP}{dr} \tag{3.22}$$

inside the convection zone. In order to make more accurate calculations, one has to take the help of *mixing length theory*, which was developed by Biermann (1948) and Vitense (1953). It is described in any standard textbook on stellar structure (see, for example, Kippenhahn and Weigert, 1990, Chapter 7). This theory assumes that upward-moving hot blobs or downward-moving cold blobs typically traverse a vertical distance called the mixing length, after which they lose their identity and mix their heat contents with their surroundings. By assuming a reasonable value of the mixing length, it is possible to calculate the small difference between the actual temperature gradient and the critical gradient, which is responsible for transporting the necessary heat flux. We shall not discuss mixing length theory in this elementary textbook.

While constructing a model of a star, one has to proceed in the following way. First one assumes that there is no convection and heat transport is entirely due to radiative transport described by (3.16). After calculating the temperature distribution on the basis of this assumption, the next step is to check if the temperature gradient obtained in this way satisfies the Schwarzschild stability condition (3.21). If it is satisfied, then it can be taken as established that the heat flux is really carried by radiative transport and the temperature gradient is given by (3.16). On the other hand, if the stability condition (3.21) is not satisfied in some regions, then the heat flux is primarily carried by convection in those regions and one has to repeat the calculation by using (3.22) instead of (3.16).

3.3   Constructing stellar models

We have already derived all the necessary equations for constructing stellar models. Let us now see how it can be done.

First of all, one has to specify the chemical composition of a star, since opacity and the nuclear energy generation rate depend on the chemical composition. The chemical composition can be given by specifying the mass fraction X_i of various elements present in the stellar material. The next step is to figure out the equation of state $P(\rho, T, X_i)$, the opacity $\chi(\rho, T, X_i)$ and the nuclear energy generation rate $\varepsilon(\rho, T, X_i)$ as functions of density, temperature and chemical composition. In §2.6, we have already discussed the opacity calculation. In the next chapter, we shall discuss how the nuclear energy generation rate is calculated. Before discussing the stellar structure models, we make a few comments about the equation of state.

The density at the centre of the Sun is more than 100 times the density of water. However, still the material there behaves like a perfect gas, because the temperature is so high that the interatomic potential energies are negligible compared to the typical kinetic energies of the particles and atoms do not get a chance to bind together to form a solid or a liquid. If we can assume the gas to be completely ionized, then the equation of state becomes particularly simple. Let X be the mass fraction of hydrogen, Y the mass fraction of helium and Z the mass fraction of other heavier elements (often referred to as 'metals' by astronomers!). The number of hydrogen atoms per unit volume is $X\rho/m_H$. Since each hydrogen atom contributes two particles (one electron and one nucleus which is a proton), there will be $2X\rho/m_H$ particles per unit volume from fully ionized hydrogen. The number density of helium atoms will be $Y\rho/4\,m_H$ and they will contribute $3Y\rho/4\,m_H$ particles. Since a heavy atom of atomic mass A approximately contributes $A/2$ particles, it is easy to see that the contribution to the number density from heavier elements is $Z\rho/2\,m_H$. Hence the number of particles per unit volume is

$$n = \left(2X + \frac{3}{4}Y + \frac{1}{2}Z\right)\frac{\rho}{m_H}$$

so that the gas pressure is given by

$$P = \frac{\kappa_B}{\mu\, m_H}\rho\, T, \tag{3.23}$$

where

$$\mu = \left(2X + \frac{3}{4}Y + \frac{1}{2}Z\right)^{-1} \tag{3.24}$$

is known as the *mean molecular weight*. We shall see that (3.23) will be quite adequate for the purpose of qualitatively understanding various properties of stars. For accurate stellar models, however, one needs to take account of the partial ionization, especially in the outer layers of the star, and should also include the radiation pressure, which becomes important for more massive stars. Finally, when the density is very high, the electron gas becomes *degenerate*, i.e. it obeys the Fermi–Dirac distribution rather than the classical Maxwellian distribution. This gives rise to what is called the *degeneracy pressure*, which will be discussed in detail in §5.2. This pressure can play a crucial role in balancing gravity when the nuclear fuel is exhausted in a star. A more complete discussion of the equation of state is postponed to §5.2.

Let us now write down all the equations for stellar structure in one place. They are

$$\frac{dM_r}{dr} = 4\pi r^2 \rho, \tag{3.25}$$

$$\frac{dP}{dr} = -\frac{GM_r}{r^2}\rho, \tag{3.26}$$

$$\frac{dL_r}{dr} = 4\pi r^2 \rho\varepsilon, \tag{3.27}$$

$$\left.\begin{aligned}\frac{dT}{dr} &= -\frac{3}{4a_Bc}\frac{\chi\rho}{T^3}\frac{L_r}{4\pi r^2}\\[2mm]\frac{dT}{dr} &= \left(1 - \frac{1}{\gamma}\right)\frac{T}{P}\frac{dP}{dr}\end{aligned}\right\}. \tag{3.28}$$

We discussed at the end of the previous section how one determines which form of (3.28) is to be used. In a typical star, the convection may take place in a certain range of radius, whereas heat is transported by radiative transfer in other regions. So, for the same stellar model, it may be necessary to use one form of (3.28) in some regions and the other form elsewhere. Once the equation of state $P(\rho, T, X_i)$, the opacity $\chi(\rho, T, X_i)$ and the nuclear energy generation rate $\varepsilon(\rho, T, X_i)$ are given, the above equations involve four independent functions of r: ρ, T, M_r and L_r. The number of independent equations is also four. It is straightforward to figure out the boundary conditions. We have the following two boundary conditions at the centre of the star

$$M_r = 0 \text{ at } r = 0, \tag{3.29}$$

$$L_r = 0 \text{ at } r = 0. \tag{3.30}$$

The radius $r = R$ of the star is the point where both ρ and T become very small compared to their values in the interior. Hence the simplest boundary conditions for them are

$$\rho = 0 \text{ at } r = R, \tag{3.31}$$

$$T = 0 \text{ at } r = R. \tag{3.32}$$

Since there are four equations involving four variables, with one boundary condition for each variable, this is clearly a mathematically well-posed problem. Unfortunately, not much progress can be made analytically unless one makes drastically simplifying assumptions. However, it is not difficult to solve the equations of stellar structure numerically.

Although it is not our aim to give a detailed discussion of the numerical methods, let us try to give an idea how one proceeds. Suppose we want to construct a model of a star with a given central density ρ_c. Taking the central temperature to have a value T_c and using the boundary conditions (3.29)–(3.30), we can start integrating (3.25)–(3.28) from $r = 0$. In general, ρ and T will not become zero at the same value of r so that it will not be possible to satisfy (3.31)–(3.32) simultaneously. We then have to try out the procedure again and again by varying the value of the central temperature T_c, until we find a combination ρ_c and T_c for which the solution would be such that ρ and T will become simultaneously zero for some particular r. We would regard that r to be the radius R of the star, and boundary conditions (3.31)–(3.32) will be satisfied. The values of M_r and L_r at $r = R$ would give us the mass and the luminosity of the star. We thus see that in principle the structure of a star with a central density ρ_c can be found this way, and such a star would have a definite mass and definite luminosity. Although the procedure outlined above gives an idea of how a stellar structure can be found, this simple procedure unfortunately does not work properly. The equation of radiative energy transfer in (3.28) has a factor T^3 in the denominator and this factor becomes very large near the surface where T is very small. This leads to a numerical instability. One can think of the alternative of starting the numerical integration from the stellar surface $r = R$. This leads to a numerical instability at the centre due to the factor r^2 in the denominator of (3.26). One possible way of getting around these difficulties is to start the numerical integrations both from $r = 0$ and $r = R$, and then match them smoothly at an intermediate point. Although this method works, it is not a particularly efficient method. A more efficient numerical algorithm was developed by Henyey, Vardya and Bodenheimer (1965) and is known as the *Henyey method*. This is a standard method widely used in solving stellar structures and is described in standard textbooks (see, for example, Kippenhahn and Weigert, 1990, §11.2).

Uniqueness of solutions?

For the sake of simplicity, let us consider stars of given uniform composition. Then the equation of state, the opacity and the nuclear energy generation rate all become functions of density and pressure alone. From the discussion of the previous section, it would seem that it will be possible to construct a unique stellar structure solution starting from a given central density ρ_c. Such a solution would correspond to a star of given mass M. Hence, at first sight, it appears that there should be a unique stellar structure solution for a star of a given mass. In fact, in the early years of stellar research, astronomers believed that the structure of a star of given mass and given chemical composition should be unique. This result was known as the *Vogt–Russell theorem* (Vogt, 1926; Russell, Dugan and Stewart, 1927). Even the otherwise careful Chandrasekhar gave a 'proof' of this theorem in his book (Chandrasekhar, 1939, pp. 252–253).

Further research showed that solving the stellar structure equations is a complicated problem and often solutions were not unique. In other words, the Vogt–Russell theorem could not be a mathematically correct result! Let us give one counter-example. Consider a star of mass $M_\odot$. Such a star can have a structure like the Sun. We shall see in Chapter 5 that it is possible for such a star to have a different configuration – the white dwarf configuration. At first sight, it may seem that this may be due to the change in chemical composition, since the Sun is expected to become a white dwarf when its nuclear fuel is exhausted, leading to a change in its chemical composition. However, we shall see in Chapter 5 that the white dwarf configuration arises when stellar matter is in a degenerate state. It should in principle be hypothetically possible to put even the solar material into a degenerate state and a white dwarf star with the solar composition is a theoretical possibility. It is thus clear that a star of mass $M_\odot$ can have at least two distinct configurations and both of these should follow from the stellar structure equations. The Vogt–Russell theorem cannot be correct.

Although the Vogt–Russell theorem is not correct from a strictly mathematical point of view, for practical purposes a normal star of a given mass M and standard composition may be taken to have a reasonably unique structure. Such a structure would correspond to a luminosity L and radius R. In other words, by solving the stellar structure equations, it should be possible to find the luminosity L and radius R of a star of mass M. If it were possible to solve the stellar structure equations (3.25)–(3.28) analytically, we could have found out how L and R are related to M. Unfortunately it is not possible to solve the stellar structure equations analytically. Only if we are allowed to make drastic assumptions and simplifications, is it possible to proceed analytically and obtain a few approximate relations amongst various quantities pertaining to a star.

3.4 Some relations amongst stellar quantities

We shall now do a few drastic things with the stellar structure equations (3.25)–(3.28) to extract some relations amongst various quantities pertaining to a star. Since some of our steps will be highly questionable in nature, we shall have to take the derived results with a degree of caution. A comparison with detailed stellar models, however, will show that we are not very much off the track. Our aim will be to find how various quantities scale with each other. We shall, therefore, ignore the constant factors in our equations. A slightly more sophisticated approach than ours is to construct what are called *homologous* stellar models, in which it is assumed that various quantities vary inside different stars in similar ways. Several standard textbooks discuss homologous stellar models, a particularly excellent account being given by Tayler (1994, pp. 110–117; see also Kippenhahn and Weigert, 1990, Chapter 20).

Let us replace the left-hand side of the hydrostatic equation (3.26) by $-P/R$, where P can be taken as the typical pressure inside the star. Replacing M_r/r^2 on the right-hand side by M/R^2, we are led to

$$\frac{P}{R} \propto \frac{M}{R^2}\rho,$$

from which

$$P \propto \frac{M^2}{R^4} \tag{3.33}$$

on taking $\rho \propto M/R^3$. The equation of state $P \propto \rho T$ would imply

$$P \propto \frac{M}{R^3}T. \tag{3.34}$$

For (3.33) and (3.34) to hold simultaneously, we must have

$$T \propto \frac{M}{R}. \tag{3.35}$$

In other words, inside temperatures of different stars should be proportional to M/R.

After subjecting (3.26) to this drastic treatment, we do a similar thing with the radiative energy transfer equation (3.28). If we assume that the radiative transfer equation holds throughout the star and further the variation of χ inside the star is not very appreciable, then we can write

$$\frac{T}{R} \propto \frac{M}{R^3 T^3} \frac{L}{R^2},$$

from which it follows that

$$L \propto \frac{(TR)^4}{M}. \tag{3.36}$$

It is seen from (3.35) that TR should be proportional to M. Substituting this in (3.36), we come to the conclusion

$$L \propto M^3. \tag{3.37}$$

This is called the *mass–luminosity relation*, which implies that a more massive star should be more luminous. Since we derived this relation by making some drastic assumptions, one may express doubts about the correctness of this relation. Figure 3.2 shows a plot of $\log L$ versus $\log M$ as obtained from detailed numerical solutions of stellar structure equations. On this figure, we superpose a dashed line with a slope corresponding to the relation (3.37). This line is not too far off from what we get from detailed stellar models. We shall present a comparison with observational data in the next section.

We saw in the previous chapter that the surface of a star behaves approximately like a blackbody. Hence, if T_{eff} is the effective surface temperature, then we must have

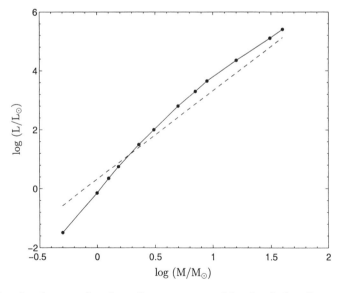

Fig. 3.2 Luminosity as a function of mass computed by detailed stellar models. The dashed line indicates the slope that would result if L varied as M^3. Adapted from Hansen and Kawaler (1994, p. 43) who use the results of Iben (1965) and Brunish and Truran (1982).

$$L = 4\pi R^2 \sigma T_{\text{eff}}^4, \tag{3.38}$$

where σ is the Stefan–Boltzmann constant. If we assume that T_{eff} is a measure of the typical interior temperature of the star (i.e. if hotter stars have hotter surface temperatures), then we can write from the above

$$L \propto R^2 T^4. \tag{3.39}$$

Since L goes as M^3 by (3.37) and RT goes as M by (3.35), it follows from (3.39) that

$$M^3 \propto M^2 T^2$$

so that

$$M \propto T^2. \tag{3.40}$$

Using (3.37) and the fact we are assuming T_{eff} to go as T, we can write

$$L \propto T_{\text{eff}}^6. \tag{3.41}$$

We thus conclude that two important observable quantities of stars – their luminosities and their effective surface temperatures – should be related as given by (3.41). A plot of luminosity versus surface temperature for a number of stars is known as the *Hertzsprung–Russell diagram*, or *HR diagram* in brief, after Hertzsprung (1911) and Russell (1913) who produced the first observational plots of this kind. For historical reasons, the convention is to plot the effective surface temperature T_{eff} increasing towards the left! We shall explain the reason for this convention in the next section, where HR diagrams of stars will be discussed in detail. Figure 3.3 shows a theoretical HR diagram constructed from detailed numerical models of stars, where the dashed line indicates the slope that we would get if the scaling relation (3.41) was strictly valid. We shall see in the next section that the scaling relation (3.41) matches the observational data also reasonably well. Thus, in spite of various crude and questionable assumptions, we have managed to derive an important scaling law which is not very far from the truth. Different points on the curve in Figure 3.3 corresponds to stars of different masses, which are also indicated.

A star lives as a normal star as long as it has got nuclear fuel to burn. Since the amount of nuclear fuel is proportional to mass and the rate at which the fuel is burnt is proportional to luminosity, the lifetime τ of a star should be given by

$$\tau \propto \frac{M}{L}. \tag{3.42}$$

Making use of (3.37), we have

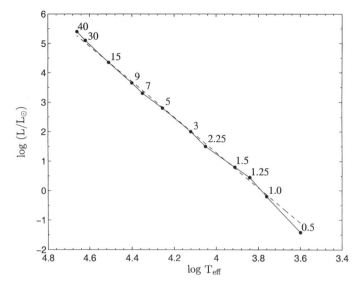

Fig. 3.3 The relation between luminosities and surface temperatures of stars as computed by detailed stellar models. The dashed line indicates the slope that would result if L varied as T_{eff}^6. The masses of stars corresponding to different points on the curve are also shown. Adapted from Hansen and Kawaler (1994, p. 40) who use the results of Iben (1965) and Brunish and Truran (1982).

$$\tau \propto M^{-2}. \tag{3.43}$$

Hence, more massive stars live for shorter times. A more massive star has more nuclear fuel to burn; but it burns this fuel at such a fast rate that it runs out of the fuel in a shorter time. This very important result that massive stars are short-lived helps us understand many aspects of observational data, as we shall see in the next two sections.

3.5 A summary of stellar observational data

In the previous section, we arrived at some theoretical conclusions about how various quantities connected with stars may be related to each other. Are these conclusions borne out by observational data? Before we can answer this question, we discuss briefly how various stellar parameters are determined.

3.5.1 Determination of stellar parameters

For any star that is not too faint, it is possible to take the spectrum. So we first discuss what we can learn from the spectrum. Then we point out what further information can be obtained if the star (i) is nearby or (ii) is in a binary system.

Stellar spectra: surface temperature, composition, stellar classification

We have seen in §2.4 that the surface of a star behaves approximately like a blackbody, the main departure from the blackbody spectrum being the spectral lines. Hence, by fitting the spectrum of a star to a blackbody spectrum, it is possible to estimate the effective surface temperature T_{eff} of the star. One of the easy things to measure of a star is its apparent magnitude in the U, B and V bands as defined in §1.4. As we pointed out in §1.4, the quantity $B - V$ is a measure of the star's colour. It is the effective surface temperature T_{eff} which determines where the peak of the spectrum will be and thereby determines the colour of the star (a hotter star being bluish and a colder star reddish). We thus expect a one-to-one correspondence between $B - V$ and T_{eff}, at least for stars of similar properties. Although we showed the theoretical HR diagram (Figure 3.3) with T_{eff} plotted on the horizontal axis, observational HR diagrams usually have the directly measurable quantity $B - V$ on their horizontal axes.

The composition of the star can be found out from its spectral lines. This is, however, not as straightforward as one may at first think. Let us explain this by considering the example of hydrogen. Since all stars are predominantly made of hydrogen, we may expect hydrogen lines to be present in the spectra of all stars. In reality, hydrogen lines are found only in stars of intermediate temperature. Hydrogen lines in the visible part of the spectrum consist of Balmer lines, which are produced due to atomic transitions to the $n = 2$ atomic state from higher states ($n = 3, 4, \ldots$). If the stellar surface temperature is too high, then hydrogen is completely ionized and such atomic transitions do not take place. On the other hand, a low surface temperature would imply that all hydrogen atoms are mostly in the ground state $n = 1$, with very few atoms occupying the states $n = 3, 4, \ldots$. Only for intermediate stellar surface temperatures, the levels $n = 3, 4, \ldots$ are well populated and appropriate atomic transitions take place to produce the Balmer lines. It was Saha (1921) who first realized that the strengths of spectral lines by themselves do not give us the composition of a stellar atmosphere. Matter of the same composition can produce very different spectra when kept at different temperatures. Saha (1920, 1921) developed his famous theory of thermal ionization and provided a satisfactory explanation why spectra of stars with different surface temperatures look different.

Around 1890, a group of astronomers at Harvard Observatory led by E. C. Pickering had developed a scheme of classifying stellar spectra in which a particular type of spectrum would be denoted by a Roman letter. Saha's work led to the realization that different spectral classes corresponded to different surface temperatures of stars. The spectral classes in the order of progressively decreasing surface temperature are O, B, A, F, G, K, M. Generations of astronomy students remembered these spectral classes with the help of the mnemonic 'Oh be a fine girl kiss me', the first letters of the successive words giving the names of spectral classes. We pointed out in §3.4 that HR diagrams are plotted with surface temperature on the horizontal axis increasing leftward. This is because HR diagrams were originally constructed by plotting spectral classes on the horizontal axis, before it was realized that spectral classes corresponded to surface temperatures.

Although spectral lines depend crucially on the surface temperature apart from composition, it is possible to carry out a sophisticated analysis of stellar spectra to determine the composition of the surface material of the star. We have indicated in §2.4.3 the basic idea behind the formation of spectral lines and discussed in §2.7 how an analysis of spectral lines can be carried on in very simple situations. Further details of spectral analysis are beyond the scope of this book. Apart from composition, spectral lines give us other crucial information. The star's velocity component along the line of sight would cause a Doppler shift of spectral lines, and by measuring this Doppler shift, the line of sight velocity of a star can be measured. Again, if the star is strongly magnetic, then one can hope to detect the Zeeman effect in the stellar spectra which would give information about the magnetic field.

Nearby stars: distance, luminosity

If a star is within a few pc, we can determine the distance of the star from a measurement of its parallax. Distances of about one hundred thousand stars have been determined by the Hipparcos astronomy satellite devoted to accurate measurements of stellar positions (Perryman *et al.*, 1995).

Once the distance to the star is known, we can find the absolute magnitude in any band by applying (1.8). The absolute magnitude in the V band, known as the *absolute visual magnitude*, is denoted by M_V and is a measure of the energy the star is giving out in visible light. A star like the Sun may be giving out most of its energy in the visible light. But stars with higher surface temperature may be giving out energy predominantly in the ultraviolet and stars with lower surface temperature in the infrared. Hence M_V does not give a correct estimate of the total luminosity of the star. If we could measure the total energy received from the star in all wavelengths and calculated the absolute magnitude from that, that would be called the *absolute bolometric magnitude*, denoted by M_{bol}. If we know the surface temperature T_{eff} of the star, then we can estimate the fraction of the emitted energy which will go in the V band. Hence, from a measurement of M_V, it is possible to infer M_{bol} which is related to the luminosity of the star. Thus, for nearby stars, once we have found out the distance, we also can infer the luminosity (or M_{bol}) from a measurement of M_V.

Binary stars: stellar mass determination

One of the fundamental parameters of a star is its mass. The mass of a star can be estimated only from the gravitational attraction it produces and we can observe the gravitational attraction only if there is a nearby object on which it acts. Luckily many stars are found in binary systems and one can determine the masses of both the stars by observing the effect of each on the other. Some binary stars are resolved through powerful telescopes. In other cases, the binary nature is inferred from indirect evidence. If one star is much dimmer than the other and the dimmer star sometimes blocks the light coming from the brighter star, then we can observe a periodic variation of brightness. Such binaries are called *eclipsing binaries*. As the two stars in a binary system move around their common centre of gravity, one star may sometimes be moving towards us and sometimes away from us, leading to a periodic variation in the Doppler shift of spectral lines. Binaries displaying such periodically varying Doppler shifts in their spectra are known as *spectroscopic binaries*.

Once the binary period and the velocities of the companions are known, it is straightforward to apply Newtonian gravitational mechanics to calculate the masses of the two companions (see, for example, Böhm-Vitense, 1989, Chapter 9). Since we can determine stellar masses only for stars in binary systems, one worry is whether stars of which we know masses constitute an unbiased statistical sample of stars. We shall discuss in §4.5 that binary stars very close to each other can transfer mass between themselves and evolve differently from isolated stars. However, if the stars in the binary system are sufficiently far away to ensure that mutual gravitational attraction does not distort their shapes significantly, then the nature of these stars would not be too different from isolated stars and they can be taken as typical representative samples in statistical studies of stars.

3.5.2 Important features of observational data

Mass–luminosity relation

If a star is both nearby and in a binary, then both its luminosity and mass can be determined. Plotting luminosities and masses of such stars, we get Figure 3.4. Our simple theoretical considerations led to (3.37), implying that luminosity should go as the cube of mass. The fact that a straight line fits the data reasonably well implies that L indeed goes as M^n, the index n being given by the slope of the straight line having value 3.7. Thus the very crude scaling arguments used in §3.4 brought us quite close to the truth.

HR diagram of nearby stars

For nearby stars, we can determine the luminosities and then plot the luminosities against surface temperatures (obtained from the spectra). As we pointed out in the previous section, the diagram obtained in this way is known as the HR diagram. Figure 3.5 shows the HR diagram of nearby stars based on the distance measurements by the Hipparcos astronomy satellite (Perryman *et al.*, 1995). It may be noted that the quantities plotted on the axes are the colour index $B - V$

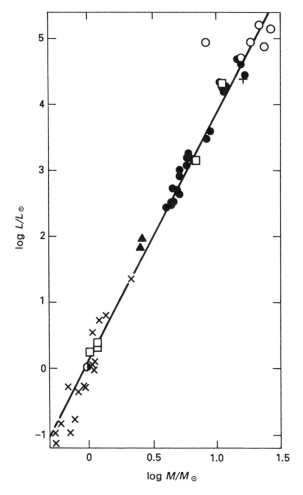

Fig. 3.4 The observational mass–luminosity relation. The different symbols correspond to different types of binaries (i.e. visual binaries are indicated by crosses, spectroscopic binaries by open squares, etc.). From Böhm-Vitense (1989, p. 87), based on the data presented by Popper (1980).

and the absolute visual magnitude M_V, which are directly measured (rather than $T_{\rm eff}$ and L which are inferred from these measurements). Stars lying on the right side of the diagram are reddish in colour, whereas stars lying on the left side are bluish in colour. HR diagrams with M_V plotted against $B - V$ are also called *colour–magnitude diagrams*. Most stars seem to lie on a diagonal strip in Figure 3.5 from the upper left corner to the lower right corner. This diagonal strip is called the *main sequence*. We shall discuss the stars outside the main sequence in the next section. The scaling laws discussed in the previous section are expected to apply to the stars on the main sequence. If we consider a median curve passing through the points on the main sequence, it will give a relation between M_V and $B - V$. As we have already discussed, $B - V$ is

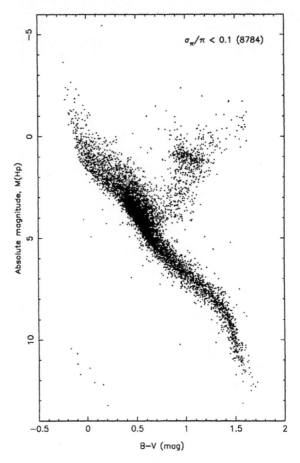

Fig. 3.5 The HR diagram (or colour–magnitude diagram) of nearby stars, constructed from the measurements by the Hipparcos astronomy satellite. From Perryman *et al.* (1995). (©European Southern Observatory. Reproduced with permission from *Astronomy and Astrophysics.*)

in turn related to T_{eff} and M_{V} is related to the absolute bolometric magnitude M_{bol}. Table 3.1 shows how these various quantities are related for stars lying on the main sequence. Figure 3.6 showing the relationship between M_{bol} and T_{eff} for main sequence stars is made from the last two columns of Table 3.1. It is clear that a straight line is a good fit. This straight line corresponds to a scaling relation $L \propto T_{\text{eff}}^{n}$ with $n = 5.6$. Our crude arguments in the last section had given a remarkably close power law index of 6 (see (3.41))!

3.6 Main sequence, red giants and white dwarfs

Although most of the data points in Figure 3.5 lie on the diagonal strip called the main sequence, there are also many data points in the upper right corner

Table 3.1 The relationship amongst colour index $B - V$, absolute visual magnitude M_V, effective surface temperature T_{eff} and absolute bolometric magnitude M_{bol} for main sequence stars. Adapted from Tayler (1994, p. 17).

$B - V$	M_V	$\log T_{eff}$	M_{bol}
0.0	0.8	4.03	0.4
0.2	2.0	3.91	1.9
0.4	2.8	3.84	2.8
0.6	4.4	3.77	4.3
0.8	5.8	3.72	5.6
1.0	6.6	3.65	6.2
1.2	7.3	3.59	6.6

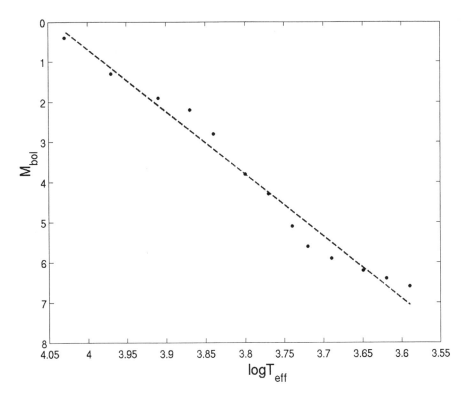

Fig. 3.6 The relation between M_{bol} and T_{eff} for stars lying on the median of the main sequence, with a best fit straight line. Based on data given by Tayler (1994, p. 17).

and a few data points in the lower left corner. The data points in the upper right correspond to stars which are red in colour and have luminosities much larger than the luminosities of red stars on the main sequence. Since unit areas of all stars with the same surface temperature give out energy at approximately the same rate (due to the Stefan–Boltzmann law of surface emission from blackbodies), the stars in the upper right corner have to be much larger in size than the red stars on the main sequence, in order to be much more luminous. This clearly follows from (3.38). The stars lying in the upper right corner of the HR diagram are, therefore, called *red giants*. The stars lying in the lower left corner of the HR diagram are bluish-white in colour and have much smaller luminosities compared to blue stars on the main sequence. By arguments similar to what we have just given, these stars have to be much smaller in size compared to bluish-white stars on the main sequence. So these stars in the lower left corner of the HR diagram are called *white dwarfs*.

We have already provided a theoretical explanation of the main sequence. The approximate scaling relation (3.41) derived in the previous section is a reasonable fit for the main sequence. We know that more luminous stars are more massive. So the upper left corner of the main sequence corresponds to more massive stars and the lower right corner to less massive stars. The mass of a star determines at which point of the main sequence the star would lie. The main sequence is essentially a sequence of stellar masses, with the mass increasing from the lower right towards the upper left. This should also be clear from the theoretical HR diagram shown in Figure 3.3, in which the masses of stars are demarcated.

Detailed explanations of red giants and white dwarfs will be provided in the next two chapters. Here we make only a few general remarks. We shall see in §4.3 that stars in the main sequence generate energy by converting hydrogen into helium. While a steady energy generation goes on in this way, the internal thermal energy balances the gravity and the structure of the star does not change much with time. This means that the position of the star in the HR diagram does not change much while hydrogen is being converted into helium in its interior. However, when hydrogen is appreciably depleted in the stellar core, the nuclear energy generation drops and is not able to balance the inward pull of gravity completely. This leads to a contraction of the core and, by the Kelvin–Helmholtz arguments presented in §3.2.2, we know that this will cause the core to heat up. Detailed calculations show that this process also dumps some heat in the surrounding layers of the star and inflates those layers. Red giant stars are believed to be caused in this way. We shall see in §4.5 that several nuclear reactions can take place in the very hot cores of red giants, leading to the production of various elements up to iron if certain conditions are satisfied. Eventually, however, all possible nuclear fuel is exhausted and the star can no longer produce thermal energy by nuclear reactions to balance gravity. What then happens? Since electrons are fermions, they obey Pauli's exclusion principle, i.e. two electrons cannot occupy the same quantum state. So electrons resist being pushed into very small volumes, once the density is sufficiently high and all the low-lying quantum states are filled. The pressure arising out of this factor, called the *electron degeneracy pressure*, will be derived in §5.2. We shall then show in §5.3 that, even in the absence of other energy sources, the electron degeneracy pressure alone can balance gravity if the mass of the star is less than the famous Chandrasekhar mass limit. White dwarfs are supposed to be very dense, dead stars in which no more nuclear reactions are taking place and gravity is balanced by the electron degeneracy pressure of the dense stellar material. The surface temperature of white dwarfs is a remnant of the heat produced in the gravitational contraction. Eventually, after the white dwarf radiates out the heat, it will become a cold dark object.

3.6.1   The ends of the main sequence. Eddington luminosity limit

The lightest stars on the main sequence at the lower right corner of the HR diagram have masses of order $0.1 M_\odot$, whereas the most massive stars at the upper left corner have masses of order $100 M_\odot$. Why do all stars have masses in this narrow range of about three orders of magnitude? Other stellar parameters like luminosity and radius vary much more.

Let us first point out what determines the lower limit of stellar mass. As we shall point out in §8.3, stars form out of the gravitational collapse of interstellar gas clouds. When a newly forming proto-star shrinks gravitationally, the Kelvin–Helmholtz theory outlined in §3.2.2 should hold and the proto-star should become hotter while it shrinks. Eventually its interior may become hot enough for nuclear reactions to start, causing the gravitational contraction to halt. Thus the proto-star becomes a real star burning nuclear fuel inside. However, if the mass of the proto-star is less than a lower limit, then the interior does not become hot enough for nuclear reactions to start, because the electron degeneracy pressure halts the gravitational contraction before the temperature can become sufficiently high. Such an object is called a *brown dwarf*. Detailed theoretical calculations suggest that $0.08 M_\odot$ is the lower limit for the mass of a star generating energy by nuclear reactions (see Exercise 5.9). A gravitationally contracting object with less mass becomes a brown dwarf. A brown dwarf will never have a surface temperature as high as that of even the least massive stars. However, after its formation, for some time a brown dwarf will be radiating away the heat produced during its gravitational contraction and can be detected. The first unambiguous detection of a brown dwarf was reported by Nakajima *et al.* (1995).

Let us now turn our attention to very massive stars, with very high temperatures inside. The radiation pressure becomes more important inside more massive stars. You will learn a historically important argument for it when you work out Exercise 5.5 in Chapter 5. Eventually the very high radiation pressure inside a massive star can make the star unstable. It is straightforward to show that a high radiation pressure can lift the outer layers of a star. The energy flux of radiation at the surface of a star with luminosity L and radius R is $L/4\pi R^2$. If χ is the opacity, then $\rho\chi$ is the absorption coefficient and the energy absorbed per unit volume per unit time is $\rho\chi(L/4\pi R^2)$. The momentum associated with this energy can be obtained by dividing this by c, which will give us the momentum absorbed per unit time in a unit volume, which is nothing but the force exerted on this unit volume. The star will be able to hold on to this outer layer of gas only if the inward force of gravity is stronger than this force exerted by radiation, i.e. if

$$\frac{GM}{R^2}\rho > \frac{L}{4\pi R^2}\frac{\rho\chi}{c},$$

from which

$$L < \frac{4\pi c\, GM}{\chi}. \tag{3.44}$$

This limit of luminosity is known as the *Eddington luminosity limit* (Eddington, 1924). Note that the radius R has cancelled out of this expression. Since we have the approximate relation (3.37) that L goes as M^3, we can write $L = \lambda M^3$. It then follows from (3.44) that the Eddington limit will be violated if the mass of the star were to be larger than $M_{\max}$ given by

$$\lambda M_{\max}^2 = \frac{4\pi c\, G}{\chi}. \tag{3.45}$$

While $M_{\max}$ given by (3.45) may be an absolute upper limit beyond which a star's outer layers would be blown off by radiation, in reality stars with mass considerably less than this $M_{\max}$ become unstable due to radiation pressure and are not able to exist (see, for example, Kippenhahn and Weigert, 1990, §22.4, §39.5).

3.6.2   HR diagrams of star clusters

Many stars are found in clusters. There are some relatively loosely bound clusters, each having a few dozens of stars. Such loosely bound clusters are called *open clusters*. Of more interest to us are the *globular clusters*, which are very tightly bound almost spherical clusters, containing of the order of 10^5 stars. Figure 3.7 shows a globular cluster. As we shall discuss in §6.1.2, the globular clusters are found around the centre of our Galaxy. From an astrophysical point of view, the main importance of a star cluster is that it gives us a group of stars which are believed to have been born at about the same time and which are at roughly the same distance from us.

    If all the stars in a cluster are at the same distance d, then we see from (1.8) that the difference between absolute magnitude and apparent magnitude

Fig. 3.7 A globular cluster of stars, photographed at Kavalur Observatory.

will be the same for all stars. Hence we can construct the HR diagram of a star cluster by plotting the apparent magnitude (instead of absolute magnitude) against $B - V$. Figure 3.8 shows such an HR diagram of a globular cluster. One can clearly see the main sequence. By using (1.8), one can easily find out the distance d of the globular cluster which will yield such values of the absolute magnitude that the main sequence of the globular cluster will coincide with the main sequence of nearby stars as seen in Figure 3.5. This is a very powerful method of determining distances to star clusters.

The overall appearance of Figure 3.8, however, is quite different from Figure 3.5. For example, in Figure 3.5 we find that the main sequence continues to values of $B - V$ less than 0.0 on the left side. On the other hand, the main sequence in Figure 3.8 seems to end at around $B - V = 0.3$. We know that main sequence stars with lower values of $B - V$ correspond to more massive stars. So the globular cluster is basically missing stars on the main sequence heavier than a certain mass. The explanation for this is not difficult to give. We have already pointed out that more massive stars have shorter lifetimes (see (3.43)). So, in a globular cluster of a certain age, stars heavier than a particular mass would have finished their lives on the main sequence. As we shall see in §4.5, stars lie on the main sequence as long as hydrogen is converted into helium. After that, a star becomes a red giant as a result of the inflation of outer layers. We see in Figure 3.8 that there is a branch of stars proceeding towards the region of red giant stars (upper right corner) from the place where the main sequence

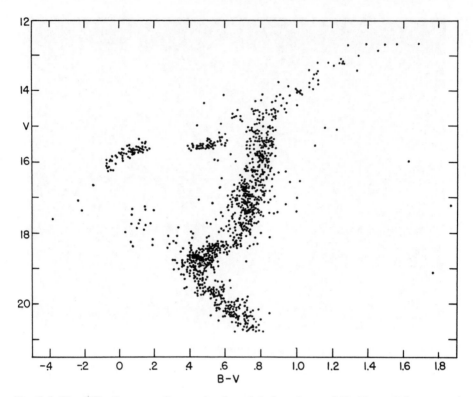

Fig. 3.8 The HR diagram of stars in the globular cluster M3. From Johnson and Sandage (1956). (ⒸAmerican Astronomical Society. Reproduced with permission from *Astrophysical Journal*.)

seems to end abruptly. Presumably these are stars in a state of transition from the main sequence to the red giant phase. Since this transition takes relatively less time compared to lifetimes of stars, the probability of coming across stars in this transitory phase is not very high in a random sample of stars. That is why we see relatively few stars between the main sequence and the red giant phase in the HR diagram of nearby stars. After finishing the red giant phase, a star may proceed towards becoming a white dwarf. Some points in Figure 3.8 seem to correspond to such stars.

The stars at the abrupt turning point of the main sequence are the stars which are just running out of hydrogen in the core. So the age of the globular cluster is essentially equal to the main sequence lifetime of these stars at the turning point. Hence, from a theoretical estimate of the lifetimes of stars, one can determine the age of a globular cluster simply by noting the turning point of the main sequence. Figure 3.9 is a composite HR diagram by superposing the HR diagrams of several star clusters. The vertical axis displays the absolute magnitude, which can be found easily after determining the distance of the cluster by matching the main sequence. The clusters with turning points lower

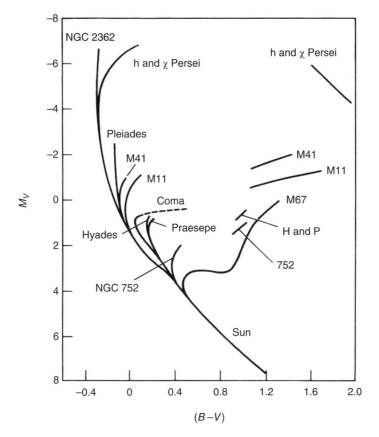

Fig. 3.9 A composite HR diagram sketching the extent of the main sequence for several star clusters. From Sandage (1957). (©American Astronomical Society. Reproduced with permission from *Astrophysical Journal*.)

down are clearly older. Detailed theoretical calculations suggest that the oldest globular clusters are about 1.5×10^{10} yr old. This poses an important constraint on cosmology, since the Universe could not be younger than this!

Exercises

3.1 Estimate the total thermal energy of the Sun from the fact that its internal temperature is of order 10^7 K. Show that this is of the same order as the rough estimate of gravitational potential energy.

3.2 If the Sun was producing its energy by slow contraction as suggested by Helmholtz and Kelvin, estimate the amount by which the radius of the Sun has to decrease every year to produce the observed luminosity.

3.3 Show that the radiation pressure at the centre of the Sun is negligible compared to the gas pressure, by estimating the ratio of the radiation pressure to the gas pressure.

3.4 The Sun has a convection zone from $0.7\,R_\odot$ to the solar surface. Find out how density, pressure and temperature vary within this convection zone by assuming that (i) equation (3.22) holds exactly inside the convection zone and (ii) the convection zone contains a very small fraction of the Sun's mass so that the gravitational field in the convection zone can be taken to be $-GM_\odot/r^2$. (According to current solar models, the convection zone contains only about 2% of solar mass.)

3.5 Using the fact that the opacity in very hot stars is provided by Thomson scattering, show that L/M has to be less than a critical value and find its numerical value. How does it compare with $L_\odot/M_\odot$? Use (3.45) to estimate the maximum mass M_{max} of a star such that the outer layers would be blown off by radiation if the mass of the star was larger. (Note that stars with such high mass actually do not exist.)

3.6 From Figures 3.5 and 3.8, estimate the distance of the globular cluster M3 from us.

3.7 Make a very rough estimate of the wavelengths at which a star of mass $9M_\odot$ and a star of mass $0.25M_\odot$ will give out maximum radiation.

4

Stellar astrophysics II: Nucleosynthesis and other advanced topics

4.1 The possibility of nuclear reactions in stars

We have seen in the previous chapter that many aspects of stellar structure can be understood without a detailed knowledge of stellar energy generation mechanisms. This is indeed fortunate because not much was known about energy generation mechanisms when Eddington was carrying out his pioneering investigations of stellar structure in the 1920s. Eddington (1920) correctly surmised that the Kelvin–Helmholtz hypothesis of energy generation by contraction (see §3.2.2) could not possibly be true and stellar energy must be produced by subatomic processes. Nuclear physics, however, was still in its infancy and details of how the stellar energy is produced could not be worked out at that time. With the rapid advances in nuclear physics within the next few years, it became possible to work out the details of energy-producing nuclear reactions inside stars. To build sufficiently detailed and realistic models of stars and stellar evolution, a good understanding of energy generation mechanisms is essential. We turn to this subject now.

Let us consider a nucleus of atomic mass A and atomic number Z. It is made of Z protons and $A - Z$ neutrons. The mass m_{nuc} of the nucleus is always found to be less than the combined mass of these protons and neutrons. It is the energy equivalent of this mass deficit which provides the *binding energy* of the nucleus and is given by

$$E_B = [Zm_{\text{p}} + (A - Z)m_{\text{n}} - m_{\text{nuc}}]c^2. \qquad (4.1)$$

To find out how tightly bound a nucleus is, we need to consider the binding energy per nucleon

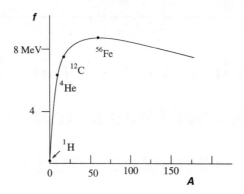

Fig. 4.1 A smooth curve showing the binding energy per nucleon, plotted against the atomic mass number.

$$f = \frac{E_B}{A}. \tag{4.2}$$

Figure 4.1 shows a plot of f for different atomic nuclei. It is seen that the intermediate-mass nuclei around iron are most tightly bound. So energy is released in two kinds of nuclear reactions: the fusion of very light nuclei into somewhat heavier nuclei or the fission of very heavy nuclei into intermediate-mass nuclei. Energy production in the interiors of stars is believed to be due to nuclear fusion. We note that f for helium is 6.6 MeV, which is about 0.007 of the mass of a nucleon. Hence, if a mass $M_\odot$ of hydrogen is fully converted into helium, the total amount of energy released will be $0.007 M_\odot c^2$. Dividing this by the solar luminosity $L_\odot$, we get an estimate of the lifetime of a star which shines by converting hydrogen into helium, i.e.

$$\tau_{\mathrm{nuc}} \approx \frac{0.007 M_\odot c^2}{L_\odot}. \tag{4.3}$$

On putting in the values of $M_\odot$ and $L_\odot$, this turns out to be

$$\tau_{\mathrm{nuc}} \approx 10^{11} \text{ yr,}$$

which is much longer than the Kelvin–Helmholtz time scale given by (3.13) and is of the same order as the age of the Universe.

All nuclei are positively charged and normally repel each other. Only when two nuclei are brought within about 10^{-15} m, can the short-range nuclear forces overcome the electrical repulsion and the nuclei can fuse. A typical internuclear potential is shown in Figure 4.2. For two nuclei with atomic numbers Z_1 and Z_2, the electrostatic potential is

$$\frac{1}{4\pi \epsilon_0} \frac{Z_1 Z_2 e^2}{r}.$$

The height of this potential at the nuclear radius $r \approx 10^{-15}$ m turns out to be about $Z_1 Z_2$ MeV. At the centre of the Sun where the temperature is of order

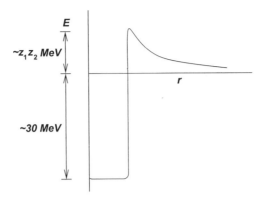

Fig. 4.2 A sketch of a typical nuclear potential.

10^7 K, the typical kinetic energy $\kappa_B T$ of a particle is about a keV, which is about 10^3 lower than the electrostatic potential barrier between nuclei. Even the centre of the Sun is not hot enough for the nuclei to overcome the mutual electrical repulsion and come close together for nuclear fusion – according to classical physics! However, one of the standard results of quantum mechanics is that a particle can tunnel through a potential barrier. While studying α-decay, Gamow (1928) calculated the probability for the α-particle to tunnel from the inside of the nucleus to the outside by penetrating the potential barrier. The same probability should hold for a particle to tunnel from the outside through the potential barrier of the nucleus. On taking account of the tunnelling probability, it was found that nuclear fusion can indeed take place in the interior of the Sun (Atkinson and Houtermans, 1929).

The basic principles for calculating the rate of any nuclear reaction inside a star will be discussed in the next section. Then in §4.3 we shall list some of the specific nuclear reactions likely to take place in stellar interiors.

4.2 Calculation of nuclear reaction rates

Suppose a nucleus having charge $Z_1 e$ can react with a nucleus having charge $Z_2 e$, their number densities per unit volume being n_1 and n_2. We want to calculate the rate of the reaction, i.e. the number of reactions taking place per unit volume per unit time.

If both types of nuclei have a Maxwellian velocity distribution, it is straightforward to show that the probability of the relative velocity between a pair being v also follows a Maxwellian distribution

$$f(v)\, dv = \left(\frac{m}{2\pi\kappa_B T}\right)^{3/2} \exp\left(-\frac{mv^2}{2\kappa_B T}\right) 4\pi v^2 dv,$$

where m is the reduced mass $m_1 m_2/(m_1 + m_2)$. In terms of the kinetic energy

$$E = \frac{1}{2} m v^2,$$

the distribution can be written as

$$f(E)\, dE = \frac{2}{\sqrt{\pi}} \frac{E^{1/2}}{(\kappa_B T)^{3/2}} \exp\left(-\frac{E}{\kappa_B T}\right) dE. \tag{4.4}$$

If $\sigma(E)$ is the reaction cross-section between the two nuclei approaching each other with energy E, then it is easy to see that the reaction rate is given by

$$r = n_1 n_2 \langle \sigma v \rangle, \tag{4.5}$$

where

$$\langle \sigma v \rangle = \int_0^\infty \sigma(E)\, v\, f(E)\, dE. \tag{4.6}$$

From (4.5) and (4.6), it should be clear that we need only the reaction cross-section $\sigma(E)$ to calculate the reaction rate. We now discuss how this cross-section can be found.

As we pointed out in §4.1, the typical particle energy in a stellar interior is much less than the height of the potential barrier sketched in Figure 4.2. Hence the cross-section $\sigma(E)$ has to depend on the probability of tunnelling through this potential barrier. The quantum mechanical tunnelling probability through such a barrier was first calculated by Gamow (1928) and is reproduced in many textbooks on nuclear physics (see, for example, Yarwood, 1958, §19.5). We write down the expression without derivation (you are asked to do the derivation in Exercise 4.1 with suitable hints). For nuclei approaching each other with energy E, the probability of tunnelling through the potential barrier is given by

$$P \propto \exp\left[-\frac{1}{2\epsilon_0 \hbar} \left(\frac{m}{2}\right)^{1/2} \frac{Z_1 Z_2 e^2}{\sqrt{E}}\right]. \tag{4.7}$$

Now, without the tunnelling probability, the reaction cross-section is expected to go as approximately λ^2, where λ is the de Broglie wavelength. Since $\lambda^2 \propto 1/E$, we can write down the cross-section including the tunnelling probability in the form

$$\sigma(E) = \frac{S(E)}{E} \exp\left[-\frac{b}{\sqrt{E}}\right], \tag{4.8}$$

where

$$b = \frac{1}{2\epsilon_0 \hbar} \left(\frac{m}{2}\right)^{1/2} Z_1 Z_2 e^2, \tag{4.9}$$

and $S(E)$ is a slowly varying function of E. It should be noted that the assumption of slow variation of $S(E)$ has its limitations. Occasionally the cross-section of a nuclear reaction may become very large for a certain energy, as sketched in

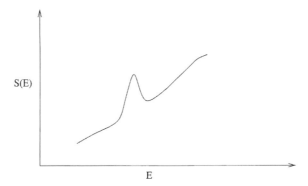

Fig. 4.3 A sketch showing the variation of a nuclear reaction cross-section with energy around a resonance.

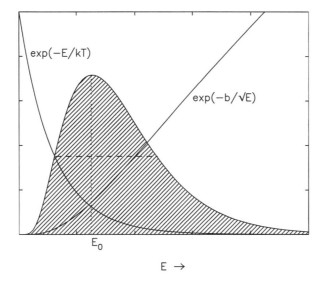

Fig. 4.4 Variation with energy of the Gamow factor, the Maxwellian factor and their product (the curve with the shading underneath).

Figure 4.3. This is called a *resonance*. Only in the absence of resonances, can we take $S(E)$ to be a slowly varying function. Usually $S(E)$ is determined from laboratory experiments.

On substituting (4.4) and (4.8) into (4.6), we finally get

$$\langle \sigma v \rangle = \frac{2^{3/2}}{\sqrt{\pi m}} \frac{1}{(\kappa_B T)^{3/2}} \int_0^\infty S(E)\, e^{-E/\kappa_B T} e^{-b/\sqrt{E}}\, dE. \tag{4.10}$$

The function $\exp(-E/\kappa_B T)$ decreases rapidly with E, whereas the other function $\exp(-b/\sqrt{E})$ increases rapidly with E, as shown in Figure 4.4. Their product has an appreciable value only for a narrow range of energy around E_0. We can replace the slowly varying function $S(E)$ by its value $S(E_0)$ at E_0 and take it outside the integral. Then the integral in (4.10) is given by

$$J = \int_0^\infty e^{g(E)} \, dE, \tag{4.11}$$

where

$$g(E) = -\frac{E}{\kappa_B T} - \frac{b}{\sqrt{E}}. \tag{4.12}$$

The value of J is given by the shaded area in Figure 4.4. From $dg/dE = 0$, we can find the value of E_0 where the function $g(E)$ is maximum, which gives

$$E_0 = \left(\frac{1}{2} b \kappa_B T\right)^{2/3} = \left[\left(\frac{m}{2}\right)^{1/2} \frac{Z_1 Z_2 e^2 \kappa_B T}{4\epsilon_0 \hbar}\right]^{2/3}. \tag{4.13}$$

Let the value of $g(E)$ at E_0 be denoted by $-\tau$, i.e.

$$\tau = -g(E_0) = 3\frac{E_0}{\kappa_B T} = 3\left[\left(\frac{m}{2\kappa_B T}\right)^{1/2} \frac{Z_1 Z_2 e^2}{4\epsilon_0 \hbar}\right]^{2/3}. \tag{4.14}$$

We can now expand $g(E)$ in a Taylor series around the point $E = E_0$, which gives

$$g(E) = g(E_0) + \left(\frac{dg}{dE}\right)_{E=E_0} (E - E_0) + \frac{1}{2}\left(\frac{d^2 g}{dE^2}\right)_{E=E_0} (E - E_0)^2 + \cdots$$

$$= -\tau - \frac{\tau}{4}\left(\frac{E}{E_0} - 1\right)^2 + \cdots$$

on calculating $d^2 g/dE^2$ from (4.12) and noting that $dg/dE = 0$ at $E = E_0$. Substituting this in (4.11), we get

$$J \approx e^{-\tau} \int_0^\infty e^{-\frac{\tau}{4}\left(\frac{E}{E_0} - 1\right)^2} \, dE. \tag{4.15}$$

Since the integrand makes significant contributions only in a narrow range of E around E_0, we can replace the lower limit by $-\infty$. Then the integral in (4.15) becomes a Gaussian integral which can be evaluated easily and gives

$$J \approx \frac{2}{3} \kappa_B T \sqrt{\pi \tau} e^{-\tau}. \tag{4.16}$$

From (4.10) (keeping in mind that τ goes as $T^{-1/3}$), we now have

$$\langle \sigma v \rangle \propto \frac{S(E_0)}{T^{2/3}} \exp\left[-3\left(\frac{e^4}{32\epsilon_0^2 \kappa_B \hbar^2} \frac{m Z_1^2 Z_2^2}{T}\right)^{1/3}\right]. \tag{4.17}$$

Once $S(E)$ for the nuclear reaction is found from laboratory experiments, the reaction rate can be obtained by substituting (4.17) into (4.5).

For calculating stellar models, we need to know the energy generation rate by the nuclear reaction. If $\Delta\mathcal{E}$ is the energy released in this nuclear reaction, then the energy generation rate per unit volume is $r\,\Delta\mathcal{E}$, with r given by (4.5). This must be equal to $\rho\varepsilon$, where ε is as defined in §3.2.3, i.e.

$$\rho\varepsilon = r\,\Delta\mathcal{E} = n_1 n_2 \langle\sigma v\rangle \Delta\mathcal{E}. \tag{4.18}$$

If X_1 and X_2 are the mass fractions of the two elements which take part in the nuclear reaction, then n_1 and n_2 should respectively be proportional to ρX_1 and ρX_2. It is then clear from (4.17) and (4.18) that the nuclear energy generation function ε should have the following functional dependence on various relevant quantities:

$$\varepsilon = C\rho\, X_1 X_2 \frac{1}{T^{2/3}} \exp\left[-3\left(\frac{e^4}{32\epsilon_0^2 \kappa_B \hbar^2}\frac{m Z_1^2 Z_2^2}{T}\right)^{1/3}\right]. \tag{4.19}$$

Once the coefficient C is estimated from the experimentally determined cross-section $S(E)$, we have the necessary input for stellar structure calculations. The function ε increases with temperature sharply because of the exponential involving temperature. Since $Z_1^2 Z_2^2/T$ appears in the exponential, it should be clear that reactions involving heavier nuclei are much less likely compared to reactions involving lighter nuclei at a given temperature. We now turn to the specific nuclear reactions which are likely to take place inside stars.

4.3   Important nuclear reactions in stellar interiors

Although nuclear reactions inside stars involve no chemical burning, it is quite customary to refer to energy generation by nuclear reactions as *nuclear burning* and the element which gets transformed in the nuclear reactions as *nuclear fuel*. As we pointed out in the previous section, one needs an experimentally determined cross-section $S(E)$ for a nuclear reaction to calculate the energy generation by that reaction in stellar interiors. As we already pointed out, typical particle energies in stellar interiors are of the order of keV. Laboratory experiments are usually done for energies of order MeV so that the Coulomb barrier does not pose a big problem and the nuclear reactions become more likely. From measurements of $S(E)$ at MeV energies, one has to extrapolate to keV energies for application to stellar interiors. For an account of the historical development of this subject, the interested reader should be referred to the Nobel Lecture by Fowler (1984), who was a pioneer in the experimental measurement of many cross-sections relevant for astrophysics. Fowler (1984) gives plots of $S(E)$ for many astrophysically relevant nuclear reactions.

In the early decades of the twentieth century, astronomers were not sure of the composition of stars. However, by the time Russell (1929) carried out an extensive spectroscopic analysis of the Sun, it had become clear that the stars are mainly made up of hydrogen. Also, it should be apparent from (4.19) that hydrogen can 'burn' at a temperature lower than the temperatures necessary to burn helium and other heavier elements with higher atomic number Z. We believe that the main-sequence stars generate their energies by burning hydrogen into helium. By the late 1930s, nuclear physics had developed sufficiently to enable physicists to come up with schemes of likely nuclear reactions inside stars.

Bethe and Critchfield (1938) proposed what is now known as the *proton–proton* or *pp chain*. The energy generation inside the Sun primarily takes place due to this chain. In the first two reactions of this chain, deuterium ^{2}H and then ^{3}He are produced as follows:

$$^1\text{H} + {}^1\text{H} \longrightarrow {}^2\text{H} + e^+ + \nu,$$
$$^2\text{H} + {}^1\text{H} \longrightarrow {}^3\text{He} + \gamma. \tag{4.20}$$

After the production of ^{3}He, the reactions can proceed through three alternative branches: *pp*1, *pp*2, *pp*3. The branch *pp*1 is by far the dominant branch for conditions corresponding to the solar interior. It involves two nuclei of ^{3}He producing a nucleus of ^{4}He:

$$pp1 : \ {}^3\text{He} + {}^3\text{He} \longrightarrow {}^4\text{He} + {}^1\text{H} + {}^1\text{H}. \tag{4.21}$$

On considering all the reactions in the *pp*1 branch, it should be clear that effectively four ^{1}H nuclei combine to form one ^{4}He nucleus. The other two branches (*pp*2 and *pp*3) start dominating only when the temperature is above 10^7 K. They require the prior existence of ^{4}He and first form ^{7}Be:

$$^3\text{He} + {}^4\text{He} \longrightarrow {}^7\text{Be} + \gamma. \tag{4.22}$$

Afterwards ^{7}Be can lead to the following two kinds of reactions:

$$pp2 : \ {}^7\text{Be} + e^- \longrightarrow {}^7\text{Li} + \nu,$$
$$^7\text{Li} + {}^1\text{H} \longrightarrow {}^4\text{He} + {}^4\text{He}. \tag{4.23}$$
$$pp3 : \ {}^7\text{Be} + {}^1\text{H} \longrightarrow {}^8\text{B} + \gamma,$$
$$^8\text{B} \longrightarrow {}^8\text{Be} + e^+ + \nu,$$
$$^8\text{Be} \longrightarrow {}^4\text{He} + {}^4\text{He}. \tag{4.24}$$

Our job is now to find the energy generation function ε for the whole chain of reactions. How can this be done? We note that the first reaction in (4.20) is mediated by the weak interaction (the emission of a neutrino is usually the signature of a reaction being mediated by the weak interaction) and is a slow reaction with a small cross-section. Even though some of the other reactions may be faster, they cannot proceed without the ^{2}H nuclei which are produced in the first slow reaction. It is thus the first reaction which determines the reaction rate in a steady state. In general, when a series of reactions will have to take place, the slowest reaction determines the rate at the steady state. However, while calculating the energy generation, it is necessary to add up the energies released in all the reactions in the chain. When all these are done carefully, the energy generation rate ε is given by

$$\varepsilon_{pp} = 2.4 \times 10^{-1} \rho X^2 \left(\frac{10^6}{T}\right)^{2/3} \exp\left[-33.8 \left(\frac{10^6}{T}\right)^{1/3}\right] \text{W kg}^{-1}, \quad (4.25)$$

when the contributions of $pp2$ and $pp3$ branches are neglected. Here X is the mass fraction of hydrogen.

If carbon, nitrogen and oxygen are already present and can act as catalysts, then hydrogen can be synthesized into helium by a completely different series of nuclear reactions. This series of reactions, known as the *CNO cycle*, was independently suggested by von Weizsäcker (1938) and Bethe (1939). The reactions in this cycle are the following:

$$^{12}\text{C} + {}^1\text{H} \longrightarrow {}^{13}\text{N} + \gamma,$$
$$^{13}\text{N} \longrightarrow {}^{13}\text{C} + e^+ + \nu,$$
$$^{13}\text{C} + {}^1\text{H} \longrightarrow {}^{14}\text{N} + \gamma,$$
$$^{14}\text{N} + {}^1\text{H} \longrightarrow {}^{15}\text{O} + \gamma,$$
$$^{15}\text{O} \longrightarrow {}^{15}\text{N} + e^+ + \nu,$$
$$^{15}\text{N} + {}^1\text{H} \longrightarrow {}^{12}\text{C} + {}^4\text{He}. \quad (4.26)$$

On adding up these reactions, the net result again is that four ^1H nuclei have combined together to make one ^4He nucleus. Again, the reaction rate in the steady state is governed by the slowest reaction in the cycle, which in this case happens to be the fourth reaction in (4.26). The energy generation rate by the CNO cycle is found to be

$$\varepsilon_{\text{CNO}} = 8.7 \times 10^{20} \rho X_{\text{CNO}} X \left(\frac{10^6}{T}\right)^{2/3} \exp\left[-152.3 \left(\frac{10^6}{T}\right)^{1/3}\right] \text{W kg}^{-1},$$

$$(4.27)$$

where X_{CNO} is the sum of the mass fractions for carbon, nitrogen and oxygen. It should be noted that both (4.25) and (4.27) are of the same form as (4.19).

The variations of ε_{pp} and ε_{CNO} as functions of T are shown in Figure 4.5 for a typical stellar composition. It should be clear from this figure that for stars like the Sun with the central temperatures of order 10^7, the *pp* chain should be the dominant energy generation mechanism. On the other hand, more massive stars with higher central temperatures generate energy predominantly by the CNO cycle.

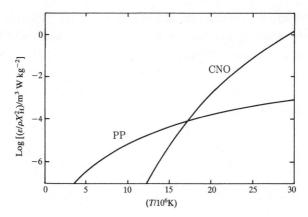

Fig. 4.5 The variation with temperature of the energy generation rate by hydrogen burning, for the two major reaction chains. From Tayler (1994, p. 92).

Apart from explaining the energy generation mechanism in stars, the other important goal of nuclear astrophysics is to explain the abundances of various elements in the Universe. As we shall see in §11.3, we believe that nuclear reactions took place in the early Universe and some significant fraction of baryonic matter was converted into helium. The helium synthesized in stars makes additions to this primordial helium. The next important question is how the heavier elements are produced. Gamow (1946) suggested that all the elements were synthesized in the early Universe. We now think that this was a wrong idea and heavier elements are synthesized in stars. Let us see how this can happen. After some helium has been synthesized from hydrogen by say *pp* chain reactions, we shall have a mixture of hydrogen and helium nuclei. Let us consider such a mixture. If heavier nuclei have to be built up from this, then the obvious first step may be either of these two reactions: (i) one hydrogen and one helium nuclei combine to produce a nucleus of mass 5; (ii) two helium nuclei combine to produce a nucleus of mass 8. However, laboratory experiments failed to discover any stable nucleus of mass 5 or 8. It became clear that these two obvious nuclear reactions could not provide the next step of synthesizing heavier nuclei. Then how are heavier nuclei produced? This problem was solved by Salpeter (1952), who suggested what is known as the *triple alpha reaction*. In this reaction, three ^{4}He nuclei combine together as follows:

$$^4\text{He} + {}^4\text{He} + {}^4\text{He} \longrightarrow {}^{12}\text{C} + \gamma. \tag{4.28}$$

Since this reaction involves three particles, it is much less likely to occur compared to reactions involving two particles. Also, the Coulomb repulsion is stronger between helium nuclei than between the nuclei involved in *pp* chain reactions, requiring a higher temperature (which should be evident from (4.19)). In the conditions prevailing in the early Universe, this reaction is found to be highly improbable and nucleosynthesis could not possibly proceed beyond helium. Inside stellar cores, however, this reaction can take place when the temperature is higher than 10^8 K. But, even then, the rate would have been too slow if the cross-section of this reaction was non-resonant. Hoyle (1954) conjectured that there must be a resonance to make the reaction rate appreciable. This resonance was almost immediately found in laboratory experiments.

Detailed calculations show that central temperatures of main-sequence stars are not high enough for the triple alpha reaction. So stars in the main sequence generate energy by the *pp* chain (less massive stars) or CNO cycle (more massive stars). When, however, hydrogen is exhausted in the core, hydrogen-burning reactions can no longer halt the inward pull of gravity. The core then starts shrinking, as we shall discuss more in §4.5. As we shall see in the next chapter, it is possible for gravity to be eventually balanced by degeneracy pressure when the core density is sufficiently high (provided the core mass does not exceed the famous Chandrasekhar limit to be derived in §5.3). However, while the core shrinks, its temperature rises by the Kelvin–Helmholtz arguments given in §3.2.2. If the star is not too massive, then its central temperature may never become high enough to start the triple alpha reaction and the star may end up as a white dwarf with a helium core. In the case of very massive stars, on the other hand, the temperature of the shrinking core may become very high for other nuclear reactions involving heavier nuclei to start. Once a new nuclear reaction is ignited, it can halt the inward pull of gravity. After carbon has been synthesized by the triple alpha process, the next heavier nuclei can be built up from carbon. There is a vast literature on the various nuclear reactions which build up heavier nuclei. We shall not get into this complex subject here. In sufficiently massive stars, it is believed that nuclear reactions can go all the way up to the most stable nucleus, iron. So such stars may eventually have an iron core, beyond which it is not possible to generate energy by nuclear reactions. All possible nuclear reactions in stellar interiors were systematically discussed by Burbidge, Burbidge, Fowler and Hoyle (1957). One important question is why we see elements heavier than iron in the Universe, or why we even see elements higher than helium in the solar system, since the Sun has not yet gone beyond the stage of synthesizing helium from hydrogen. These issues will be discussed in §4.7.

4.4 Detailed stellar models and experimental confirmation

We explained in §3.3 how detailed stellar models are calculated. One of the important inputs in a stellar model calculation is the nuclear energy generation rate. We have seen in §4.2 and §4.3 how this rate can be determined. So we now understand in principle how a stellar model is constructed. The equation of state $P(\rho, T, X_i)$, the opacity $\chi(\rho, T, X_i)$ and the nuclear energy generation rate $\varepsilon(\rho, T, X_i)$ all depend on the chemical composition of the star. So we

Table 4.1 The standard solar model. The density ρ is in kg m^{-3}. Adapted from Bahcall and Ulrich (1988).

$R/R_\odot$	$M_r/M_\odot$	$L_r/L_\odot$	T	ρ
0.000	0.000	0.000	1.56e+7	1.48e+5
0.053	0.014	0.106	1.48e+7	1.23e+5
0.103	0.081	0.466	1.30e+7	8.40e+4
0.151	0.192	0.777	1.11e+7	5.61e+4
0.201	0.340	0.939	9.31e+6	3.51e+4
0.252	0.490	0.989	7.86e+6	2.09e+4
0.302	0.620	0.999	6.70e+6	1.20e+4
0.426	0.830	1.001	4.73e+6	2.96e+3
0.543	0.924	1.001	3.53e+6	8.42e+2
0.691	0.974	1.000	2.38e+6	2.05e+2
0.822	0.993	1.000	1.19e+6	6.42e+1
0.909	0.999	1.000	5.25e+5	1.87e+1
1.000	1.000	1.000	5.77e+3	0.00e+0

need to specify the composition, keeping in mind that the composition changes continously due to nuclear reactions – at least in the core where these reactions take place. To construct the model of a star of a definite mass, usually an initial uniform composition is assumed and first a stellar model is calculated on the basis of it. This model would correspond to a star of this mass when it is just born. Then one finds out how the composition of the core will change due to nuclear reactions after some time. A stellar model calculated with this changed composition corresponds to the star some time after it is born. By constructing successive models with changed compositions, one finds how the star evolves with time. While hydrogen is being converted into helium in the core of a star, the overall structure of the star is found not to change much and the star lies on the main sequence in the HR diagram. Only when hydrogen is depleted sufficiently in the core, drastic changes in the overall characteristics of the star start taking place. We shall discuss these in the next section.

The age of the solar system is estimated by such methods as the analysis of radioactive nuclei with long half-lives in the old rocks and meteorites. We believe the Sun to be about 4.6×10^9 yr old. So a standard solar model is constructed by first solving the stellar structure equations by assuming that the present composition of the solar surface was initially the composition of the whole Sun and then by advancing this model through 4.6×10^9 yr. Table 4.1 presents the standard solar model. Before discussing how this standard solar model has been beautifully confirmed by recent experiments, we turn to some other important points.

We showed in §3.4 that many properties of stars can be understood without solving the stellar structure equations in detail. Now we want to discuss some important results which follow from detailed stellar structure models. We saw in §3.4 that more massive stars are more luminous and hotter, i.e. both their surfaces and central regions are hotter than surfaces and central regions respectively of less massive stars. It should be clear from Figure 4.5 that the CNO cycle must be the main hydrogen burning process for more massive stars, whereas the *pp* chain is the main hydrogen burning process for less massive stars (up to stars slightly heavier than the Sun). From the exponential factors in (4.25) and (4.27), it follows that ε_{CNO} is a much more rapidly increasing function of temperature than ε_{pp}. As a result, the CNO cycle in the core of a massive star tries to create a steep temperature gradient. A steep temperature gradient is likely to violate the Schwarzschild stability condition (3.21), giving rise to convection. Detailed calculations show that the massive stars have convective cores, whereas the cores of less massive stars are stable against convection. In the case of less massive stars, the temperature in the outer layers just below the surface is less than the temperature in the outer layers of more massive stars. A look at Kramers's law (2.79) and Figure 2.8 should convince the reader that the opacity should be higher in the outer layers of less massive stars. If the energy flux were to be carried by radiative transfer, it follows from (3.16) that the temperature gradient will have to be steep if the opacity was high. Again we expect the Schwarzschild condition (3.21) to be violated and the energy flux to be carried by convection in the regions where opacity is high. To sum up, more massive stars have convective cores surrounded by stable envelopes, whereas less massive stars have convective envelopes surrounding stable cores.

It follows from the standard solar model that the Sun has a stable core up to a radius of about $0.7 R_\odot$, beyond which the temperature gradient is unstable and heat is transported by convection. This theoretical conclusion is corroborated by high-quality photographs of the solar surface like the one in Figure 4.6. This photograph really gives the impression that we are looking at the top of a layer of convecting fluid. Since the upcoming hot gases are brighter and the downgoing cold gases are darker, we get the granular pattern which changes in a few minutes.

One of the main triumphs of stellar structure theory is that it can account for various properties of the stars on the main sequence (mass–luminosity relation and colour–magnitude relation). We saw in §3.4 that even fairly crude arguments based on the stellar structure equations give us a reasonable idea how these relations arise. However, apart from explaining these relations, stellar structure theory has led to very detailed stellar models constructed by many theorists over the years. Is there some way to test if these detailed theoretical stellar models are indeed close to reality? In other words, do densities, temperatures and pressures vary in the interiors of stars exactly in accordance with these theoretical stellar models? Two recent developments described below give us confidence that the standard solar model describes the interior of the Sun

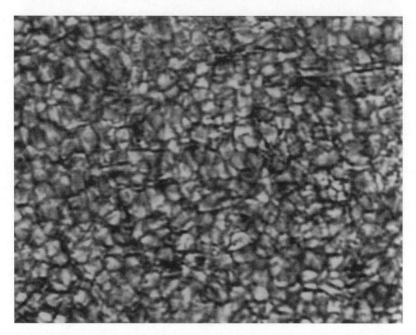

Fig. 4.6 A photograph of the solar surface showing the granulation pattern due to convection (photographed with the Vacuum Tower Telescope of the Kippenheuer Institut located in Tenerife). Courtesy: W. Schmidt.

extremely well and probably the same is true for theoretical models of stars with other masses.

4.4.1 Helioseismology

This subject, which is the study of solar oscillations, began when Leighton, Noyes and Simon (1962) discovered that the surface of the Sun is continuously oscillating with periods of the order of a few minutes. We know that an air column in a pipe vibrates only at some eigenfrequencies. A careful analysis of the solar oscillations revealed the existence of many discrete frequencies. It became clear that the observed oscillations are essentially superpositions of many modes with discrete eigenfrequencies. By now several thousands of eigenfrequencies have been measured very accurately.

The eigenfrequencies of an air column depend on the length of the column and the sound speed inside it, since sound waves travel back and forth inside the column to set up the standing modes. Similarly, the eigenmodes of the Sun are caused by sound waves (we would call them 'sound waves' even though their frequencies are usually outside the audible range) which interfere constructively after passing through and around the Sun. Since different modes go up to different depths in the interior of the Sun, the analysis of many modes together

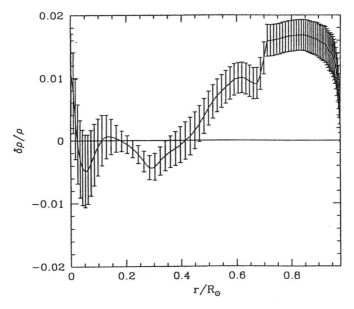

Fig. 4.7 The difference between the density inferred from helioseismology and the density calculated from the standard solar model (divided by the density), as a function of the solar radius. From Chitre and Antia (1999). (©Indian Academy of Sciences. Reproduced with permission from *Current Science*.)

tells us how the sound speed varies with depth in the interior of the Sun. The sound speed is given by

$$c_\mathrm{S} = \sqrt{\frac{\gamma P}{\rho}} \tag{4.29}$$

(see §8.3). Once sound speeds at different depths are inferred from helioseismology, one can use (4.29) to determine the density as a function of depth inside the Sun. Figure 4.7 shows how the density, inferred from helioseismology and calculated from the standard model, differ from each other. The difference is considerably less than 2% at all depths. Thus helioseismology has verified the standard solar model to a very high degree of accuracy.

4.4.2 Solar neutrino experiments

We believe that energy inside stars is produced by nuclear fusion, because that is the most satisfactory theoretical idea we have been able to come up with. However, we saw in §3.4 and §3.5 that many aspects of observational data can be explained to a reasonable extent without any detailed knowledge of the energy generation process. So, can we have an independent experimental check that nuclear reactions are really taking place inside stars? The nuclear reactions taking place in the interior of the Sun are listed in (4.20)–(4.24). It may be noted that a neutrino is a by-product in many of the reactions. Since neutrinos interact with matter only through the weak interaction, most of the neutrinos created at the centre of the Sun would come out without interacting with the material of the Sun at all. Thus, at the Earth, we expect a flux of neutrinos directly coming from the centre of the Sun. Detecting this flux of neutrinos is a sure way of confirming that nuclear reactions are indeed taking place in the centre of the Sun. In the 1960s the famous first solar neutrino experiment began (Davis, Harmer and Hoffman, 1968). The flux of neutrinos was detected, but the experimentally measured flux was found to be about one-third of what was theoretically predicted. Let us take a more detailed look at the solar neutrino experiment.

We take stock of the nuclear reactions which produce neutrinos. In the first reaction of (4.23), ^{7}Be gives rise to a neutrino besides a nucleus ^{7}Li. Since there are only two end products, the conservations of momentum and energy easily show that each of the product particles should have a specific value of energy. Actually, the ^{7}Be neutrino can have two discrete energies: 0.38 MeV and 0.86 MeV. There are two other important reactions producing neutrinos: (i) the first reaction of (4.20) and (ii) the second reaction of (4.24). We would refer to these neutrinos as *pp* neutrinos and ^{8}B neutrinos respectively. In both these cases, the neutrino is one of the three end products. So it is possible for the neutrino to have a distribution of energy. The *pp* neutrinos have energy in the range 0–0.4 MeV, whereas the ^{8}B neutrinos have the energy range 0–15 MeV. Figure 4.8 shows the spectrum of neutrinos theoretically predicted by the standard solar model. It should be noted that the vertical axis is logarithmic and the flux of ^{8}B neutrinos is several orders smaller than the flux of *pp* neutrinos. The flux of ^{8}B neutrinos depends sensitively on the solar model, since these neutrinos are produced in a reaction in the *pp*3 branch. This branch becomes more important if the temperature is higher. In a different solar model with the central temperature lower than what is predicted by the standard model, the ^{8}B neutrino flux can be considerably less. On the other hand, the *pp* neutrinos come from the main nuclear reaction. The luminosity of the Sun fixes the number of reactions taking place per unit time and determines the *pp* neutrino flux. The value of this flux, therefore, is independent of the solar model used.

Since neutrinos interact so weakly with matter, it is not easy to detect them. The pioneering experiment of Davis used the following reaction

$$^{37}\text{Cl} + \nu \rightarrow {}^{37}\text{Ar} + e^-, \tag{4.30}$$

for which the threshold neutrino energy is 0.814 MeV, as indicated in Figure 4.8. So only the ^{8}B neutrinos can produce this reaction. A huge tank of the cleaning fluid C_2Cl_4 was placed deep underground in a gold mine, to cut down the disturbances expected at the terrestrial surface. Neutrinos are the only particles which penetrated to this depth and occasionally interacted with a ^{37}Cl nucleus to produce ^{37}Ar. Since ^{37}Ar is radioactive, one could estimate the number

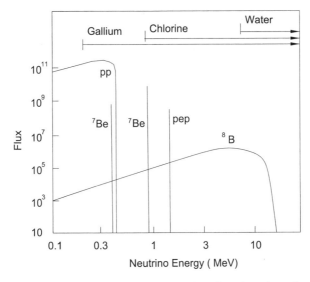

Fig. 4.8 The expected spectrum of the solar neutrino flux, based on the standard solar model. The detection ranges of the different experiments are indicated. Adapted from Bahcall (1999).

of ^{37}Ar nuclei produced from the number of radioactive decays and thereby find the solar neutrino flux. A convenient unit to express the neutrino flux measurement is SNU (Solar Neutrino Units), defined as 10^{-36} interactions per target atom per second. The chlorine experiment of Davis has fixed the flux to a value 2.56 ± 0.23 SNU on the basis of 25 years of operation, whereas the theoretical value predicted by the standard solar model is 7.7 ± 1.2 SNU (Bahcall, 1999). For many years, Davis's experiment was the only solar neutrino experiment and one possible explanation of the discrepancy was that the central temperature of the Sun could be less than what is predicted by the standard solar model. Other neutrino experiments were planned to settle this question.

Two experiments in Japan – Kamiokande and SuperKamiokande – used pure water, in which neutrinos with energy above 7 MeV can scatter electrons to high velocities which produce Cerenkov radiation. Again only the ^{8}B neutrinos could be detected and the flux was found to be half of what was theoretically predicted. However, using an array of Cerenkov detectors, it was possible to ascertain the direction from which the neutrinos were coming and to show for the first time that neutrinos were really coming from the Sun (Hirata *et al.*, 1990).

It became of utmost importance to detect the low-energy *pp* neutrinos, since the predicted theoretical flux is independent of the solar model. Low-energy neutrinos induce the following reaction in gallium

$$^{71}\mathrm{Ga} + \nu \rightarrow {}^{71}\mathrm{Ge} + e^{-}. \tag{4.31}$$

Hence one can use gallium as a detector of *pp* neutrinos. Two experiments using gallium – GALLEX (Anselmann *et al.*, 1995) and SAGE (Abdurashitov *et al.*, 1996) – have given the rate of 73 ± 5 SNU, whereas the theoretical prediction from the standard solar model is 129 ± 8 SNU (Bahcall, 1999). Although there is virtually no uncertainty in the predicted flux of *pp* neutrinos, gallium detectors detect neutrinos from other reactions as well, giving rise to the uncertainty in the theoretical flux.

The results of the various solar neutrino experiments taken together left little doubt that something must be happening to a part of the solar neutrino flux so that a part is not detected by the detectors on the Earth. We know that there are three kinds of neutrinos: the electron neutrino, the muon neutrino and the tau neutrino. If neutrinos have non-zero mass, then it can be shown that it is possible for one type of neutrino to get spontaneously converted into other types. Such neutrino oscillations have been confirmed recently by the Sudbury Neutrino Observatory (Ahmad *et al.*, 2002). The nuclear reactions in the Sun produce electron neutrinos and all the solar neutrino experiments also detect electron neutrinos only. Presumably, during the flight from the Sun to the Earth, some of the electron neutrinos get converted into the other types and are not detected in the solar neutrino experiments. This is now believed to be the reason why the measured flux is less than what is theoretically predicted.

4.5 Stellar evolution

We pointed out in §4.3 that a main-sequence star is expected to generate energy steadily as long as hydrogen in the core is converted into helium. The luminosity or the surface temperature of the star does not change much during this phase when it lies on the main sequence. Eventually, the hydrogen in the core of the star is exhausted. What then happens to the star? This is the central question of stellar evolution. Unfortunately, the only way of answering this question is through very detailed numerical computations. Nothing much can be done analytically or on the basis of general arguments. Stellar evolution is a very important topic for the professional astrophysicist and very large numbers of detailed computations have been done by many groups on this subject. The picture which emerges from these computations is quite complicated in its details. A star evolves through many very different stages. Also, stars of different masses evolve very differently. To a physicist who is not interested in very detailed astrophysical phenomenology, stellar evolution often appears to be a messy, confusing and unattractive subject. Since the emphasis of this book has been on those astrophysical topics which are of interest to physicists, we refrain from giving a detailed account of stellar evolution. There is another reason for not getting into the details of the subject. The author of this book does not claim to have any particular insight into this subject. Instead of reading an account by this author, the readers will do much better to read the excellent reviews of Iben (1967, 1974) or the relevant chapters from the book by Kippenhahn and Weigert (1990, Ch. 31–34). The groups of Iben and Kippenhahn have been responsible for some of the most thorough calculations of stellar evolution in the last few decades. For a relatively non-technical but superbly written account of the subject, see Tayler (1994, Ch. 6). We describe below only some of the salient features of stellar evolution.

Once hydrogen is exhausted in the stellar core, not enough energy is generated there to balance the inward pull of gravity. As a result, the core of the star starts shrinking and the gravitational potential energy released in the process generates heat, as suggested in the Kelvin–Helmholtz theory (§3.2.2). We know from the Kelvin–Helmholtz theory that the core would get hotter in this process. This has two important consequences.

(1) Heavier elements undergo nuclear fusion at higher temperatures, since a stronger Coulomb barrier has to be overcome. When the core becomes sufficiently hot, helium starts burning to produce carbon, halting the Kelvin–Helmholtz contraction. When helium is exhausted, the same cycle repeats, until the core becomes hot enough for the next nuclear fuel to burn. In very heavy stars, the core eventually ends up being iron, which has the most strongly bound nuclei. On the other hand, for light stars, the core temperature may never become high enough (before the electron degeneracy pressure halts the gravitational contraction) even for helium burning, so that the core remains a helium core. Very massive stars go through a complicated phase when different nuclear fuels burn in different spherical shells of the star with different temperatures.

(2) The excess heat produced in the Kelvin–Helmholtz contraction of the core inflates the outer layers of the star. Hence the star can bloat up to a huge size, while its luminosity does not change that much, so that its surface temperature drops. This causes the position of the star in the HR diagram to move away from the main sequence and follow the trajectory shown in Figure 4.9. Thus the star ends up being a red giant. Detailed computations show that the trajectories of massive stars can be even more complicated than the trajectory sketched in this figure, since the position of the star in the HR diagram changes whenever a new nuclear fuel is ignited in the core of the star. Figure 4.10 shows theoretical trajectories of stars of different masses based on detailed computations. Whenever a new nuclear fuel is ignited, there is a tendency of the trajectory proceeding back towards the main sequence. It may be noted that the trajectories never move towards the right of the HR diagram beyond a certain regime, since it was shown by Hayashi (1961) that stellar models lying too far on the right side of the HR diagram would be unstable and there would be a forbidden region there within which no stars can lie.

Eventually all nuclear fuels in the core that could be ignited at the prevailing conditions are exhausted and no more nuclear energy is produced to halt the gravitational contraction. If the mass of the core is less than a critical mass,

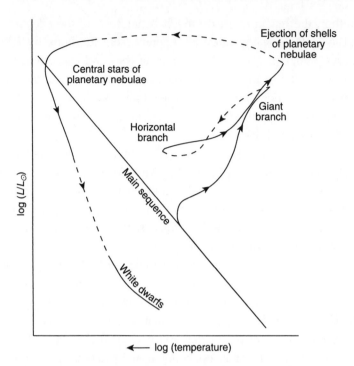

Fig. 4.9 A schematic trajectory of a star in the HR diagram. From Longair (1994, p. 31), after a figure of Mihalas and Binney (1981).

then gravity inside it can be balanced by electron degeneracy pressure (to be discussed in §5.2) when the density rises to sufficiently high values. Then the core may stop shrinking any more. In such a situation, the bloated envelope of the star also cannot persist too long. There are various mass loss mechanisms by which a large part of the outer envelope may be lost – either steadily or more violently, as we discuss in the next two sections. Any remaining part of the envelope may again settle on the core, so that we finally may have a compact star, which has a hot white surface initially and then gradually cools. Figure 4.9 shows the trajectory of the star as it evolves to become a white dwarf. Most stellar evolution codes fail to predict very reliable trajectories in the HR diagram in this phase, because many aspects of the theory are still rather ill-understood. As we shall see in §5.3, the mass limit of white dwarfs is about $1.4M_\odot$. However, considerably more massive stars also eventually may end up as white dwarfs by losing a large part of the mass. If the final mass remains larger than this mass limit, then the other possible final configurations are neutron stars and black holes, to be discussed in the next chapter.

4.5.1   Evolution in binary systems

A significant fraction of all stars are estimated to be in binary systems. If it is a close binary with the two stars very near each other, then their evolutions

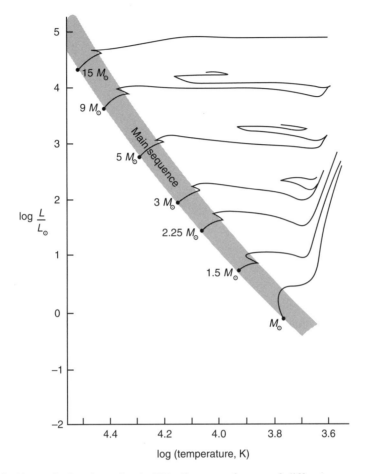

Fig. 4.10 Theoretical trajectories in HR diagram of stars of different masses, based on detailed computations. From Longair (1994, p. 33), after a figure of Mihalas and Binney (1981).

can differ in important ways from the evolution of isolated stars. We shall see in §5.5 and §5.6 that the topic of binary evolution is of great significance in understanding many astrophysical phenomena. We, therefore, make some brief remarks on the binary evolution problem.

The two stars in a binary system revolve around their common centre of mass, with an angular velocity denoted by Ω. In a frame of reference rotating with Ω, the two stars will be at rest. The force acting on a particle at rest in this frame will be the gravitational attractions of the two stars plus the centrifugal force. The effective potential will be given by

$$\Phi = -\frac{GM_1}{r_1} - \frac{GM_2}{r_2} - \frac{1}{2}\Omega^2 s^2, \tag{4.32}$$

where r_1 and r_2 are the distances of the particle from the centres of the two stars, whereas s is the distance from the rotation axis passing through the centre

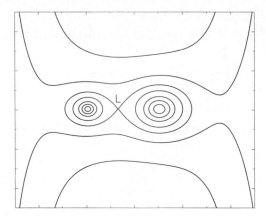

Fig. 4.11 Equipotential surfaces of two stars rotating around a common centre of mass, in the rotating frame of reference.

of mass. Figure 4.11 shows some of the equipotential surfaces in a typical case. The surface of a star should be an equipotential surface, if we want to ensure that there are no unbalanced horizontal forces at the stellar surface. Each of the stars should extend up to some equipotential surface.

We notice in Figure 4.11 that equipotential surfaces near any one of the stars go around that star alone. On the other hand, the equipotential surfaces far away surround both the stars. There is a critical surface made with the equipotential surfaces around the two stars touching at a point L. This point is called the *inner Lagrange point*, whereas the critical surface is known as the *Roche lobe*. When one of the stars becomes a red giant, its surface may bloat up to the Roche lobe, after which the gas from the surface should start falling into the other star through the inner Lagrange point. We shall discuss some consequences of binary mass transfer in §5.5 and §5.6. Such a mass transfer can lead to varieties of complicated situations. The more massive star of the binary finishes its life on the main sequence first and becomes a red giant. If it succeeds in transferring a significant amount of mass to the other star, then this other star may become more massive and may start evolving faster.

4.6 Mass loss from stars. Stellar winds

When a star becomes a red giant, the gravitational attraction at its inflated surface becomes much smaller than that at an ordinary stellar surface. This reduces the star's ability to hold on to the material on its surface and the surface material may keep escaping. Even in the case of an ordinary star like the Sun, material is continuously escaping from its corona in the form of a flow known as the *solar wind*. Let us consider why this happens.

Although the temperature of the solar surface is about 6000 K, the corona has a much higher temperature of the order of a million degrees. An elementary discussion of some possible reasons for this high temperature of the corona will be given in §8.9. Before the discovery of the solar wind, the corona was believed to be in static equilibrium. We reproduce below a simple but famous derivation by Parker (1958), showing that a hydrostatic solution of the corona leads to inconsistencies. Although the corona appears fairly non-spherical, one can try to construct a first approximate model of the corona by assuming spherical symmetry so that quantities like density and pressure can be regarded as functions of radius r alone. Since the corona has very little mass, the gravitational field in the corona can be regarded as an inverse-square field created by the mass of the Sun $M_\odot$. The hydrostatic equation (3.2) applied to the corona becomes

$$\frac{dP}{dr} = -\frac{GM_\odot}{r^2}\frac{\mu\, m_H}{\kappa_B}\frac{P}{T}, \tag{4.33}$$

where we have made use of (3.23) to eliminate ρ.

Without bothering about what heats the corona, we shall assume that the heat is produced in the lower layers of the corona so that outer regions of the corona can be modelled by taking a boundary condition that $T = T_0$ at some radius $r = r_0$ near the base of the corona. This is somewhat like calculating the temperature distribution in a metal rod with one end heated in a furnace. If the temperature of the furnace is given as a boundary condition, then the problem can be solved without knowing whether the furnace is heated by charcoal, gas or electricity. In the tenuous gas of the corona, conduction is the main mode of heat transport. In steady state, we expect that the same heat flux will pass through successive spherical surfaces in the outer corona. Let us consider a spherical surface at radius r. The heat flux through unit area of this surface is given by $K(dT/dR)$, where K is the thermal conductivity. Hence the heat flux through the whole spherical surface is

$$4\pi r^2 K\frac{dT}{dr},$$

which should be a constant for different r. It follows from the kinetic theory of plasmas that the thermal conductivity K of a plasma goes as the 5/2 power of temperature (see, for example, Choudhuri, 1998, §13.5). Hence we have

$$r^2 T^{5/2}\frac{dT}{dr} = \text{constant}, \tag{4.34}$$

of which the solution is

$$T = T_0\left(\frac{r_0}{r}\right)^{2/7} \tag{4.35}$$

satisfying the boundary conditions that $T = T_0$ at $r = r_0$ and $T = 0$ at infinity. Substituting for T in (4.33) from (4.35), we get

$$\frac{dP}{P} = -\frac{GM_\odot \mu m_{\mathrm{H}}}{\kappa_{\mathrm{B}} T_0 r_0^{2/7}} \frac{dr}{r^{12/7}},$$

of which the solution satisfying $P = P_0$ at $r = r_0$ is

$$P = P_0 \exp\left[\frac{7 G M_\odot \mu m_{\mathrm{H}}}{5 \kappa_{\mathrm{B}} T_0 r_0} \left\{\left(\frac{r_0}{r}\right)^{5/7} - 1\right\}\right]. \qquad (4.36)$$

The surprising thing to note is that the pressure has a non-zero asymptotic value as r goes to infinity. It is not possible to obtain a solution of the problem such that both P and T are zero at infinity. The asymptotic value of P at infinity is much larger than the typical value of the pressure of the interstellar medium.

What is the significance of this non-zero pressure at infinity? Parker (1958) concluded correctly that the hot solar corona could be in static equilibrium only if some appropriate pressure is applied at infinity to stop it from expanding. Since there is nothing to contain the corona by applying the necessary pressure, Parker (1958) suggested that the outer parts of the corona must be expanding in the form of solar wind. The solar wind was detected from spacecraft observations just a few years after Parker's bold prediction. Parker (1958) worked out a detailed hydrodynamic model of the solar wind as well, which we shall not discuss in this book. If the Sun were surrounded by a gas cloud with pressure larger than the pressure at infinity needed to maintain hydrostatic equilibrium, then we would have an inflow of gas into the Sun. Such a process is called *accretion*. The theory of spherical accretion was worked out by Bondi (1952). It is basically the reverse of a spherical wind (see Choudhuri, 1998, §6.8).

The solar wind is an example of what is called a thermally driven wind. It is caused by the high temperature of the corona, which makes it difficult for gravity to hold on to the gas. There are other mechanisms of driving winds. We pointed out in §3.6.1 that the radiation force in the outer atmosphere of a very massive star may become comparable to gravity. This may cause a radiatively driven wind. If a star is rotating very fast, that may lead to a centrifugally driven wind. The Sun loses only about $10^{-14} M_\odot$ yr^{-1} due to the solar wind. Because of the weak gravity at the surface of a red giant, often red giants have much stronger winds. It is possible for a star to lose a significant fraction of its mass while passing through the red giant phase. A dramatic confirmation of mass loss comes from the observations of what are called *planetary nebulae*. Figure 4.12 shows a planetary nebula. Through the low-resolution telescopes of earlier times, a planetary nebula looked somewhat like a planet. We now know that a planetary nebula is essentially the outer shell of a star which has been blown off. At the centre of a planetary nebula, we usually find the hot core of the star, which is eventually expected to become a white dwarf. There is an even more violent mass loss mechanism, a *supernova*, which we discuss now.

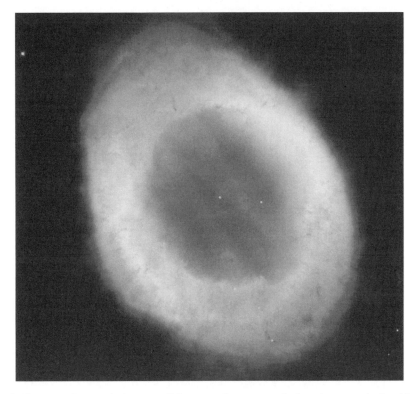

Fig. 4.12 The Ring Nebula, a well-known planetary nebula, photographed with the Hubble Space Telescope. Courtesy: NASA and Space Telescope Science Institute.

4.7 Supernovae

Chinese astronomers recorded that in the year 1054 a star in the Taurus constellation became so bright that it was visible during daytime. Figure 4.13 shows what a modern telescope finds in that spot of the sky. We see a luminous gas shell, known as the Crab Nebula because of its crab-like appearance. By comparing photographs taken at intervals of a few years, one easily finds that the shell is increasing in size and a simple backward extrapolation suggests that this shell must have started from a very small size around 1054. Presumably, what the Chinese astronomers recorded was the explosion of a star which created today's Crab Nebula. Statistical estimates suggest that there should be about 30 such supernova explosions in our Galaxy in every 1000 years. However, we are able to see only a very small fraction of our Galaxy in visible light, as we shall discuss in §6.1.3. Tycho and Kepler carefully studied two supernovae in our Galaxy seen in the years 1572 and 1604 respectively. No supernova has been observed in our Galaxy after the invention of the telescope! However, a supernova was seen in 1987 in the Large Magellanic Cloud, which is a companion to our Galaxy at a distance of about 55 kpc. Christened as SN 1987A, this

Fig. 4.13 The Crab Nebula, the remnant of the supernova seen from the Earth in 1054. Photographed with the Hubble Space Telescope. Courtesy: NASA, ESA, J. Hester and A. Loll.

was the most thoroughly studied supernova in the history of astronomy and has considerably increased our knowledge about supernovae. The energy involved in a typical supernova explosion is estimated to be about 10^{45} J.

By studying many supernovae, astronomers have concluded that supernovae can be divided into two types: Type I supernovae and Type II supernovae, which have certain different characteristics. These two classes are divided into some subclasses, but we shall not get into those details here. Amongst Type I supernovae, we shall confine our attention to the subclass Type Ia. All Type Ia supernovae appear almost identical. They reach exactly the same maximum intrinsic luminosity and afterwards their luminosities also decrease in exactly the same way. On the other hand, the Type II supernovae show some variations from one supernova to the other. We summarize below our current ideas of the physical mechanisms which trigger these two types of supernovae. The readers should be warned to take these theoretical ideas as provisional and not yet completely established.

Any model of Type Ia supernovae should explain why they always look almost identical. Suppose a white dwarf is in a close binary system. When its companion becomes a red giant, it is possible for a mass transfer to take place onto the white dwarf, as discussed in §4.5.1. We shall show in §5.3 that the maximum mass which a white dwarf can have is the Chandrasekhar mass of about $1.4 M_\odot$, beyond which it is not possible for electron degeneracy pressure to balance gravity. Suppose the mass transfer increases the mass of the white dwarf just beyond the Chandrasekhar mass. Then gravity cannot be balanced any more and the white dwarf star may have a catastrophic explosion, which probably disrupts the star completely without leaving any remnant behind. If all the Type Ia supernovae are produced in this way, by the explosions of white dwarfs of identical mass under identical conditions, then it is certainly expected that all these supernovae should appear identical.

Type II supernovae are believed to take place in much more massive stars. This is inferred from the fact that they usually take place in regions where star formation has taken place recently and massive stars, which are short-lived, are found only in such regions. When the core of the massive star completely runs out of all nuclear fuels, it starts shrinking until the core density becomes comparable to the density inside an atomic nucleus ($\approx 10^{17}$ kg m^{-3}). We shall show in §5.4 that the neutron degeneracy pressure may balance gravity at such densities. When this happens, the rapidly shrinking core suddenly stops shrinking any more. The surrounding material falling inward with the core gets bounced back when the collapse of the core is suddenly halted. Presumably the Type II supernova is caused by the explosive bouncing off of the envelope surrounding the newly formed neutron star core.

The variation of the supernova luminosity with time is called its *light curve*. Figure 4.14 shows the light curve of SN 1987A, which was a Type II supernova. A large portion of the light curve appears like an exponential decay (note that the vertical axis in Figure 4.14 is logarithmic) with a half-life of about 77 days. Now ^{56}Co, which is a radioactive isotope of cobalt, decays into ^{56}Fe with a half-life of 77.1 days. It is believed that copious amounts of ^{56}Co are produced in a Type II supernova and it is the decay of this which is responsible for the light curve.

We pointed out in §4.3 that nuclear reactions in the interiors of very massive stars may convert the core into iron, which is the maximally bound nucleus, so that no more nuclear burning is possible after its formation. How are the heavier elements produced then? Our current view is that the elements heavier than iron are synthesized in Type II supernovae, in a way suggested in a classic paper by Burbidge, Burbidge, Fowler and Hoyle (1957). Let us summarize the main ideas below. It is possible for an electron and a proton to combine to form a neutron:

$$p + e^- \rightarrow n + \nu. \tag{4.37}$$

However, since the mass of a neutron is more than the combined mass of a proton and an electron, this reaction cannot proceed unless some extra energy is

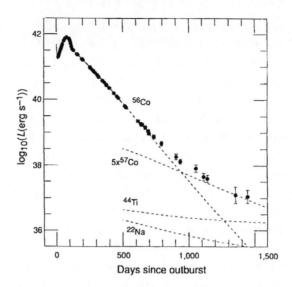

Fig. 4.14 The light curve of SN 1987A. The dashed lines indicate how the number densities of the radioactively decaying nuclei ^{56}Co and ^{57}Co would decline with time. From Chevalier (1992). (©Nature Publishing Group. Reproduced with permission from *Nature*.)

supplied. In a supernova explosion, electrons suddenly become highly energetic and it becomes possible for the above reaction to proceed, producing large numbers of neutrons and neutrinos. Now consider a nucleus of mass A and charge Z. The electrostatic repulsion of a heavy nucleus is much stronger than that of a light nucleus. So another charged particle cannot easily come near the heavy nucleus. But the uncharged neutron can come close and get absorbed by it, increasing the mass of the nucleus to $A + 1$. It is well known that nuclei too massive for their charge Z tend to be unstable to β-decay. If the nucleus emits a β-particle, we end up with a nucleus of mass $A + 1$ and charge $Z + 1$ starting from a nucleus of mass A and charge Z. Heavier nuclei can be built up in this way.

Our solar system has many elements heavier than iron which, as far as our present understanding goes, could only be synthesized in a supernova. Presumably there was a very massive star in our neighbourhood before the solar system formed. This massive star must have ended its life in a supernova and the debris of this supernova with heavy elements got mixed with interstellar gas, out of which the solar system formed.

Our preceding discussion suggests that many neutrinos should be produced by reaction (4.37) when the core collapses violently to trigger a Type II supernova. Evidence for this was found when 20 neutrinos from SN 1987A were detected by two experiments – one of them being Kamiokande which we have discussed in §4.4.2 in connection with solar neutrinos. The flux estimated from these neutrinos suggests that a very major portion of the gravitational potential energy lost (in the core collapse to produce a neutron star) must be carried away by the neutrinos. The arrival times of the neutrinos were spread over 12 s. If all the neutrinos were emitted at the same time and had zero mass, then they would have all travelled at speed c and should have arrived simultaneously. On the other hand, if the neutrinos had mass, then the less energetic neutrinos would have travelled slightly slower and one gets an upper bound of 20 eV for the neutrino mass from the observed spread in arrival times. The reader is asked to work this out in Exercise 4.6. This is an upper bound, since it is possible that the neutrinos were emitted at slightly different times and then travelled at the same speed.

4.8 Stellar rotation and magnetic fields

In our discussion of stellar structure, we have assumed spherical symmetry. There are two factors which could cause departures from spherical symmetry of a star – rotation and magnetic field. We know quite a lot about the rotation and magnetic field of our nearest star – the Sun. Within the last few years, our knowledge about the rotation and magnetic field of other stars has also increased. For normal stars, the effect of rotation or magnetic field is usually not enough to cause appreciable departures from spherical symmetry, which is the case for the Sun. Even when stellar rotation or the stellar magnetic field may not be important from the point of view of stellar structure, they are certainly intriguing astrophysical effects which can have many other consequences. Before leaving the subject of stellar astrophysics, here we provide a brief summary of what we know about solar rotation and magnetic fields.

Solar rotation

It was known for a long time that the Sun does not rotate like a solid body. The equator of the Sun rotates faster than the pole, taking about 25 days to go around the rotation axis, whereas a point near the pole would take more than 30 days to go around. It has now become possible to map the distribution of angular velocity in the interior of the Sun with the help of helioseismology, which was introduced briefly in §4.4.1. We basically measure the eigenfrequencies of many modes of oscillation in the Sun. Because of the spherical geometry, we expect that the velocity associated with a normal mode must be of the form

$$\mathbf{v}(t, r, \theta, \phi) = \exp(-i \, \omega_{nlm} t) \, \xi_{nlm}(r) Y_{lm}(\theta, \phi), \qquad (4.38)$$

where $Y_{lm}(\theta, \phi)$ is a spherical harmonic. If the Sun were non-rotating, it can be shown that ω_{nlm} would be independent of m. In other words, the eigenfunctions with the same n and l, but different m, would have the same frequencies. But

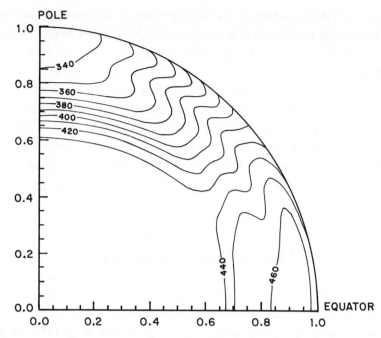

Fig. 4.15 The contours of constant angular velocity inside the Sun, as obtained by helioseismology. The contours are marked with rotation frequency in nHz. It may be noted that frequencies of 340 nHz and 450 nHz correspond respectively to rotation periods of 34.0 days and 25.7 days. Courtesy: J. Christensen-Dalsgaard and M. J. Thomson.

rotation causes frequencies with different m to be split (Gough, 1978). We point out the analogy from atomic physics that the energy levels of the hydrogen atom for different m are degenerate in the absence of a magnetic field. But a magnetic field lifts this degeneracy and splits the levels. In exactly the same way, the rotation of the Sun lifts the degeneracy of eigenfrequencies with different m. The amount of splitting of a mode depends basically on the angular velocity in the region where the mode has the largest amplitude. By studying the splittings of different modes having the largest amplitudes in different regions of the Sun, one can then obtain a map of how the angular velocity varies in the interior of the Sun. Figure 4.15 shows a map giving the distribution of angular velocity in the interior of the Sun. The Sun has a convection zone from $0.7 R_\odot$ to $R_\odot$, within which the variations of angular velocity are confined, with a radial gradient of angular velocity at the bottom of the convection zone.

Solar magnetic fields

It has been known in the Western world from the time of Galileo that the Sun often has dark spots on the surface. Hale (1908) discovered Zeeman splitting in the spectra of sunspots, thereby concluding that sunspots are regions of

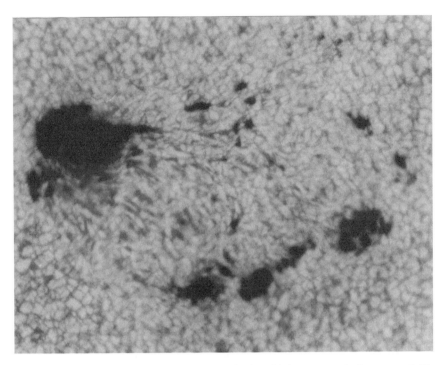

Fig. 4.16 A newly formed bipolar sunspot pair, in which one spot is fragmented. From Zwaan (1985). (©Springer. Reproduced with permission from *Solar Physics*.)

concentrated magnetic field of order 0.3 T. This is the first time that somebody conclusively established the existence of magnetic fields outside the Earth's environment. Often one finds two large sunspots lying side by side at nearly the same solar latitude. Figure 4.16 shows a sunspot pair in which one sunspot is actually broken into several fragments, which is often the case. Hale *et al.* (1919) discovered that two sunspots in such a pair have opposite polarities, making up a magnetic bipole. They also found that these magnetic bipoles are oriented in opposite directions in the two hemispheres. Figure 4.17 is a magnetogram image of the whole solar disk, where regions of positive polarity are indicated by white and regions of negative polarity by black, the regions without appreciable magnetic field being represented in grey. One notes that most bipolar magnetic regions are roughly aligned parallel to the solar equator. In the magnetic bipolar regions in the northern hemisphere, one finds the positive polarity (white) to appear on the right side of the negative polarity (black). This is reversed in the southern hemisphere, where white appears to the left of black. We shall discuss in §8.6 how one theoretically explains these remarkable observations.

Even before it was realized that sunspots are regions of strong magnetic fields, it was discovered that the number of sunspots on the solar surface increases and decreases in a cyclic fashion, with a period of about 11 years.

Fig. 4.17 A magnetogram picture of the full solar disk. The regions with positive and negative polarities are respectively shown in white and black, with grey indicating regions where the magnetic field is weak. Courtesy: K. Harvey.

There is a phase in the cycle when not many sunspots are seen. Then sunspots start appearing at around 40° latitude. As time goes on, newer sunspots tend to appear at lower and lower latitudes. This is clearly seen in the so-called *butterfly diagram* first introduced by Maunder (1904). Figure 4.18 shows a butterfly diagram in which the horizontal axis is time. At any particular time, those ranges of latitude (vertical axis) are marked where sunspots appear. The butterfly pattern results from the equatorward shift of the latitude zones where sunspots are seen. Eventually one finds only very few sunspots near the equator. Then the next cycle begins with sunspots appearing again around 40° latitude. It is found that the polarities of bipolar sunspots get reversed from one 11-year cycle to the next. In other words, if we had taken a magnetogram exactly like Figure 4.17 about 11 years before or after the time when Figure 4.17 was produced, then we would see black on the right side in the northern hemisphere and white on the right side in the southern. It thus implies that the period of the

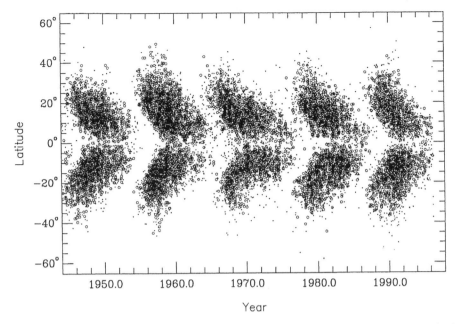

Fig. 4.18 The butterfly diagram showing the distribution of sunspots in latitude (vertical axis) at different times (horizontal axis). Courtesy: K. Harvey.

solar cycle is actually 22 years, if we want the magnetic field to come back to the initial configuration.

Astronomers have got evidence that many other stars have large starspots and also magnetic cycles like the Sun. Why do stars have magnetic fields at all and what gives rise to the cyclic behaviour of the magnetic fields? In §8.7 we shall give a qualitative introduction to the complex subject known as *dynamo theory* which seeks to answer this question.

4.9 Extrasolar planets

The study of planetary motions played a key role in the historical development of astronomy. The study of physical characteristics of planets, however, has now become a branch of science quite distinct from astrophysics and is usually referred to as *planetary science*. In this book, we do not get into a discussion of planetary science, since the methods and concepts used in planetary science are quite different from those used in modern astrophysics. But there is one question connected with planets which has always excited astronomers: do other Sun-like stars also have planets? The direct detection of a planet even around a nearby star is still extremely difficult with today's observing techniques. The best chance of discovering extrasolar planets is through indirect methods. For example, if a sufficiently heavy planet (like Jupiter or heavier) revolves around a star in a nearby orbit, its gravitational attraction would make the star also go in a circular or elliptical orbit around the common centre of mass. This would make the radial velocity of the star with respect to us vary periodically with time, which can be detected from the Doppler shifts of the star's spectral lines. While there have been several claims in the past for the discovery of extrasolar planets, Mayor and Queloz (1995) are credited with the first discovery which is accepted by astronomers to be genuine and which ushered in an era of many subsequent discoveries of extrasolar planets in very rapid succession. There are a few hundred confirmed detections at the time of writing this book and the list is growing rapidly.

As we pointed out in §3.6.1 and shall discuss in more detail in §8.3, stars form due to gravitational collapse of gas clouds in the interstellar medium. Planetary systems are also believed to form as a part of this star formation process. So presumably the planets can throw some light as to how stars form. Astrophysicists have yet to figure out what clues the recently discovered extrasolar planets give of the star formation process.

Exercises

4.1 Consider a nucleus of charge $Z_1 e$ approaching another nucleus of charge $Z_2 e$ with the energy of relative motion equal to E. According to classical physics, the nuclei should not be able to come closer than a distance r_1 given by

$$E = \frac{1}{4\pi \epsilon_0} \frac{Z_1 Z_2 e^2}{r_1}.$$

Using the WKB approximation of quantum mechanics, show that the tunnelling probability of the two nuclei coming within the range of nuclear forces is given by

$$P \propto \exp\left[-2 \int_{r_0}^{r_1} \left\{ \frac{2m}{\hbar^2} \left(\frac{1}{4\pi \epsilon_0} \frac{Z_1 Z_2 e^2}{r} - E \right) \right\}^{1/2} dr \right],$$

where m is the reduced mass and $r = r_0$ is the inner edge of the potential barrier at the nuclear surface. You can easily work out this integral by substituting $r = r_1 \cos^2 \theta$ and assuming $r_1/r_0 \gg 1$. Show that the final result is (4.7).

4.2 For two protons, show that the argument of the exponential given in the nuclear energy generation rate expression (4.19) becomes what is given in (4.25).

4.3 According to current solar models, the centre of the Sun has a temperature of about 1.56×10^7 K, a density of about 1.48×10^5 kg m^{-3} and a chemical composition given by $X_H = 0.64$, $X_{He} = 0.34$, $X_{CNO} = 0.015$. Estimate the amount of energy that is generated per unit volume at the centre of the Sun due to the *pp* chain and the CNO cycle.

4.4 Make a very rough estimate of the time that an acoustic wave propagating radially inward in the Sun would take to go from one end of the Sun to the other end.

4.5 Near the orbit of the Earth, the solar wind has a velocity of about 400 km s^{-1} and contains about 10 protons per cm^3. Assuming that the solar wind always had these characteristics during the Sun's lifetime of 4.5×10^9 yr, estimate the fraction of mass the Sun would have lost in the solar wind during its lifetime.

4.6 Neutrinos from Supernova 1987A which reached the Earth travelling a distance of 55 kpc were found to have energies in the range 6–39 MeV. If the spread of 12 s in arrival times was caused by neutrinos of different energies travelling at different speeds, show that the neutrino mass cannot be much more than about 20 eV.

5

End states of stellar collapse

5.1 Introduction

We have seen in the previous two chapters that the gravitational attraction inside a normal star is balanced by the thermal pressure caused by the thermonuclear reactions taking place in the stellar interior. Eventually, however, the nuclear fuel of the star is exhausted and there is no further source of thermal pressure to balance gravity. We have pointed out in §4.5 that such a star keeps on contracting – unless some kind of pressure other than thermal pressure is eventually able to balance gravity again. The aim of this chapter is to discuss the possible end configurations of stars which have no nuclear fuel left in them.

We have to make use of one very important property of Fermi particles. In a unit cell of volume h^3 in the six-dimensional position-momentum phase space, there cannot be more than two Fermi particles (one with spin up and the other with spin down). The electrons inside the stellar matter make up a Fermi gas, and when the density inside the contracting star becomes sufficiently high, this electron gas becomes 'degenerate'. This means that the theoretical limit of two particles per unit cell of phase space is almost reached. We shall show in §5.2 that such a degenerate Fermi gas exerts what is known as the *degeneracy pressure*. White dwarf stars discussed in §3.6 are believed to represent stellar configurations in which the inward pull of gravity is balanced by the degeneracy pressure of the electron gas. The structure of white dwarfs is discussed in §5.3, where we derive the famous result that a white dwarf configuration is possible only if the mass of the star is less than the Chandrasekhar mass limit of about $1.4 M_{\odot}$.

Another possible end configuration of stars is the neutron star configuration. As we shall show in §5.4, at very high densities electrons are forced to combine with the nuclei to produce matter primarily consisting of neutrons. Since neutrons are also Fermi particles, a gas of neutrons also exerts degeneracy pressure. A neutron star is a stellar configuration in which gravity is balanced by the neutron degeneracy pressure. Since the equation of state of matter at the very high densities prevailing inside neutron stars is not accurately known, the structure of neutron stars is not understood as well as the structure of white dwarfs. Neutron stars also have an upper limit of mass like white dwarfs. But this mass limit is not known very precisely due to the uncertainties in our knowledge of the equation of state. Most comprehensive calculations suggest that this mass limit is not more than $2M_\odot$.

Although neutron stars were theoretically postulated in the 1930s soon after the discovery of the neutron, they remained a theoretical curiosity for more than three decades. Pulsars were discovered in 1968 and were quickly identified to be rotating neutron stars. The very important field of observational investigations of neutron stars is summarized in §5.5–5.6.

The initial mass of a star does not necessarily have to be less than the mass limit of white dwarfs or neutron stars for the star to end up into one of these configurations. We have pointed out in §4.6–4.7 that a star can lose a considerable part of its mass during the late phases of evolution – in the form of a steady wind during the red giant phase, or through more drastic ejection mechanisms like the shedding of the outer shell as a planetary nebula or a supernova explosion. From statistical studies of various kinds of stars, it is inferred that stars less massive than about $4M_\odot$ eventually become white dwarfs, whereas stars with initial masses in the range $4M_\odot$ to $10M_\odot$ are believed to end up as neutron stars, typically after undergoing a supernova explosion (see, for example, Shapiro and Teukolsky, 1983, §1.3). Stars with initial masses more than $10M_\odot$ probably cannot shed enough mass to become white dwarfs or neutron stars. They have to go on contracting until the gravitational attraction is so strong that even light cannot escape. The physics of this *black hole* configuration will be discussed in §13.3. However, we shall make some comments on the observational evidence for black holes in §5.6.

5.2 Degeneracy pressure of a Fermi gas

The pressure in a gas arises from the random motions of the particles constituting the gas. If $4\pi f(p)p^2 dp$ is the number of particles having momentum between p and $p + dp$ (assuming the distribution function to be isotropic), whereas v is the velocity of a particle having momentum p, then the pressure P of the gas is given by a standard expression in kinetic theory

$$P = \frac{1}{3} \int vpf(p)\, 4\pi p^2 \, dp. \tag{5.1}$$

The reader should be able to derive it easily by considering a unit area on the wall of the gas container, figuring out the distribution of particles hitting this area in unit time and keeping in mind that the momentum changes in the elastic collisions provide the pressure (Exercise 5.1).

For an ordinary gas, on substituting the Maxwellian distribution in (5.1), the pressure is found to be given by $nk_B T$, where n is the number of particles per unit volume (Exercise 5.1). The pressure of stellar material containing different types of particles is given by (3.23). It is clear that this pressure, which arises out of thermal motions of particles, should go to zero at $T = 0$ – provided we assume the validity of classical physics. However, when a gas of Fermi particles is compressed to very high density, many of the particles are forced to remain in non-zero momentum states even at $T = 0$, thereby giving rise to the degeneracy pressure. When stellar matter is compressed, electrons become degenerate much before protons and other nuclei. The reason behind this is quite simple. If the kinetic energy $p^2/2m$ is equally partitioned amongst different types of particles, the lighter electrons are expected to have smaller momenta. Hence they occupy a much smaller volume of the momentum space and consequently their number density in this region of momentum space is higher than the corresponding number density of heavier particles. At a density which makes electrons degenerate, the heavier particles still remain non-degenerate (i.e. their phase space occupancy remains well below the theoretical limit). Electrons which occupy real space volume V and have momenta in the range d^3p in momentum space have $2\,V\,d^3p/h^3$ states in phase space available to them (two being due to the two spin states). If d^3p corresponds to the shell between p and $p + dp$, then the number of states per unit volume within this shell is clearly $8\pi p^2\,dp/h^3$. The occupancies of these states are given by the Fermi–Dirac statistics (see, for example, Pathria, 1996, Chapter 8). To make life simple, we shall neglect the finite-temperature effects and assume that all states below the Fermi momentum p_F are occupied, whereas all states above p_F are unoccupied. Then the number density n_e of electrons is given by

$$n_e = \int_0^{p_F} \frac{8\pi}{h^3} p^2\,dp = \frac{8\pi}{3h^3} p_F^3. \tag{5.2}$$

If all states between p and $p + dp$ are occupied, then $8\pi p^2 dp/h^3$ must equal $4\pi f(p)p^2 dp$, implying that $f(p)$ in (5.1) should be $2/h^3$ if $p < p_F$ and 0 if $p > p_F$. Hence

$$P = \frac{8\pi}{3h^3} \int_0^{p_F} v\,p^3\,dp. \tag{5.3}$$

We now use the relativistic expression that the momentum of a particle is given by $p = m\gamma v$, where γ is the Lorentz factor (see, for example, Jackson, 1999, §11.5). Then

$$v = \frac{p}{m\gamma} = \frac{pc^2}{E} = \frac{pc^2}{\sqrt{p^2c^2 + m^2c^4}}. \tag{5.4}$$

On using (5.3) and (5.4), the pressure due to the degenerate electron gas is finally given by

$$P = \frac{8\pi}{3h^3} \int_0^{p_F} \frac{p^4 c^2}{\sqrt{p^2 c^2 + m_e^2 c^4}} dp. \tag{5.5}$$

Our aim is to derive an equation of state connecting the pressure and density. Protons and other heavier nuclei present in the stellar material contribute to density, but not to pressure because they are non-degenerate. Let us first find out the relation between the density ρ and the electron number density n_e. If X is the hydrogen mass fraction, then the number density of hydrogen atoms (which are ionized and no longer exist in atomic form) is $X\rho/m_H$. These atoms contribute $X\rho/m_H$ electrons per unit volume. A helium atom has atomic mass 4 and contributes two electrons, i.e. the number of electrons contributed is 0.5 per atomic mass unit. For heavier atoms also, the number of electrons contributed is usually very close to 0.5 per atomic mass unit. In other words, for helium and atoms heavier than helium, the number of electrons is half the number of nucleons. In a unit volume of stellar matter, these atoms provide a mass $(1 - X)\rho$, which corresponds to $(1 - X)\rho/m_H$ nucleons. There are $(1 - X)\rho/2m_H$ corresponding electrons. Hence the electron number density is given by

$$n_e = \frac{X\rho}{m_H} + \frac{(1 - X)\rho}{2m_H} = \frac{\rho}{2m_H}(1 + X).$$

We write this in the form

$$n_e = \frac{\rho}{\mu_e m_H}, \tag{5.6}$$

where μ_e is the mean molecular weight of electrons given by

$$\mu_e = \frac{2}{1 + X}. \tag{5.7}$$

From (5.2) and (5.6), it follows that the Fermi momentum p_F is given by

$$p_F = \left(\frac{3h^3 \rho}{8\pi \mu_e m_H} \right)^{1/3}. \tag{5.8}$$

On evaluating the integral (5.5) with this expression of p_F, we get the equation of state relating P and ρ. Here we shall only consider the two extreme cases of the electrons being non-relativistic and fully relativistic. The reader is asked to work out the general case in Exercise 5.2.

When the electrons are non-relativistic, we can write

$$\sqrt{p^2 c^2 + m_e^2 c^4} \approx m_e c^2$$

so that (5.5) gives

$$P = \frac{8\pi}{15h^3 m_e} p_F^5.$$

On substituting from (5.8), we have

$$P = K_1 \rho^{5/3}, \tag{5.9}$$

where K_1 is given by

$$K_1 = \frac{3^{2/3}}{20\pi^{2/3}} \frac{h^2}{m_e m_H^{5/3} \mu_e^{5/3}} = \frac{1.00 \times 10^7}{\mu_e^{5/3}} \tag{5.10}$$

if we use SI units. The above non-relativistic equation of state for degenerate electrons was derived by Fowler (1926) who was the first person to realize that gravity inside a white dwarf must be balanced by electron degeneracy pressure.

When the electrons are fully relativistic, we can write

$$\sqrt{p^2 c^2 + m_e^2 c^4} \approx pc$$

so that (5.5) gives

$$P = \frac{2\pi c}{3h^3} p_F^4.$$

On substituting from (5.8), we have

$$P = K_2 \rho^{4/3}, \tag{5.11}$$

where K_2 is given by

$$K_2 = \frac{3^{1/3}}{8\pi^{1/3}} \frac{hc}{m_H^{4/3} \mu_e^{4/3}} = \frac{1.24 \times 10^{10}}{\mu_e^{4/3}} \tag{5.12}$$

if we use SI units.

We now have (5.9) and (5.11) giving the two extreme limits of the equation of state of degenerate stellar matter, whereas the ideal gas equation of state is given by (3.23). One important question is: which equation of state should be used when? For a particular combination of ρ and T, one of the expressions (3.23), (5.9) or (5.11) would be the most appropriate. In other words, if we make a diagram by plotting T versus ρ, regimes where different expressions of pressure should be used would correspond to different regions in this diagram. On a boundary between two such regions in the T versus ρ plot, the two different expressions for pressure valid on the two sides of the boundary should give the same value. Figure 5.1 shows the diagram constructed in this way, indicating the regions of validity of the ideal gas equation of state (3.23), the non-relativistic degenerate equation of state (5.9) and the relativistic equation of state (5.11). Blackbody radiation at temperature T exerts pressure $(1/3)a_B T^4$

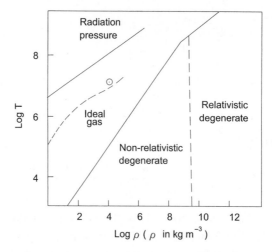

Fig. 5.1 Different regions in a density–temperature plot in which different equations of state hold. The dashed line indicates the run of density and temperature in the interior of the Sun. Adapted from Kippenhahn and Weigert (1990, p. 130).

and this also has to be included in a complete treatment. Figure 5.1 also indicates the region where the radiation pressure is going to be the dominant pressure. The dashed line corresponds to the run of temperature and density inside the Sun, indicating that the ideal gas equation of state is completely adequate in dealing with stars like the Sun.

5.3 Structure of white dwarfs. Chandrasekhar mass limit

It should be clear from the previous section that the equation of state of degenerate matter relates pressure with density (i.e. it does not involve temperature). Suppose we now want to calculate the structure of a star entirely made of degenerate matter (such as a white dwarf). The equations (3.25) and (3.26) alone suffice to formulate the problem completely if P is known as a function of ρ alone. Out of the three unknown variables ρ, P and M_r appearing in these two equations, one is no longer independent and the other two can be obtained by solving these two equations. The remaining two equations of stellar structure, (3.27) and (3.28), become redundant. Constructing the model of a star made of degenerate matter is, therefore, a mathematically simpler and cleaner problem than the problem of constructing the model of a normal star. We turn to this problem now. We can easily combine (3.25) and (3.26) into one single equation by eliminating M_r:

$$\frac{1}{r^2}\frac{d}{dr}\left(\frac{r^2}{\rho}\frac{dP}{dr}\right) = -4\pi G\rho. \tag{5.13}$$

Given an equation of state of the form $P(\rho)$, we can easily integrate (5.13).

The two limiting equations of state (5.9) and (5.11) are both of the form

$$P = K\rho^{(1+\frac{1}{n})} \tag{5.14}$$

with n equal to 3/2 and 3 respectively for the non-relativistic and fully relativistic cases. A relation like (5.14) between density and pressure is called a *polytropic relation*. We now write the density inside the star in the form

$$\rho = \rho_c \theta^n, \tag{5.15}$$

where ρ_c is the density at the centre of the star and θ is a new dimensionless variable which clearly has to have the value 1 at the centre. Substituting (5.15) into (5.14), we get

$$P = K\rho_c^{\frac{n+1}{n}} \theta^{n+1}. \tag{5.16}$$

We also introduce another dimensionless variable ξ through

$$r = a\xi, \tag{5.17}$$

where a defined as

$$a = \left[\frac{(n+1)K\rho_c^{\frac{1-n}{n}}}{4\pi G} \right]^{1/2} \tag{5.18}$$

has the dimension of length. On using (5.15), (5.16), (5.17) and (5.18), the basic structure equation (5.13) reduces to

$$\frac{1}{\xi^2} \frac{d}{d\xi} \left(\xi^2 \frac{d\theta}{d\xi} \right) = -\theta^n, \tag{5.19}$$

which is known as the *Lane–Emden equation* (Homer Lane, 1869; Emden, 1907). If the material inside a star satisfies the polytropic relation, the structure of the star can be found by solving the Lane–Emden equation. Since this is a second-order equation, we need two boundary conditions to integrate it. One boundary condition is obviously

$$\theta(\xi = 0) = 1. \tag{5.20}$$

The other boundary condition comes from the consideration that we do not want a cusp in the density at the centre of the star, which implies

$$\left(\frac{d\theta}{d\xi} \right)_{\xi=0} = 0. \tag{5.21}$$

It may be noted that the polytropic relation and the Lane–Emden equation played an important role in the history of stellar structure research. Some of the early pioneers like Emden (1907) and Eddington (1926) tried to obtain insights into the structures of stars by assuming the polytropic relation (5.14) to hold inside stars and then by solving the Lane–Emden equation. We know that such an approach gives only a very approximate model of a normal star. For understanding the structures of white dwarfs, however, this is the standard approach.

We now need to solve the Lane–Emden equation (5.19) subject to the two boundary conditions (5.20) and (5.21). It is possible to find analytical solutions if n has the values 0, 1 or 5 (see Exercise 5.4). Solving the Lane–Emden equation numerically for other values of n is fairly straightforward. If n is less than 5, then θ falls to zero for a finite value of ξ, which we denote by ξ_1. We interpret this as the surface of the star, where density and pressure as given by (5.15) and (5.16) have to go to zero.

We now want to show that it is possible to draw important conclusions without actually solving the Lane–Emden equation. Suppose we have a group of stars made up of matter satisfying the polytropic equation of state (5.14) with a particular value of n. Different stars in this group will have different values of ρ_c. We expect that a star with a particular value of ρ_c will have a particular value of radius R and a particular value of mass M. We now want to find out how ρ_c, R and M are related to each other amongst the stars in our group. If ξ_1 is the value of ξ where θ goes to zero, then the physical radius of the star is given by

$$R = a\xi_1.$$

Looking at the expression of a as given by (5.18), we conclude that

$$R \propto \rho_c^{\frac{1-n}{2n}}, \tag{5.22}$$

since all the other quantities appearing in the expression of R are the same for all members in our group of stars. The mass of the star is given by

$$M = \int_0^R 4\pi r^2 \rho \, dr = 4\pi a^3 \rho_c \int_0^{\xi_1} \xi^2 \theta^n d\xi. \tag{5.23}$$

Again the integral $\int_0^{\xi_1} \xi^2 \theta^n d\xi$ is going to be the same for all the members in our group of stars. Noting the dependence of a on ρ_c, we find

$$M \propto \left(\rho_c^{\frac{1-n}{2n}}\right)^3 \rho_c,$$

i.e.

$$M \propto \rho_c^{\frac{3-n}{2n}}. \tag{5.24}$$

We have noted that $n = 3/2$ substituted into (5.14) gives the non-relativistic equation of state (5.9). On putting $n = 3/2$ in (5.22) and (5.24), we get

$$R \propto \rho_c^{-1/6}, \quad M \propto \rho_c^{1/2},$$

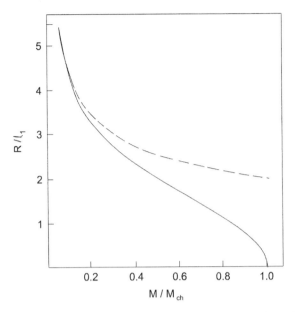

Fig. 5.2 The variation of radius with mass for white dwarfs. The solid curve corresponds to the full solution, where the dashed curve is obtained by using the non-relativistic equation of state (5.9). This figure is adapted from Chandrasekhar (1984), where the unit of radius l_1 used on the vertical axis is defined.

which combine to give

$$R \propto M^{-1/3}. \tag{5.25}$$

This is the very important mass–radius relation of white dwarfs within which matter satisfies the non-relativistic equation of state (5.9). The dashed line in Figure 5.2 shows how radius varies with mass when (5.9) is used to solve the structure of the white dwarf. It is clear that white dwarfs of increasing mass are smaller in size.

We now need to consider the case of the relativistic equation of state (5.11), which follows from (5.14) on taking $n = 3$. A very surprising result is that the mass M becomes independent of ρ_c on substituting $n = 3$ in (5.24). In other words, the mass of a star obeying the relativistic equation of state (5.11) has a fixed value and can be obtained from (5.23). On multiplying (5.19) by ξ^2 and integrating from $\xi = 0$ to $\xi = \xi_1$, we get

$$\int_0^{\xi_1} \xi^2 \theta^n d\xi = -\xi_1^2 \left(\frac{d\theta}{d\xi} \right)_{\xi=\xi_1}. \tag{5.26}$$

The integral in (5.23) can thus be replaced by $|\xi_1^2 \theta'(\xi_1)|$, where the prime denotes differentiation with respect to ξ. Additionally, we substitute the expression of a as given by (5.18) into (5.23) and then put the expression (5.12) in the place of K. This finally gives

$$M_{Ch} = \frac{\sqrt{6}}{32\pi} \left(\frac{hc}{G}\right)^{3/2} \left(\frac{2}{\mu_e}\right)^2 \frac{\xi_1^2 |\theta'(\xi_1)|}{m_H^2}. \tag{5.27}$$

On solving the Lane–Emden equation numerically for $n = 3$, we find $\xi_1^2|\theta'(\xi_1)| = 2.018$. Substituting the values of other quantities in (5.27), we find

$$M_{Ch} = 1.46 \left(\frac{2}{\mu_e}\right)^2 M_{\odot}. \tag{5.28}$$

We have come to the surprising conclusion that only this fixed value of mass is possible if the stellar material satisfies the relativistic equation of state (5.11) exactly. This fixed mass M_{Ch} is taken as the unit of mass on the horizontal axis of Figure 5.2. To understand what is happening, we have to consider the full equation of state following from (5.5) instead of considering the non-relativistic and fully relativistic limits. Using this equation of state in (5.13), one can find out the variation of radius with mass. This problem was worked out numerically by Chandrasekhar (1935). The solid curve in Figure 5.2 indicates the results we get on using the full equation of state. For white dwarfs of smaller masses (which also have larger sizes), the interior density is not so high and the non-relativistic limit of the equation of state holds. Hence the solid curve coincides with the non-relativistic dashed curve on the left side of the figure. For increasing masses and larger interior densities, the Fermi momentum p_F starts becoming larger as seen from (5.8). When $p_{FC} \approx m_e c^2$, the relativistic effects become important and the dashed curve deviates from the solid curve. On comparing (5.9) and (5.11), we find that the relativistic effects make the equation of state 'less stiff' or 'softer', i.e. the pressure does not rise with density as rapidly as in the non-relativistic case. This is basically due to the fact that the speeds of particles saturate at c and the pressure, which results from the random motions of particles, cannot increase with density as rapidly as it was increasing before the saturation. Matter with a softer equation of state is less efficient in counteracting gravity. As a result, we find that the solid curve is below the dashed curve, which implies that the radius of a white dwarf of given mass is less when the complete equation of state (which is softer than the non-relativistic one) is used. Eventually, as we move towards the right side of the figure, the radius becomes too small and the interior density becomes too high so that the relativistic limit of the equation of state is approached. The mass M_{Ch} corresponding to the relativistic limit of the equation of state is the limiting mass for which the radius goes to zero. This is the celebrated *Chandrasekhar mass limit* (Chandrasekhar, 1931). It is not possible for white dwarfs to have larger masses.

White dwarfs usually form from the cores of stars in which hydrogen has been completely burnt out to produce helium (and higher elements in some circumstances). If the hydrogen mass fraction $X \approx 0$, then it follows from (5.7) that $\mu_e \approx 2$. Hence (5.28) implies that the Chandrasekhar mass limit should be around $1.4 M_\odot$. It is seen in Figure 5.1 that the equation of state starts becoming relativistic when the density is of order $10^9 \, \text{kg m}^{-3}$. This can be taken as the typical density inside a white dwarf. If the mass is of order $10^{30} \, \text{kg}$, then the radius has to be about $10^7 \, \text{m} \approx 10^4 \, \text{km}$. This is indeed the typical size of a white dwarf.

5.4 The neutron drip and neutron stars

Just as the degeneracy pressure of electrons supports a white dwarf against gravity, the degeneracy pressure of neutrons supports a neutron star. According to astrophysical folklore, on hearing of the discovery of the neutron in Cavendish Laboratory (Chadwick, 1932), Landau immediately suggested that there can be stars primarily made up of neutrons. Unlike protons, neutrons are electrically neutral and hence many neutrons can be brought together without being disrupted by electrostatic repulsion. However, neutrons are known to decay according to the reaction

$$n \rightarrow p + e + \bar{\nu} \tag{5.29}$$

with a half-life of about 13 minutes. A reverse reaction is also in principle possible:

$$p + e \rightarrow n + \nu. \tag{5.30}$$

Since the neutron mass is more than the combined mass of a proton and an electron, the reaction (5.30) can take place only if some energy is supplied to make up for this mass deficit. Therefore, under ordinary laboratory circumstances, (5.30) is an unlikely reaction and free neutrons decay away following (5.29).

When matter is compressed to very high densities, things change drastically. For simplicity, let us assume that the highly compressed matter consists of electrons, protons and neutrons (i.e. we do not include the possibility that nuclei form). As we already pointed out in §5.2, the electrons become degenerate with the rise of density while the other heavier particles still remain non-degenerate. Suppose we want to put an additional electron in a region of high density. We know that all the levels are filled up to the Fermi momentum p_F, which is related to the number density n_e of electrons by (5.2). Let

$$E_F = \sqrt{p_F^2 c^2 + m_e^2 c^4}$$

be the Fermi energy associated with this Fermi momentum p_F. Unless an energy $E_F - m_e c^2$ is added to an electron, it is not possible to put the electron in the region of high density, since all the lower energy states are filled. Consider the situation when this excess energy required becomes equal to or larger than $(m_n - m_p - m_e)c^2$, the amount by which the neutron mass exceeds the sum of the proton mass and the electron mass. In this situation, it will be energetically favourable for the electron to combine with a proton to produce a neutron, in accordance with (5.30), rather than to exist as a free electron (assuming that neutrons are non-degenerate and a neutron can be created at the lowest energy state). The condition for this critical situation is

$$\sqrt{p_{F,c}^2 c^2 + m_e^2 c^4} - m_e c^2 = (m_n - m_p - m_e)c^2,$$

where $p_{F,c}$ is the critical Fermi momentum. From this

$$m_e c^2 \left(1 + \frac{p_{F,c}^2}{m_e^2 c^2}\right)^{1/2} = Qc^2,$$

where $Q = m_n - m_p$. This equation can be cast in the following form to give the critical Fermi momentum:

$$p_{F,c} = m_e c \left[\left(\frac{Q}{m_e}\right)^2 - 1\right]^{1/2}. \tag{5.31}$$

Since the Fermi momentum increases with density, we expect the Fermi momentum to be less than $p_{F,c}$ when the density is below a critical density. In this situation, free electrons are energetically favoured and we do not expect any neutrons to be present, since they would decay away following (5.29). The critical density, at which the Fermi momentum becomes equal to $p_{F,c}$, can be obtained by putting the values of fundamental constants in (5.31) to get $p_{F,c}$, then obtaining n_e with the help of (5.2) and multiplying n_e by $m_p + m_e$ (since only protons and electrons are present below the critical density). This gives

$$\rho_c = 1.2 \times 10^{10} \text{ kg m}^{-3}. \tag{5.32}$$

When the density is made higher than this, the electrons start combining with protons to give neutrons. This phenomenon is called the *neutron drip*. At densities well above the critical density, matter would mainly consist of neutrons. These neutrons do not decay according to (5.29) which is now completely suppressed, since there are no free states for the product electron to occupy (below the very high Fermi level).

We presented above a simplified calculation of neutron drip without considering the possible formation of nuclei. When the existence of nuclei is taken into account, the calculation becomes much harder. The interested reader may look at Shapiro and Teukolsky (1983, §2.6) for a derivation. On making various reasonable assumptions, the more realistic value of the critical density for neutron drip is found to be 3.2×10^{14} kg m^{-3}. Strictly speaking, the term 'neutron drip' refers to neutrons getting out of nuclei when the density is raised above the critical density.

If a stellar core is compressed by some means to densities higher than what is needed for the neutron drip, the core will essentially consist of neutrons.

Since neutrons are Fermi particles like electrons and obey the Pauli exclusion principle, neutrons also can give rise to a degeneracy pressure. While deriving the degeneracy pressure due to electrons in §5.2, we had used the Fermi–Dirac statistics, which tacitly assumes that the particles are non-interacting. This is not that bad an assumption for the electron gas inside a white dwarf. However, when neutrons are packed to densities close to the density inside an atomic nucleus (which is the case in the interiors of neutron stars), the neighbouring neutrons interact with each other through nuclear forces and it is no longer justified to treat them as non-interacting particles. Hence finding an accurate equation of state for matter at such high densities is very difficult and the subject is still not on a very firm footing. Like the Chandrasekhar limit of white dwarfs, neutron stars also have a mass limit. However, this mass limit is not known very accurately due to the uncertainty in our knowledge of the equation of state. One can get an absolute theoretical limit by demanding that the equation of state cannot be so stiff that the speed of sound is larger than the speed of light (Rhoades and Ruffini, 1974). While this absolute theoretical limit of neutron star mass is $3.2M_\odot$, it is generally believed that the actual mass limit is somewhat less than this and most likely around $2M_\odot$.

Detailed calculations suggest that a neutron star typically has a radius of order 10 km and internal density close to 10^{18} kg m^{-3}. We have pointed out in §1.5 that general relativity can be neglected if the factor $2GM/c^2r$ is small compared to 1. For a neutron star of mass $M_\odot$ and radius 10 km, this factor is as large as 0.3. Hence general relativistic effects cannot be neglected in a rigorous calculation. The hydrostatic equations (3.25) and (3.26) have to be modified when general relativity is included, as shown by Oppenheimer and Volkoff (1939). It is beyond the scope of this elementary textbook to discuss these relativistic corrections.

Neutron stars remained a theorist's curiosity for many years. Baade and Zwicky (1934) made a remarkable suggestion that a neutron star may form in a supernova explosion. When a star of mass $M_\odot$ collapses to a radius of 10 km, the gravitational potential energy lost is of order 10^{46} J, which is tantalizingly close to the energy output of a supernova. If the gravitational energy lost in the collapse of the inner core to form a neutron star is somehow dumped into the outer layers of the star, then the outer layers can explode with this energy. Nobody took this idea seriously until a dramatic confirmation of this idea came in the late 1960s, as we discuss in the next section.

5.5 Pulsars

A definitive observational confirmation for the existence of neutron stars came when Hewish *et al.* (1968) discovered radio sources which were giving out radio pulses at intervals of typically a second. The signal from such a source

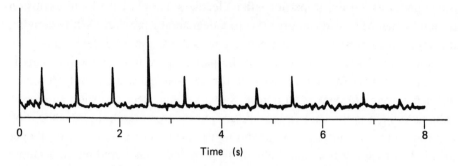

Fig. 5.3 Radio signals from the pulsar PSR 0329 + 54, which has a period of 0.714 s. Note that different pulses are not identical and some pulses are even missing.

called a *pulsar* is shown in Figure 5.3. Soon after the discovery, Gold (1968) identified pulsars as rotating neutron stars. The pulse period must be due to some physical mechanism like rotation or oscillation. Theoretical estimates of oscillation periods of white dwarfs or neutron stars show that they do not match the observed pulsar periods (oscillation periods of normal stars are much longer). If the pulsar period has to be identified with the rotation period of some object, one has to make sure that the centrifugal force is not stronger than gravity, i.e.

$$\Omega^2 r < \frac{GM}{r^2},$$

which implies

$$\Omega < (G\rho)^{1/2}. \tag{5.33}$$

A rotation period of 1 s demands that the rotating object should have a density higher than 10^{11} kg m^{-3} if it is not to be disrupted by the centrifugal force. The pulsars with shortest periods could not be rotating white dwarfs (which have densities of order 10^9 kg m^{-3}). The only possibility is that the pulsars are rotating neutron stars.

When pulsars were found near the centres of Crab and Vela supernova remnants, the idea of Baade and Zwicky (1934) that neutron stars are born in supernova explosions got dramatic support. However, only a few clear pulsar and supernova remnant associations are known. Most of these cases are for supernova remnants which are not very old (less than 10^5 yr). One possibility is that many of the supernova explosions may be somewhat asymmetric and the neutron stars may be born with a net momentum. So they move away from the centres of the supernova remnants and are found associated with the remnants only if not too much time elapsed since the explosion. The other possibilities are: many supernovae may not produce neutron stars, or the neutron stars may not be visible to us as pulsars.

This brings us to a central question: why do rotating neutron stars become visible as pulsars? Presumably the radio emission is produced at the magnetic poles of the neutron star by complicated plasma processes which we shall not discuss in this book. Very often the magnetic axis is inclined with respect to the rotation axis. When the magnetic pole gets turned towards the observer during a rotation period, the observer receives the radio pulse. The duty cycle of a typical pulsar (i.e. the fraction of time during which the radio signal is received) is less than 10%.

From where does the pulsar get the energy which is radiated away? The rotational kinetic energy of the neutron star is believed to be the ultimate source of energy. As this energy source is tapped, the neutron star rotation slows down. The periods of all pulsars keep on increasing very slowly as a result of this. The typical period increase rate is $\dot{P} \approx 10^{-15}$ s s^{-1}. This gives the pulsar lifetime $P/\dot{P}$, which is of order 10^7 yr. After a neutron star has existed as a pulsar for time of the order of 10^7 yr, presumably its rotation becomes so slow that it can no longer act as a pulsar. From the period increase rate of the Crab pulsar, one can calculate the rotational kinetic energy loss rate (by making some reasonable assumptions about mass and radius to get the moment of inertia). This energy loss rate is about 6×10^{31} W (see Exercise 5.8) and turns out to be approximately the same as the rate of total energy emission from the whole Crab Nebula, which is several orders of magnitude larger than the energy given out in the radio pulses. It thus seems that the energy for powering the whole Crab Nebula ultimately comes from the rotational kinetic energy of the pulsar.

A rapidly rotating object like a pulsar is expected to be somewhat flattened near the poles. As the rotation slows down, the pulsar tries to take up a more spherical shape. Since the crust of a neutron star is believed to be solid, the shape of the neutron star cannot change continuously. When sufficient stress builds up due to the slowing down of the neutron star, the crust suddenly breaks and the neutron star is able to take up a less flattened shape, causing a decrease in the moment of inertia because more material is brought near the rotation axis. When this happens, the moment of inertia changes abruptly and the angular velocity increases suddenly to conserve the angular momentum, leading to a decrease in pulsar period. Such sudden decreases of pulsar periods have been observed and are known as *glitches*. Apart from these occasional sudden glitches, pulsar periods steadily keep on increasing. Figure 5.4 shows the variation of the period of a pulsar with time. Four glitches can be seen in this figure.

Standard textbooks of classical electrodynamics usually derive the expression for energy loss from an oscillating dipole (see, for example, Jackson, 1999, §9.2). It is instructive to show that the analogous expression for the energy loss

Fig. 5.4 Variation of the period of pulsar PSR 0833 − 45 from late 1968 to mid-1980. Four glitches are seen. From Downs (1981). (©American Astronomical Society. Reproduced with permission from *Astrophysical Journal*.)

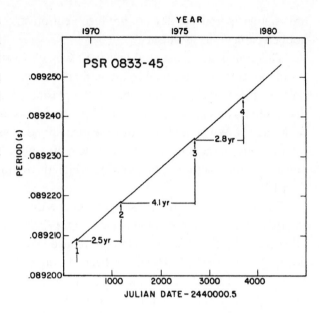

rate from an oscillating magnetic dipole **m** is

$$\dot{E} = -\frac{\mu_0}{6\pi c^3}|\ddot{\mathbf{m}}|^2.$$

If the variation of **m** arises due to a magnetic dipole rotating about an axis with an inclination α, then it follows that

$$\dot{E} = -\frac{\mu_0 \Omega^4 \sin^2 \alpha}{6\pi c^3}|\mathbf{m}|^2, \tag{5.34}$$

where $|\mathbf{m}|$ is the amplitude of the magnetic dipole and Ω is the angular velocity of rotation. A simple way of modelling the emission from a pulsar is to treat it as a rotating magnetic dipole. If the magnetic field of the pulsar is of dipole nature, then the magnetic field at the pole is given by

$$B_{\mathrm{p}} = \frac{\mu_0|\mathbf{m}|}{2\pi R^3},$$

where R is the radius of the neutron star. Writing $2\pi B_{\mathrm{p}} R^3/\mu_0$ for $|\mathbf{m}|$ in (5.34), we get

$$\dot{E} = -\frac{2\pi B_{\mathrm{p}}^2 R^6 \Omega^4 \sin^2 \alpha}{3\mu_0 c^3}.$$

If this energy comes from the rotational kinetic energy $\frac{1}{2}I\Omega^2$ (where I is the moment of inertia), then we must have

$$I\dot{\Omega} = -\frac{2\pi B_{\mathrm{p}}^2 R^6 \Omega^3 \sin^2 \alpha}{3\mu_0 c^3}. \tag{5.35}$$

Once Ω and $\dot{\Omega}$ of a pulsar have been determined, one can use (5.35) to obtain the pulsar magnetic field B_p by putting reasonable values of I and R. For the Crab pulsar, this yields

$$B_p \approx 5 \times 10^8 \text{ T},$$

if we take $\sin \alpha \approx 1$. The magnetic fields of pulsars are the strongest magnetic fields known to mankind. A possible reason for these very strong magnetic fields will be pointed out in §8.5. It is true that the assumption used in deriving (5.35), namely that the pulsar is a rotating magnetic dipole sitting in a vacuum, is approximate. It was shown by Goldreich and Julian (1969) that a rotating neutron star should be surrounded by a magnetosphere filled with plasma. However, even on purely dimensional grounds, we expect something like (5.35) to hold at least approximately.

5.5.1 The binary pulsar and testing general relativity

We now discuss a very intriguing object which was first discovered by Hulse and Taylor (1975). They found a pulsar with a mean period of 0.059 s. However, the actual value of the period was found to vary above and below this mean value periodically, with a period of about 8 hours. The most obvious explanation is that the pulsar is orbiting around an unseen companion and the variation in the pulsar period is due to the Doppler effect. When the pulsar is moving towards us, its period is observed to decrease, whereas when the pulsar is moving away, the period increases. One can determine the masses of both the pulsar and the unseen binary companion by analysing the various orbit parameters (see, for example, Shapiro and Teukolsky, 1983, §16.5). Both the masses are found to be close to $1.4 M_\odot$. The unseen companion seems to have exactly the mass beyond which the white dwarf configuration is impossible. The unseen companion is very likely to be another neutron star.

We thus have a remarkable system in which two neutron stars are orbiting around each other, one of them acting as a pulsar. The orbit is found to be highly eccentric, the eccentricity being 0.62. According to general relativity, such an object would emit gravitational radiation, just as an orbiting charge would emit electromagnetic radiation according to classical electrodynamics. As the system loses energy in the gravitational radiation, the two neutron stars should come closer and the orbital period should decrease. Careful general relativistic calculations suggest a value $\dot{P}_{orb} = -2.40 \times 10^{-12}$ for the orbital period change. The measured value $(-2.30 \pm 0.22) \times 10^{-12}$ is in very good agreement. This provides a test of general relativity to a high degree of precision and provides an indirect confirmation of the existence of gravitational radiation (to be discussed in §13.5), which astronomers have yet to detect directly from any astronomical system.

5.5.2   Statistics of millisecond and binary pulsars

Backer *et al.* (1982) discovered a pulsar with a period of 1.56 ms, which was considerably shorter than the period of any pulsar known at that time. The pulsar with the second shortest period known at that time, the Crab pulsar, had a period of 33.1 ms. Subsequently several other pulsars with periods less than 10 ms were discovered. One striking feature is that a majority of them were found in binary systems. After measuring the period variation $\dot{P}$ of these millisecond pulsars, their magnetic fields could be estimated by applying (5.35). Most of the millisecond pulsars were found to have magnetic fields around 10^4 T, considerably less than the typical magnetic fields of ordinary pulsars (around 10^8 T). Figure 5.5 is a plot of magnetic field B against pulsar period of P. A pulsar with known values of B and P is represented by a point in this figure. Pulsars in binary systems are indicated by small circles. The ordinary pulsars are towards the upper right part of the figure, whereas the millisecond pulsars are towards the lower left. While very few of the ordinary pulsars are in binary systems, many of the millisecond pulsars are found in binaries. It is clear that the ordinary pulsars and the millisecond pulsars make two very distinct

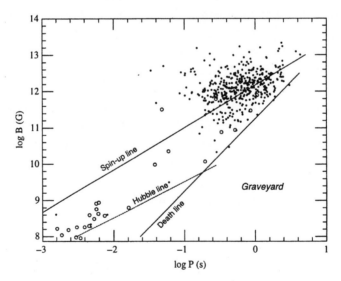

Fig. 5.5 The periods and magnetic fields (in $G = 10^{-4}$ T) of different pulsars. Pulsars in binary systems are indicated by circles. See text for explanations of *death line* and *Hubble line*. After a neutron star is spun up by binary accretion, it is expected to end up slightly below a line denoted as the *spin-up line*. The arguments for calculating this line are not given here. From Deshpande, Ramachandran and Srinivasan (1995), based on the pulsar parameters provided by Taylor, Manchester and Lyne (1993). (©Indian Academy of Sciences. Reproduced with permission from *Journal of Astrophysics and Astronomy*.)

population groups. If a neutron star is rotating too slowly or has a too weak magnetic field, then presumably it would not act as a pulsar. The line marked *death line* in Figure 5.5 is a line beyond which a neutron star no longer acts as a pulsar. One can obtain this death line by theoretical arguments based on the physics of pulsar magnetospheres, which we shall not discuss here. As a pulsar becomes older, its period becomes longer and it follows a trajectory moving towards the right in Figure 5.5. Eventually it crosses the death line and is no longer visible as a pulsar. The age of a pulsar is approximately given by $P/\dot{P}$. The *Hubble line* in Figure 5.5 indicates a line below which the age of a pulsar would be larger than the Hubble time (which is the approximate age of the Universe, to be introduced in §9.3).

What is the relation of millisecond pulsars with ordinary pulsars? The fact that millisecond pulsars are usually found in binary systems (those which are found single probably had the binaries disrupted at some stage) has led to a unified scenario in the last few years. When a neutron star is born, it is expected to have values of rotation period P and magnetic field B typical of an ordinary pulsar. Suppose the neutron star is in a binary system. At some stage, the binary companion may become a red giant and fill up the Roche lobe. As discussed in §4.5.1, this would lead to a transfer of mass from the inflated companion star to the neutron star. The binary X-ray sources to be discussed in §5.6 are believed to be neutron stars accreting matter from inflated binary companions. Because of the orbital motion of the companion, the matter accreting onto the neutron star from its companion will carry a considerable amount of angular momentum. This is expected to increase the angular velocity of the accreting neutron star, leading to a decrease in rotation period. Eventually, when the red giant phase of the companion star is over (it may become a white dwarf or another neutron star), the neutron star which has been spun up by accreting matter with angular momentum becomes visible as a millisecond pulsar with a short period P. It is necessary to provide an explanation of the magnetic field decrease as well. Various alternative theoretical ideas have been suggested. One idea is that the accreted material on the neutron star covers up and buries the magnetic field so efficiently that very little magnetic field is present at the surface. A numerical simulation of this idea by Konar and Choudhuri (2004) shows that this is possible and the surface magnetic field decreases exactly by a factor which is consistent with observational data.

5.6 Binary X-ray sources. Accretion disks

A second kind of evidence for the existence of neutron stars started coming at about the same time when pulsars were discovered. Giacconi *et al.* (1962) discovered several celestial X-ray sources with the help of Geiger counters sent aboard a rocket. After the satellite Uhuru devoted exclusively to X-ray astronomy was launched, these X-ray sources could be studied in more detail. Most of these sources were found to be in the galactic plane, indicating that they are galactic objects. It became possible to identify the optical counterparts of some of these X-sources. The optical counterparts were invariably binary stellar systems. Something must be happening in these binaries to produce the X-rays.

Suppose we drop a mass m from a height h in a gravitational field g. The gravitational potential energy mgh is first converted into kinetic energy, and then, on hitting the ground, this energy is transformed into other forms such as heat and sound. Ordinarily, in this process, a very small fraction of the rest mass energy mc^2 is released. If, however, the mass m is dropped from infinity to a star of mass M and radius R, then the gravitational energy lost is

$$\frac{GM}{R}m = \frac{GM}{c^2 R}mc^2.$$

For a typical neutron star of mass $1 M_\odot$ and radius $10\,\text{km}$, the factor $GM/c^2 R$ turns out to be about 0.15. Hence the loss of gravitational energy may be a very appreciable fraction of the rest mass energy, making such an infall of matter into the deep gravitational well of a compact object like a neutron star a tremendously efficient process for energy release.

We have pointed out in §4.5.1 that there can be mass transfer between the two stars in a binary system. Suppose one member of a binary is a compact object like a neutron star or a black hole, whereas the other member is a star which has filled up the Roche lobe. Then the compact star will accrete matter from its companion. The accreted matter loses a large amount of gravitational potential energy while falling towards the compact star and this energy presumably is radiated away. This seems to be the likely mechanism by which most of the X-ray sources are powered. We pointed out in §5.5.2 that millisecond pulsars are believed to be neutron stars spun up by the deposition of angular momentum in a binary mass transfer process. The X-ray binary sources are basically such systems caught in the act of such mass transfer. A millisecond pulsar is a possible end product after the mass transfer is over.

Since the accreting material carries angular momentum, it is unlikely to fall radially inward, but is expected to move inward slowly in the form of a disk as shown in Figure 5.6. Such a disk is called an *accretion disk*. A particular parcel of gas will follow a spiral path. A classic investigation of the accretion disk physics is due to Shakura and Sunyaev (1973). Here we merely point out some of the salient features. Just as the planets move in nearly circular orbits around the Sun, a parcel of gas in an accretion disk also moves in a nearly circular

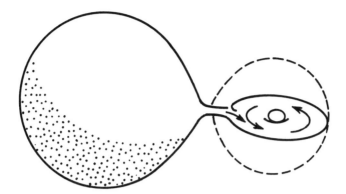

Fig. 5.6 Sketch of an accretion disk in a binary stellar system.

orbit. Balancing gravity by centrifugal force, we can easily find that the angular velocity at a distance r is given by

$$\Omega = \left(\frac{GM}{r^3}\right)^{1/2}. \tag{5.36}$$

The angular velocities of planets around the Sun indeed vary as $r^{-3/2}$, leading to Kepler's third law of planetary motion. Hence a circular motion satisfying (5.36) is often called *Keplerian motion* in astronomical jargon. If there was no viscosity in the accretion disk, then parcels of gas could forever move in Keplerian orbits, just as planets seem to move forever around the Sun. However, the viscous drag between adjacent layers of gas moving with different angular velocities causes material to spiral inward continuously in the inner regions of the disk. As material spirals inward in the accretion disk losing gravitational potential energy, this energy is radiated away from the disk.

   If the accreting material falls on a compact object of mass M and radius R, then a parcel of unit mass loses energy $-GM/R$ in falling onto that object and this energy is radiated away. If $\dot{M}$ is the mass accretion rate, then we expect the resultant luminosity to be

$$L = \frac{GM\dot{M}}{R}. \tag{5.37}$$

It is clear that the accretion rate $\dot{M}$ determines how luminous the source will be. If the accretion rate is too high and the source is too luminous, then the outward force on matter due to radiation pressure may be more than the inward pull due to gravity. We have discussed this in §3.6.1 and came to the conclusion that the luminosity cannot exceed the Eddington luminosity. Otherwise, matter will be blown outward reducing the accretion rate until the accretion rate adjusts to such a value that the luminosity does not exceed the Eddington luminosity. On the basis of such arguments, we expect the luminosity of the brightest accreting objects to be close to the Eddington luminosity. It is Thomson scattering which is the main source of opacity in accreting matter. Using the expression (2.84) for opacity due to Thomson scattering, we find the Eddington luminosity from (3.44) as follows:

$$L_{Edd} = \frac{4\pi c \, GM m_H}{\sigma_T} = 1.3 \times 10^{31} \left(\frac{M}{M_\odot}\right) W, \qquad (5.38)$$

on putting values of various quantities. It is quite remarkable that the brightest X-ray sources are found to have luminosities close to 10^{31} W on the lower side. If the luminosity as given by (5.37) is equal to the Eddington luminosity given by (5.38), then we find that the accretion rate is given by

$$\dot{M} = 1.5 \times 10^{-8} M_\odot \, yr^{-1} \qquad (5.39)$$

on taking $M = M_\odot$, $R = 10$ km. This is the typical accretion rate in binary systems. Suppose the luminosity is emitted thermally from the neutron star surface where the accreting material falls. The temperature T of this region can be found from

$$L = 4\pi R^2 \sigma T^4.$$

On taking $L = 10^{31}$ W and $R = 10$ km, the temperature is found to be about 2×10^7 K. Blackbody radiation at this temperature peaks in the X-ray part of the

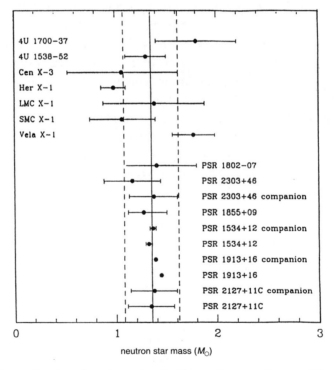

Fig. 5.7 Mass estimates of neutron stars in binary X-ray systems and in binary pulsars. From Longair (1994, p. 114) who credits J. Taylor for the figure. (©Cambridge University Press.)

spectrum. Thus the theoretical model of accretion onto neutron stars gives a very natural explanation of how the X-rays arise. For accretion onto white dwarfs, the temperature would be much less and the radiation would not predominantly be in the X-rays.

As we pointed out in §3.5.1, the mass of a star can be determined if it is in a binary system. The masses of many neutron stars in binary X-ray sources and binary pulsars have been determined. Figure 5.7 shows the masses of several neutron stars which could be determined with reasonable accuracy. All the masses are presumably below the upper mass limit of neutron stars (which is not accurately known). However, there are a few binary X-ray sources with accreting objects which possibly have masses higher than $3M_\odot$. The best-studied of these objects is Cygnus X-1. It shows variabilities in luminosity in different time scales. The central accreting object is believed to be a black hole rather than a neutron star, since its estimated mass is well above what would be the neutron star mass limit based on any reasonable equation of state.

Exercises

5.1 Derive the general expression (5.1) for pressure in a gas by considering a unit area on the wall of the gas container, figuring out the distribution of particles hitting this area in unit time and keeping in mind that the momentum changes in the elastic collisions provide the pressure. From the expression of the Maxwellian distribution given by (2.27), figure out $f(p)$ for that distribution and show that the pressure exerted is $n\kappa_B T$.

5.2 Work out the integral in (5.5) by substituting $p = m_e c \sinh\theta$ and show that the general expression for the electron degeneracy pressure given by (5.5) is equal to

$$P = \frac{\pi m_e^4 c^5}{3h^3} f(x),$$

where

$$f(x) = x(2x^2 - 3)\sqrt{x^2 + 1} + 3\sinh^{-1} x$$

and $x = p_F/m_e c$. Evaluate $f(x)$ numerically for various values of x and use these numerical values to make a plot of $\log P$ against $\log \rho$. Indicate regions of the plot corresponding to the two limiting equations (5.9) and (5.11).

5.3 Carry out all the algebraic and numerical steps to obtain (5.9)–(5.12). Then produce Figure 5.1 yourself by following the procedure mentioned in the text. Justify this procedure by careful arguments.

5.4 Consider the Lane–Emden equation

$$\frac{1}{\xi^2}\frac{d}{d\xi}\left(\xi^2\frac{d\theta}{d\xi}\right) = -\theta^n$$

to be solved with the boundary conditions

$$\theta = 1, \frac{d\theta}{d\xi} = 0$$

at $\xi = 0$. Obtain analytical solutions for the cases $n = 0$ and $n = 1$.

[Hint: To solve for $n = 1$, first substitute

$$\theta = \frac{\chi}{\xi},$$

where χ is a new variable. Then show that this substitution transforms the Lane–Emden equation to

$$\frac{d^2\chi}{d\xi^2} = -\frac{\chi^n}{\xi^{n-1}}.]$$

5.5 Consider a star in which gas pressure and radiation pressure are both important (i.e. the total pressure is the sum of the two). If the gas pressure given by (3.23) is equal to a constant fraction β of the total pressure everywhere inside the star, then show that the total pressure has to be related to the density in the following way

$$P_{\text{tot}} = \left(\frac{3\kappa_B^4}{a_B\mu^4 m_H^4}\right)^{1/3}\left(\frac{1-\beta}{\beta^4}\right)^{1/3}\rho^{4/3}.$$

Now consider several stars with different masses having the same composition (i.e. the same μ). Assuming that inside each of these stars the gas pressure is everywhere a constant fraction β of the total pressure (but β has different values for different stars), show that β inside a star would be related to its mass M by an equation of the form

$$\frac{1-\beta}{\beta^4} = CM^2,$$

where C is a constant which you have to evaluate. Show that β is smaller for larger M, implying that radiation pressure is increasingly more important inside more massive stars. This is a historically important argument first given by Eddington (1926, §84).

5.6 Those of you who are proficient in doing numerical calculations with computers can use the equation of state derived in Exercise 5.2 to solve the structure equation (5.13). On solving the equation with a particular value ρ_c of central density, you will get a model of a star with mass M and radius R. Plot R as a function of M and show that R falls to zero when M is equal to the Chandrasekhar mass. If you can do all these, then you have repeated the calculation for which Chandrasekhar won the Nobel Prize!

5.7 The Sun has a rotation period of about 27 days. If the Sun collapsed to become a white dwarf conserving its angular momentum, what would be the expected rotation period? What would be the rotation period if the Sun collapsed to become a neutron star?

5.8 The Crab pulsar has period $P = 0.033$ s and characteristic slowing time $P/\dot{P} = 2.5 \times 10^3$ yr. Estimate the energy loss rate and the magnetic field by using (5.35).

5.9 Determine the constant of proportionality in the mass–radius relation (5.25), using the fact that $\xi_1 = 3.65$ and $\xi_1^2|\theta'(\xi_1)| = 2.71$ for $n = 3/2$. We pointed out in §3.6.1 that the limiting mass of a brown dwarf is $0.08 M_\odot$. Assuming that gravity is balanced by the electron degeneracy pressure, estimate the radius of this limiting brown dwarf. If the brown dwarf formed by gravitational collapse from a much larger size, estimate the thermal energy acquired by the brown dwarf during its formation. Assuming the brown dwarf to have a uniform temperature (which is not too bad an assumption because the thermal conductivity of degenerate matter is high), estimate its temperature. Note that the temperature has to be higher than 10^7 for nuclear burning to start.

6

Our Galaxy and its interstellar matter

6.1 The shape and size of our Galaxy

When we look around at the night sky, we find that the stars are not distributed very uniformly. There is a faint band of light – the Milky Way – going around the celestial sphere in a great circle. Even a moderate telescope reveals that the Milky Way is a collection of innumerable faint stars. Herschel (1785) offered an explanation of the Milky Way by suggesting that we are near the centre of a flat disk-like stellar system. When we look in the plane of the disk, we see many more stars than what we see in the other directions, thus producing the band of the Milky Way. After the development of photography, it became much easier to record distributions of stars in different directions. In the beginning of the twentieth century, Kapteyn attempted to put Herschel's view on a firm footing, by undertaking a huge programme of counting stars in different directions and measuring their proper motions with a view of estimating distances. From a painstaking statistical analysis of these data, it was inferred that we are at the centre of an oblate stellar disk with a thickness of a few hundreds of pc and a disk radius of about a few kpc (Kapteyn and van Rhijn, 1920; Kapteyn, 1922). This model is usually referred to as the *Kapteyn Universe*, since it was believed at that time that this was the whole Universe! Before we discuss how the Kapteyn Universe was demolished by Shapley's work and what is still our accepted view of our Galaxy got established, we want to say a few words about star count analysis.

6.1.1 Some basics of star count analysis

We shall not discuss here the details of how the statistical analysis of star count data is carried out. The interested readers may look at Chapter 4 of Mihalas and Binney (1981). We just present some elementary considerations by assuming that the space around us is free of any absorbing material. Suppose we are surrounded by identical stars of absolute magnitude M distributed in space with a uniform density. We want to find the number $N(m)$ of stars which appear brighter than apparent magnitude m. It should be clear from (1.8) that a star would have apparent magnitude m if it is located at a distance

$$d = (10)^{1+0.2(m-M)} \text{ pc.} \tag{6.1}$$

All stars within a sphere of size $(4/3)\pi d^3$ around us would appear brighter than the magnitude m. The number $N(m)$ of such stars, which is clearly proportional to d^3, can be written as

$$N(m) = C_1 10^{0.6m}, \tag{6.2}$$

where C_1 is a constant. So, if we find that the observed $N(m)$ obeys (6.2) up to a certain value of m, then we can conclude that stars are distributed uniformly up to the distance d corresponding to that m as given by (6.1). If the observed $N(m)$ falls below what is theoretically expected from (6.2) beyond a certain m, then we know that we are reaching the edge of the system at the distance corresponding to that m. Checking whether the observed $N(m)$ for a certain type of stars agrees with (6.2) is a powerful test for finding if those stars are distributed uniformly around us. This test can also be applied to study the distribution of galaxies around our Galaxy.

If we had an infinite Universe uniformly populated with stars, then it can be easily shown that the brightness of the sky would have been infinite – a result known as the *Olbers paradox* (Olbers, 1826). The differential star count $A(m)$ (defined such that the number of stars having apparent magnitude between m and $m + dm$ is $A(m)dm$) is obviously given by

$$A(m) = \frac{dN(m)}{dm} = C_2 10^{0.6m}, \tag{6.3}$$

where $C_2 = 0.6 C_1 \ln 10$. From (1.6), we know that the light received by us from the star of apparent magnitude m can be written as

$$l(m) = l_0 10^{-0.4m}. \tag{6.4}$$

Hence the light received by us from stars with apparent magnitudes between m and $m + dm$ is

$$l(m) A(m) dm = l_0 C_2 10^{0.2m} dm$$

on substituting from (6.3) and (6.4). The total light received from all stars brighter than m is then given by

$$\mathcal{L} = \int_{-\infty}^{m} l(m') A(m') dm' = l_0 C_2 \int_{-\infty}^{m} 10^{0.2m'} dm' = K 10^{0.2m}, \tag{6.5}$$

where

$$K = \frac{l_0 C_2}{0.2 \ln 10} = 3 l_0 C_1.$$

It is clear from (6.5) that $\mathcal{L}$ diverges exponentially with m as we include fainter stars at greater distances which have increasingly larger values of m. Because of the finite size of our Galaxy, we can get around the Olbers paradox for stars in the Galaxy. However, we encounter this paradox again when we consider light received by all the galaxies outside our Galaxy. The resolution of this paradox for galaxies will be discussed in §14.4.1.

Our elementary discussion of star count analysis has been based on the assumption that all stars are alike. It is not very difficult to extend this discussion for a distribution of stars with different intrinsic properties (see Exercise 6.1). Often one tries to count only stars of a particular spectral type which have the absolute magnitudes lying in a narrow range. By obtaining the distribution function $N(m)$ for these stars in different directions and by comparing it with the result (6.2) for uniform distribution, it is in principle possible to determine the distances in different directions where these stars are under-abundant or over-abundant, thereby generating a map of the density distribution of these stars. Usually a particular telescope has a limit of apparent magnitude m to which it can go. Intrinsically faint stars (with large M) reach the apparent magnitude m at a relatively short distance, whereas intrinsically bright stars (with smaller M) have this magnitude at a larger distance, as can be easily seen from (6.1). Hence the telescope will show intrinsically bright stars at large distances where intrinsically faint stars are no longer visible. If we do a statistical analysis of the data taken by this telescope without properly taking account of this fact, then we may end up with the conclusion that intrinsically bright stars are more abundant at large distances compared to intrinsically faint stars. This is called the *Malmquist bias* (Malmquist, 1924). In any statistical analysis involving objects of different intrinsic luminosity, care has to be taken to avoid this bias.

6.1.2   Shapley's model

Even before the detailed papers on the Kapteyn Universe were published (Kapteyn and van Rhijn, 1920; Kapteyn, 1922), a serious rival to this model arose. In §3.6.2 we have discussed globular clusters, which are compact spherical clusters of typically about 10^5 stars. Shapley (1918) noted that most of the globular clusters are found around the constellation Sagittarius in the sky. Shapley (1919) suggested that the centre of our Galaxy must be in the direction of this constellation and the globular clusters must be distributed symmetrically around this centre. Figure 6.1 shows an edge-on view based on our modern perception of what the Galaxy would look like. The Galaxy has a thin disk with a spheroidal bulge around its centre. The Sun is located in an outlying region

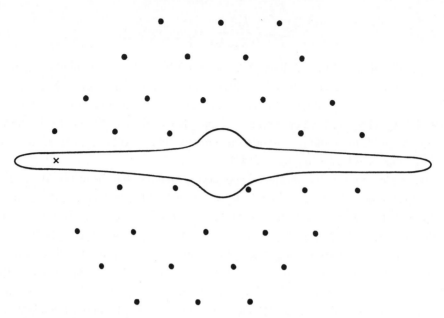

Fig. 6.1 A schematic edge-on view of our Galaxy. The position of the Sun is indicated by ×.

of this disk indicated by × in Figure 6.1 far away from the centre. About 200 globular clusters make up a roughly spherical halo around the galactic centre.

To establish the size of the Galaxy, we need to know the distances of the globular clusters from us. For measuring distances of reasonably faraway stellar systems, two kinds of stars with periodically varying luminosity – Cepheid variables and RR Lyrae stars – have proved very useful. Leavitt (1912) discovered that there was a relation between the period and the apparent luminosity of Cepheid variables in the Small Magellanic Cloud (which we now know to be a galaxy not far from our Galaxy), the brighter ones having longer periods. Since all the Cepheid variables in the Small Magellanic Cloud are approximately at the same distance from us, there must be a relation between period and absolute luminosity of these stars. The period–luminosity relation of Cepheid variables was established later when the distances (and hence absolute luminosities) of some Cepheid variables could be determined (essentially by studying Cepheid variables in star clusters within our Galaxy of which the distances could be estimated). So, if you measure the period of a Cepheid variable, you can infer its absolute luminosity and, by comparing with the apparent luminosity, you can then find the distance. In other words, a measurement of the period of a Cepheid variable leads to a determination of its distance. Initially it was thought that the Cepheid variables and RR Lyrae stars obey exactly the same period–luminosity relation, leading to erroneous estimates of some distances. Finally Baade (1954) showed that a Cepheid variable is somewhat brighter than an RR Lyrae star with the same period, necessitating the revision of many extragalactic distances.

Shapley (1919) used the RR Lyrae stars in some globular clusters to estimate their distances. From these measurements, he concluded that the galactic centre is situated at a distance of 15 kpc from us. The current best estimate for this distance is about 8 kpc (see §7.4.1 of Binney and Merrifield, 1998). The disk of the Galaxy has a thickness of the order of 500 pc. The actual estimate of the thickness depends on the kinds of stars we use to find this thickness. The bright O and B stars are usually found rather close to the mid-plane of the disk, such that one gets a lower value of the thickness of the disk on using these stars to find the thickness (the number densities of these stars fall with a scale height of about 50 pc from the mid-plane). On the other hand, stars of the other types can be found at greater distances from the mid-plane, their densities falling with more typical scale heights of order 200 or 300 pc (Gilmore and Reid, 1983). Since O and B stars are short-lived, they are statistically younger than other stars. So, presumably, as the stars grow older, they can acquire larger random velocities, enabling them to rise further from the mid-plane against gravity. We shall discuss this more in §7.6.2. Although we now know many more details not known in Shapley's time, our present view of the Galaxy is still essentially what Shapley surmised.

While Shapley was establishing the size and shape of our Galaxy, a fierce debate was going on whether some of the nebulous objects seen in the sky are outside our Galaxy or are inside it. Shapley (1921) believed that they are inside. However, this question was settled very soon by Hubble (1922) by studying Cepheid variables in some of these nebulae and by demonstrating from the distance estimates that they must be independent stellar systems outside our Galaxy. We shall discuss external galaxies in Chapter 9. Some of these have beautiful spiral structures. Figure 6.2 shows the Andromeda Galaxy, which is the nearest large spiral galaxy. We believe that our Galaxy and the Andromeda Galaxy are very similar in size, shape and appearance. If we go outside our Galaxy and look at it, it would probably appear very similar to Figure 6.2.

6.1.3   Interstellar extinction and reddening

The main reason why the Sun was put in the centre of the Kapteyn Universe is that the Milky Way looks reasonably symmetric around us. If the Sun is actually at the edge of our Galaxy, then why does the Milky Way look so symmetric? If the interstellar space has some obscuring material, then we would not be able to see too far into the galactic disk and our view of the disk would be symmetric, even though the disk may actually extend much more in one direction than in the other.

Fig. 6.2 The Andromeda Galaxy M31. Courtesy: Robert Gendler.

A clear proof of the existence of interstellar obscuration was provided by Trumpler (1930), who made a statistical study of open clusters, which are typically loosely bound clusters of a few dozen stars. Unlike globular clusters many of which are found away from the galactic disk, the open clusters mostly lie in the disk of our Galaxy. Assuming that the open clusters are statistically of the same size, one can estimate the distance from the angular size. Trumpler (1930) found that the stars in more distant open clusters appeared dimmer than what is expected from a simple inverse-square fall in intensity, clearly indicating that the starlight coming from distant clusters has undergone some attenuation. A more detailed discussion of the interstellar medium will be taken up in §6.5 and §6.6. Here we just mention that the interstellar medium contains particles of dust mixed with gas. It is the dust particles which are responsible for the absorption of starlight.

We had written down (1.8) assuming that there was no interstellar absorption and intensity fell by a simple inverse-square law. In the presence of interstellar absorption, (1.8) should be modified to

$$m = M + 5\log_{10} d - 5 + A_\lambda, \tag{6.6}$$

where A_λ gives the dimming caused by the interstellar dust. Since dimming implies an increase of the apparent magnitude m, it should be clear that A_λ has to be positive. For visible light coming from stars in the galactic plane, a rough rule of thumb for the dimming term is

$$A_V \approx 1.5\,d, \tag{6.7}$$

if d is measured in kpc. In other words, the amount of dimming of visible light with distance is approximately equal to 1.5 magnitude kpc^{-1} in the galactic plane. The subscript V in (6.7) implies that we are considering the extinction A_λ in the V band introduced in §1.4.

Since the dust particles absorb more light at the shorter wavelengths (on the bluer side), distant stars appear redder. We saw in §1.4 that the redness of a star is given by $(B - V)$. As starlight passes through interstellar matter, its redness measure $(B - V)$ keeps increasing. The change in it is denoted by $E(B - V)$ and the rule of thumb for this in the galactic plane is

$$E(B - V) \approx 0.5\,d. \tag{6.8}$$

Again d has to be in kpc. Since both A_λ and $E(B - V)$ depend linearly on the distance d, their ratio $A_\lambda/E(B - V)$ is independent of d and is a measure of interstellar extinction as a function of wavelength λ. Figure 6.3 plots the

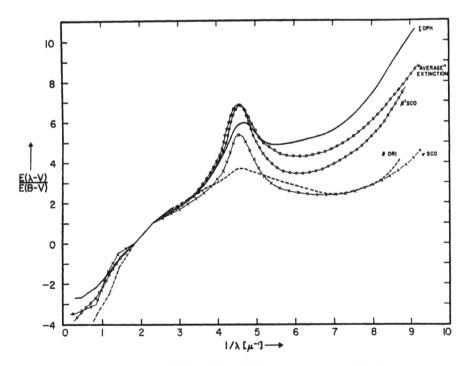

Fig. 6.3 A plot of $E(\lambda - V)/E(B - V)$, which is a measure of light extinction by interstellar dust, as a function of λ^{-1} in the directions of a few stars. An 'average' extinction curve is also indicated. From Bless and Savage (1972). (©American Astronomical Society. Reproduced with permission from *Astrophysical Journal*.)

related quantity $E(\lambda - V)/E(B - V)$, which is also a measure of interstellar extinction, as a function of inverse wavelength in the directions of a few stars. It is seen that there is an extinction peak around 2200 Å, which is usually interpreted to be due to graphite present in the dust. Apart from this peak, a straight line would not be a too bad fit for the absorption curve shown in Figure 6.3. This implies that interstellar absorption roughly goes as λ^{-1}, which is a much weaker dependence than the dependence λ^{-4} expected from Rayleigh scattering by molecules (see §2.6.1).

The existence of interstellar extinction and reddening makes the star count analysis more complicated than what it would have been in the absence of interstellar matter. For example, the expression (6.2) for $N(m)$ was obtained by assuming no absorption. There are, however, systematic methods of handling the effects of interstellar matter in star count analysis, which we shall not discuss here. Luckily interstellar dust is confined in a layer of thickness of about ± 150 pc around the mid-plane of the Galaxy, close to which we lie. So, when we look in directions away from the galactic plane, our view is not impaired by interstellar extinction or reddening. It was known for a long time that external galaxies could not be seen in a narrow zone near the galactic plane. This is known as the *zone of avoidance*.

6.1.4 Galactic coordinates

We have introduced the widely used equatorial system of celestial coordinates in §1.3. While presenting many galactic observations, it is often useful to introduce galactic coordinates. The galactic latitude b of an object is its angular distance from the galactic plane, which is taken as the equator in this system. The galactic longitude l is measured from the direction of the galactic centre, which is taken to be at $l = 0°, b = 0°$.

6.2 Galactic rotation

The gravitational field at a point inside or near the Galaxy is expected to be directed towards the galactic centre. How is this gravitational field balanced, to ensure that there is not a general fall of everything towards the galactic centre? There are basically two ways of balancing gravity. A star may move in a circular orbit such that the centrifugal force balances gravity (as in the case of planets in the solar system). The other way of balancing gravity is through random motions. Lindblad (1927) was the first to recognize that our Galaxy must be having two subsystems. Most of the stars in the disk move in roughly circular orbits around the galactic centre and constitute one subsystem. From the nearly spherical shape of the halo of globular clusters, Lindblad (1927) guessed that this must be a non-rotating subsystem in which gravity is balanced by random motions. At any particular instant of time, a globular cluster may be falling towards the galactic centre. Eventually, however, this globular cluster will come out on the other side of the Galaxy because of the kinetic energy it gains in falling towards the galactic centre. Although some individual globular clusters may be falling towards the galactic centre and the others may be moving away, the overall statistical appearance of the system of globular clusters should not change with time.

The Sun, in its orbit around the galactic centre, would circle around the non-rotating subsystem of globular clusters. The line of sight component of the relative velocity of the Sun with respect to a globular cluster can be determined by measuring the Doppler shifts of lines in the spectra of stars in this cluster. From the statistical analysis of such measurements for many globular clusters, it is possible to estimate the speed with which the Sun is going around the galactic centre, if we assume that the system of globular clusters has zero net rotation around the galactic centre. The best value for the speed of the Sun around the galactic centre, usually denoted by Θ_0, is about $\Theta_0 = 220\,\mathrm{km\ s^{-1}}$. If the Sun is located at a distance of $R_0 = 8\,\mathrm{kpc}$ from the galactic centre, then the period of revolution of the Sun around the galactic centre is

$$P_{\mathrm{rev}} = \frac{2\pi R_0}{\Theta_0} \approx 2 \times 10^8\,\mathrm{yr}. \tag{6.9}$$

Since the age of the Galaxy is believed to be of order 10^{10} yr (as we shall see in §9.3), the Sun had time to make not more than 50 rounds about the galactic centre. The approximate mass M of the Galaxy inside the solar orbit can be estimated by balancing the gravitational and centrifugal forces:

$$\frac{GM}{R_0^2} \approx \frac{\Theta_0^2}{R_0}.$$

The gravitational field would have been given by GM/R_0^2 exactly if the mass inside the solar orbit were distributed in a spherically symmetric manner. On substituting the estimated values of R_0 and Θ_0 in the above approximate equation, we find M to be of order $10^{11}\,M_\odot$.

As the gravitational field of the Galaxy is expected to fall off with distance, stars further out in the disk will have to move around the galactic centre with slower speeds. In other words, the disk of the Galaxy should have *differential rotation*. Oort (1927) carried out a classic analysis to show how this can be demonstrated by studying the motions of stars in the solar neighbourhood. We now present this analysis, based on the simplifying assumption that all stars move in exactly circular orbits. This assumption, of course, is not strictly true and §6.3 is devoted to looking at the consequences of small departures from exactly circular orbits.

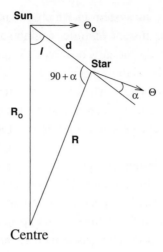

Fig. 6.4 A sketch indicating the Sun and a star going around the galactic centre.

Figure 6.4 shows the Sun at a distance R_0 from the galactic centre moving with speed Θ_0 in a circular orbit. We consider a star at a distance d from the Sun at galactic longitude l. As indicated in Figure 6.4, this star is at a distance R from the galactic centre moving with circular speed Θ. Let us consider the triangle made up by the lines R_0, R and d. If α is the angle made by the direction of the star's velocity Θ with d, then it follows from Figure 6.4 that the angle opposite to R_0 in our triangle is $90° + \alpha$. From the standard trigonometric properties of a triangle, we have

$$\frac{R}{\sin l} = \frac{R_0}{\cos \alpha} \qquad (6.10)$$

and

$$R_0 \cos l = d + R \sin \alpha. \qquad (6.11)$$

The relative radial velocity of the star (along the line of sight) with respect to the Sun is

$$v_R = \Theta \cos \alpha - \Theta_0 \sin l = \left(\frac{\Theta}{R} R_0 - \Theta_0 \right) \sin l$$

on making use of (6.10). Writing the angular velocities of the star and the Sun as

$$\omega = \frac{\Theta}{R}, \quad \omega_0 = \frac{\Theta_0}{R_0}, \qquad (6.12)$$

we get

$$v_R = (\omega - \omega_0) R_0 \sin l. \qquad (6.13)$$

The tangential velocity of the star with respect to the Sun is

$$v_T = \Theta \sin\alpha - \Theta_0 \cos l = \Theta \frac{R_0 \cos l - d}{R} - \Theta_0 \cos l$$

on substituting for $\sin\alpha$ from (6.11). On making use of (6.12), this gives

$$v_T = (\omega - \omega_0)R_0 \cos l - \omega d. \qquad (6.14)$$

The very important expressions (6.13) and (6.14) give general expressions of radial and tangential velocities of stars in the galactic disk moving in circular orbits around the galactic centre. They can even be applied to stars far away from the Sun.

We now consider stars in the solar neighbourhood for which $d \ll R_0$. For such stars, we approximately have

$$R_0 - R = d\cos l. \qquad (6.15)$$

We also can write

$$(\omega - \omega_0) = \left(\frac{d\omega}{dR}\right)_{R_0} (R - R_0) = \left[\frac{1}{R_0}\left(\frac{d\Theta}{dR}\right)_{R_0} - \frac{\Theta_0}{R_0^2}\right](R - R_0)$$

on substituting from (6.12) for ω. Using (6.15), we get

$$(\omega - \omega_0) = \left[\frac{\Theta_0}{R_0} - \left(\frac{d\Theta}{dR}\right)_{R_0}\right]\frac{d}{R_0}\cos l. \qquad (6.16)$$

Substituting (6.16), we get from (6.13) that

$$v_R = \frac{1}{2}\left[\frac{\Theta_0}{R_0} - \left(\frac{d\Theta}{dR}\right)_{R_0}\right] d \sin 2l, \qquad (6.17)$$

whereas (6.14) gives

$$v_T = \left[\frac{\Theta_0}{R_0} - \left(\frac{d\Theta}{dR}\right)_{R_0}\right] d \cos^2 l - \frac{\Theta}{R}d.$$

Since $\cos^2 l = \frac{1}{2}(\cos 2l + 1)$, we get

$$v_T = \frac{1}{2}\left[\frac{\Theta_0}{R_0} - \left(\frac{d\Theta}{dR}\right)_{R_0}\right] d\cos 2l - \frac{1}{2}\left[\frac{\Theta_0}{R_0} + \left(\frac{d\Theta}{dR}\right)_{R_0}\right] d. \qquad (6.18)$$

We can finally write (6.17) and (6.18) in the form

$$v_R = A d \sin 2l, \tag{6.19}$$

$$v_T = A d \cos 2l + B d, \tag{6.20}$$

where

$$A = \frac{1}{2} \left[\frac{\Theta_0}{R_0} - \left(\frac{d\Theta}{dR} \right)_{R_0} \right] = -\frac{1}{2} R_0 \left(\frac{d\omega}{dR} \right)_{R_0} \tag{6.21}$$

and

$$B = -\frac{1}{2} \left[\frac{\Theta_0}{R_0} + \left(\frac{d\Theta}{dR} \right)_{R_0} \right] \tag{6.22}$$

are known as *Oort constants*.

The radial velocity v_R of a star can be easily determined from the Doppler shifts of spectral lines. Suppose we measure v_R of many stars located at approximately the same distance d in the galactic plane. From (6.19), we expect v_R to vary as $\sin 2l$ with the galactic longitude of the star. Joy (1939) was one of the first astronomers to carry out such an analysis. Figure 6.5 from Joy (1939) shows radial velocities of four groups of Cepheid variables of which distances could be estimated from periods, the members of each group lying at a fixed distance d. We clearly see a sinusoidal variation in v_R with the galactic longitude. Data points not lying exactly on the fitted curves indicate that stars do not move in precise circular orbits. Since the amplitude of the oscillation is Ad, we can find the Oort constant A if we know d. To determine the other Oort constant B, we need to find the tangential velocity v_T of many nearby stars with respect to some non-rotating frame (such as the frame provided by extragalactic objects). The determination of B is more difficult than the determination of A and has been discussed by Mihalas and Binney (1981). Before quoting the best modern values of A and B, let us discuss the unit in which we should be expressing A and B. It should be clear from (6.19) and (6.20) that both A and B are obtained by dividing velocities by distances. Since stellar velocities are usually expressed in km s^{-1}, whereas galactic distances are expressed in kpc, it has been the convention to express A and B in units of km s^{-1} kpc^{-1}. Even though we could use a conversion factor between kpc and km to express A and B in s^{-1} (the dimensions of A and B are of inverse time), we follow the standard convention. The most reliable modern values of the Oort constants are obtained from the proper motion measurements by the Hipparcos astronomy satellite (our Figure 3.5 showing the HR diagram of nearby stars is also based on data from this satellite). They are

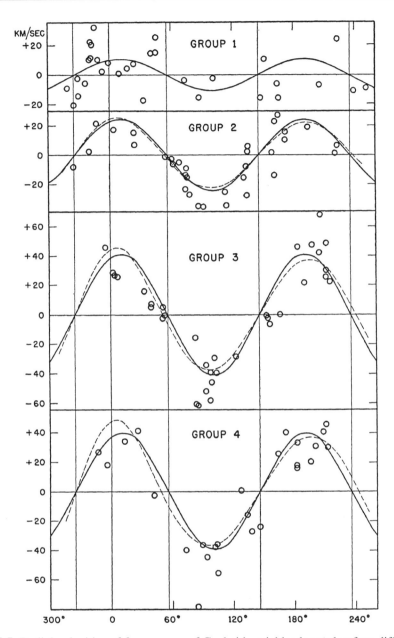

Fig. 6.5 Radial velocities of four groups of Cepheid variables located at four different distances. Note that the galactic coordinates indicated refer to the old system and not the presently used system (in which the galactic centre is taken as zero). From Joy (1939). (©American Astronomical Society. Reproduced with permission from *Astrophysical Journal*.)

$$A = 14.8 \pm 0.8 \text{ km s}^{-1}\text{kpc}^{-1}, \tag{6.23}$$

$$B = -12.4 \pm 0.6 \text{ km s}^{-1}\text{kpc}^{-1} \tag{6.24}$$

as given by Feast and Whitelock (1997).

6.3 Nearly circular orbits of stars

We assumed in §6.2 that all stars in the galactic disk move exactly in circular orbits. In reality, however, we do not expect most stars to move in exactly circular orbits, just as planets in the solar system do not move in exactly circular orbits. We know that a planet moves in an ellipse, which is the orbit in a gravitational field falling as the inverse square of distance from the central mass. Since the mass of the Galaxy is not concentrated in a central region but distributed all over the Galaxy, we expect that the gravitational field will not follow a simple inverse-square law and the orbits in the galactic disk will not be simple ellipses. We now want to find out the orbit of a star by assuming that the departure from a circular orbit is small.

Let $\Theta_{\mathrm{circ}}(r)$ be the speed which a star will need to move in a circular orbit at a distance r from the galactic centre. If f_r is the gravitational force at this distance r, then we must have

$$f_r = -\frac{\Theta_{\mathrm{circ}}^2}{r}. \tag{6.25}$$

Let $\Theta_0 = \Theta_{\mathrm{circ}}(R_0)$ be the circular speed where the Sun is located, at a distance R_0 from the galactic centre. We can think of a frame of reference at the Sun's position moving with speed Θ_0 in a circular orbit around the galactic centre. This frame of reference is known as the *local standard of rest* (LSR). If a star has a small velocity with respect to the LSR, then the orbit of the star can be found by determining its movements with respect to the LSR by using a perturbation technique.

6.3.1 The epicycle theory

We consider a star moving with speed Θ_0 in a circular orbit at distance R_0. Suppose the star is suddenly given a small kick in the radial direction. According to classical mechanics, its subsequent motion will be governed by the following equations

$$\ddot{r} - r\dot{\theta}^2 = f_r,$$

$$r^2\dot{\theta} = \mathrm{constant}$$

(see, for example, Goldstein, 1980, §3–2). Since the speed Θ in the θ direction is given by $\Theta = r\dot{\theta}$, the above two equations can be written as

$$\ddot{r} = \frac{\Theta^2}{r} - \frac{\Theta_{\mathrm{circ}}^2}{r}, \tag{6.26}$$

$$r\,\Theta = R_0\Theta_0, \tag{6.27}$$

using (6.25) to substitute for f_r and noting that the angular momentum of the star remains $R_0\Theta_0$, which did not change when we gave the star a radial kick. We now write

$$r = R_0 + \xi = R_0 \left(1 + \frac{\xi}{R_0} \right) \tag{6.28}$$

and assume that $\xi \ll R_0$, since the star will not move too far away from the circular orbit $r = R_0$ after receiving the small radial kick. We shall neglect the quadratic and higher powers of ξ in our discussion. Using (6.27), we clearly have

$$\frac{\Theta^2}{r} = \frac{R_0^2 \Theta_0^2}{r^3} \approx \frac{\Theta_0^2}{R_0} \left(1 - \frac{3\xi}{R_0} \right) \tag{6.29}$$

on making use of (6.28) and keeping only the linear term in ξ. We can write

$$\Theta_{\mathrm{circ}}(r) \approx \Theta_{\mathrm{circ}}(R_0) + \left(\frac{d\Theta}{dr} \right)_{R_0} \xi \approx \Theta_0 - (A + B)\xi, \tag{6.30}$$

where A and B are the Oort constants defined through (6.21) and (6.22). Then it follows

$$\frac{\Theta_{\mathrm{circ}}^2}{r} = \frac{\Theta_0^2 \left[1 - \frac{(A+B)}{\Theta_0}\xi \right]^2}{R_0 \left(1 + \frac{\xi}{R_0} \right)} \approx \frac{\Theta_0^2}{R_0} \left[1 - \frac{2(A+B)}{\Theta_0}\xi - \frac{\xi}{R_0} \right]. \tag{6.31}$$

On noting that $\ddot{r} = \ddot{\xi}$ and on making use of (6.29) and (6.31), we can write down (6.26) in the following approximate form

$$\ddot{\xi} = 2\frac{\Theta_0}{R_0}(A + B)\xi - 2\frac{\Theta_0^2}{R_0^2}\xi.$$

On substituting

$$\frac{\Theta_0}{R_0} = A - B,$$

this leads to

$$\ddot{\xi} = 4B(A - B)\xi,$$

which can be written as

$$\ddot{\xi} + \kappa^2\xi = 0, \tag{6.32}$$

where

$$\kappa = \sqrt{-4B(A - B)} \tag{6.33}$$

is a real quantity because B is negative. It is clear that there will be a simple harmonic motion of the star in the radial direction with respect to the circular orbit $r = R_0$. The radial velocity $\Pi = \dot{r}$ with respect to the LSR also should vary in a simple harmonic fashion and can be written as

$$\Pi = \Pi_0 \cos \kappa t \tag{6.34}$$

so that the displacement should be

$$\xi = \frac{\Pi_0}{\kappa} \sin \kappa t. \tag{6.35}$$

Now we look at the motion in the θ direction. From the constancy of the angular momentum $r^2 \dot{\theta}$, we have

$$\dot{\theta} = \frac{R_0 \Theta_0}{r^2} \approx \frac{\Theta_0}{R_0} \left(1 - \frac{2\xi}{R_0} \right).$$

Since the first term Θ_0 / R_0 corresponds to the motion of the LSR, the part corresponding to the motion of the star with respect to the LSR is approximately given by

$$\Delta\dot{\theta} = -\frac{2\Theta_0 \xi}{R_0^2}.$$

This translates into a linear velocity which, in the linear order in ξ, is

$$\Delta\Theta = (R_0 + \xi)\Delta\dot{\theta} = -\frac{2\Theta_0 \xi}{R_0} = -\frac{2\Pi_0 \Theta_0}{\kappa R_0} \sin \kappa t \tag{6.36}$$

on substituting from (6.35). The corresponding displacement is

$$\eta = \frac{2\Pi_0 \Theta_0}{\kappa^2 R_0} \cos \kappa t.$$

Since it follows from (6.21), (6.22) and (6.33) that

$$\frac{\Theta_0}{\kappa^2 R_0} = \frac{(A - B)}{-4B(A - B)} = \frac{1}{-4B},$$

we get

$$\eta = \frac{\Pi_0}{-2B} \cos \kappa t. \tag{6.37}$$

It should be clear from (6.35) and (6.37) that the star moves in an ellipse with respect to the LSR, while the LSR is revolving around the galactic centre, as shown in Figure 6.6. The ancient Greek astronomers Hipparchus (2nd century BC) and Ptolemy (2nd century AD) ascribed motions of a similar kind to planets in their geocentric theory. Borrowing a term from ancient astronomy, we call such motions *epicyclic*. The elliptical path of the star with respect to the LSR is called an *epicycle*. It follows from (6.35) and (6.37) that the ratio

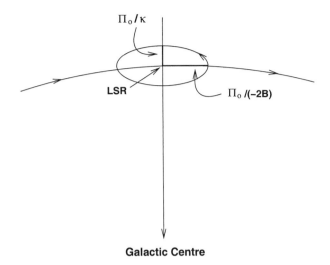

Fig. 6.6 A sketch showing the epicyclic motion of a star around the LSR.

of the semimajor axis (in the θ direction) to the semiminor axis (in the r direction) is

$$\frac{\Pi_0/2|B|}{\Pi_0/\kappa} = \sqrt{\frac{A-B}{|B|}}$$

on substituting for κ from (6.33). Putting values of A and B as given by (6.23) and (6.24), this ratio turns out to be 1.48. So the ellipse is elongated in the tangential direction. The period of oscillation in the epicycle is related to the revolution period in the following way

$$\frac{P_{\text{osc}}}{P_{\text{rev}}} = \frac{2\pi/\kappa}{2\pi R_0/\Theta_0} = \frac{A-B}{\sqrt{-4B(A-B)}} = \frac{1}{2}\sqrt{\frac{A-B}{-B}}. \qquad (6.38)$$

On putting the values of A and B, this ratio of periods is found to be 0.74 for stars in the solar neighbourhood. Since this ratio is not in general a rational number for a star at an arbitrary distance from the galactic centre, the orbit of the star will not close.

6.3.2   The solar motion

A star in the solar neighbourhood would not in general be at rest in the LSR, but would move in an epicycle with respect to the LSR. Is the Sun at rest in the LSR? Unless it is an unusual accident, we expect the answer to be 'no'. The motion of the Sun with respect to the LSR at the present epoch is called the *solar motion*. This motion can be found out by studying the motions of the stars in the solar neighbourhood and by assuming that these stars do not have any net drift in the radial direction or perpendicular to the galactic plane. This implies

$$\langle \Pi \rangle = 0, \quad \langle Z \rangle = 0. \tag{6.39}$$

Here Z is the component of velocity perpendicular to the galactic plane and $\langle \ldots \rangle$ implies averaging over stars in the solar neighbourhood. We have shown in §6.3.1 that Π for a particular star varies sinusoidally. The reader is asked in Exercise 6.4 to show the same for Z. So it is no wonder that their averages will be zero.

Let $(\Pi_\odot, \Theta_\odot - \Theta_0, Z_\odot)$ be the components of solar motion. From the Doppler shifts of spectral lines, we can find the line of sight velocity of a star with respect to the Sun, whereas the proper motion gives the velocity perpendicular to the line of sight. Combining these measurements, one can find out the components $\Pi - \Pi_\odot$ and $Z - Z_\odot$ of relative velocity, and then their averages over the stars in the solar neighbourhood. Because of (6.39), we have

$$\langle \Pi - \Pi_\odot \rangle = \langle \Pi \rangle - \Pi_\odot = -\Pi_\odot \tag{6.40}$$

and similarly

$$\langle Z - Z_\odot \rangle = -Z_\odot. \tag{6.41}$$

Thus these averages give us the components of solar motion, which are found to be

$$\Pi_\odot = -10.0 \pm 0.4 \text{ km s}^{-1}, \quad Z_\odot = 7.2 \pm 0.4 \text{ km s}^{-1}. \tag{6.42}$$

Since the LSR itself does not have any Π or Z velocities, it is relatively easy to find the components of solar motion with respect to LSR in these directions. Now let us consider the θ direction. We certainly have

$$\langle \Theta - \Theta_\odot \rangle = -(\Theta_\odot - \langle \Theta \rangle), \tag{6.43}$$

which can be found out from the measurements of stellar velocities with respect to the Sun. Now, if $\langle \Theta \rangle$ is equal to the velocity $\Theta_0 = \Theta_{\text{circ}}(R_0)$ of the LSR, then $\Theta_\odot - \langle \Theta \rangle$ would give the solar motion with respect to the LSR. But is it true that $\langle \Theta \rangle = \Theta_0$? From the epicycle theory presented in §6.3.1, especially (6.36), it would seem that a star would simply oscillate forward and backward with respect to the LSR and $\langle \Theta \rangle$ averaged over many stars in the solar neighbourhood would give Θ_0. However, this result is a consequence of the assumption of linearity. If we go beyond the linear theory, then we find that the centre of the epicycle, known as the *guiding centre*, moves slower than the LSR. The reason is not difficult to find if we look at Figure 6.6. Because of the curvature of the path of the guiding centre, the length of the epicycle path on the outside (i.e. away from the galactic centre) is larger than the length of the epicycle path on the inside. From (6.27) the velocity Θ is less when the star is in the outer part of the epicycle. Hence the star covers a longer path with a slower speed and the average Θ of the star should actually be less than Θ_0. We missed this effect in

§6.3.1, since the curvature of the guiding centre path was not taken into account in the linear theory. As we shall discuss in §7.6.2, things can be even more complicated when we consider the fact that the guiding centres of different stars in the solar neighbourhood may lie at different distances from the galactic centre. We see from (6.43) that the solar motion in the θ direction is given by

$$\Theta_\odot - \Theta_0 = -\langle(\Theta - \Theta_\odot)\rangle + \langle\Theta\rangle - \Theta_0. \tag{6.44}$$

Thus, apart from $\langle(\Theta - \Theta_\odot)\rangle$, which is found from the observations of stellar motions in the solar neighbourhood, we need to know how $\langle\Theta\rangle$ differs from the velocity of LSR Θ_0 to find out the solar motion in the θ direction. We shall see in §7.6.2 how $\langle\Theta\rangle - \Theta_0$ can be found. Here let us only quote the final result that

$$\Theta_\odot - \Theta_0 = 5.2 \pm 0.6 \text{ km s}^{-1} \tag{6.45}$$

according to the current available data. The values quoted in (6.42) and (6.45) are from Binney and Merrifield (1998, §10.3.1).

The amplitude of the solar motion is of order 10 km s^{-1}. The typical random velocity of a star in the solar neighbourhood is also of this order. Then the amplitude of oscillation in the radial direction, which is given by Π_0/κ according to (6.35), should be of order 1 kpc on taking Π_0 to be of order 10 km s^{-1}.

6.3.3   The Schwarzschild velocity ellipsoid

A measurement of the velocities of stars in the solar neighbourhood may give the impression that the distribution of velocities is random. In reality, however, most of the stars are moving in their epicyclic orbits. Since the typical amplitude of radial oscillation is of order 1 kpc, stars with guiding centres lying within a band of width 1 kpc on either side of R_0 (the distance of the Sun from the galactic centre) can come into the solar neighbourhood during their epicyclic motions. Let the guiding centre of a star be at a distance $r = R_g$ from the galactic centre. We now want to apply the theory of §6.3.1 by assuming that a radial displacement ξ has brought the star into the solar neighbourhood, i.e.

$$R_0 = R_g + \xi$$

so that

$$\Theta_{\text{circ}}(R_0) = \Theta_{\text{circ}}(R_g) + \left(\frac{d\Theta}{dr}\right)\xi. \tag{6.46}$$

In the place of (6.27), we have

$$R_0 \Theta(R_0) = R_g \Theta_{\text{circ}}(R_g) = (R_0 - \xi)\Theta_{\text{circ}}(R_g),$$

where $\Theta(R_0)$ is the tangential speed of the star when it comes near the Sun and is clearly equal to

$$\Theta(R_0) = \Theta_{\text{circ}}(R_g)\left(1 - \frac{\xi}{R_0}\right). \tag{6.47}$$

The relative tangential speed of the star with respect to the LSR is

$$\Theta(R_0) - \Theta_{\text{circ}}(R_0) = -\left[\frac{\Theta_{\text{circ}}(R_g)}{R_0} + \frac{d\Theta}{dr}\right]\xi$$

on substituting from (6.46) and (6.47). To the order in which we ignore terms quadratic in ξ, the term within the square bracket should be equal to $-2B$, as seen from (6.22). Then

$$\Theta(R_0) - \Theta_{\text{circ}}(R_0) = 2B\xi = \frac{2B}{\kappa}\Pi_0 \sin \kappa t \tag{6.48}$$

on making use of (6.35). Since the radial speed of the star is given by $\Pi_0 \cos \kappa t$ according to (6.34), we should have

$$\frac{\langle|\Pi|\rangle}{\langle|\Delta\Theta|\rangle} = \frac{\kappa}{-2B} = \sqrt{\frac{A - B}{-B}} \tag{6.49}$$

on using (6.33).

It was proposed by Schwarzschild (1907) that the stars in the solar neighbourhood would have an ellipsoidal distribution in the velocity space. In other words, the number of stars with velocity components lying between Π and $\Pi + d\Pi$, Θ and $\Theta + d\Theta$, Z and $Z + dZ$ should be

$$f(\Pi, \Theta, Z)\,d\Pi\,d\Theta\,dZ = C\exp\left[-\frac{\Pi^2}{\sigma_\Pi^2} - \frac{(\Theta - \Theta_0)^2}{\sigma_\Theta^2} - \frac{Z^2}{\sigma_Z^2}\right]d\Pi\,d\Theta\,dZ. \tag{6.50}$$

It should be clear that the random velocities in the r and θ directions are $\langle|\Pi|\rangle = \sigma_\Pi$ and $\langle|\Delta\Theta|\rangle = \sigma_\Theta$. It then follows from (6.49) that

$$\frac{\sigma_\Pi}{\sigma_\Theta} = \sqrt{\frac{A - B}{-B}},$$

which has the numerical value 1.48. We thus have the beautiful result that even the ellipticity of the Schwarzschild velocity ellipsoid depends on the Oort constants. Figure 6.7 shows the random velocities in different directions for stars of different colours in the solar neighbourhood. It is seen that redder stars tend to have more random velocities. However, the ratio of σ_Π to σ_Θ turns out to be not very different from 1.48 for stars of any colour.

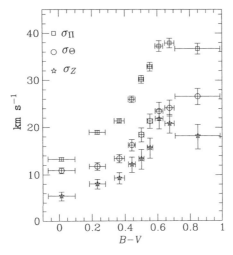

Fig. 6.7 The velocity dispersions for stars of different colours in the solar neighbourhood. From Dehnen and Binney (1998). (©Royal Astronomical Society. Reproduced with permission from *Monthly Notices of Royal Astronomical Society*.)

6.4 Stellar populations

We have already pointed out in §6.2 that our Galaxy contains two subsystems. One subsystem consists of stars in the disk which revolve around the galactic centre in nearly circular orbits. We shall see in §6.5 that the interstellar matter also revolves around the galactic centre with these stars and belongs to this subsystem. The other subsystem contains the globular clusters which have no systematic rotation around the galactic centre. Apart from globular clusters, the spheroidal component of our Galaxy (i.e. the bulge around the galactic centre seen in Figure 6.1) belongs to this subsystem. For stars in the spheroidal component also, gravity is balanced by random motions, since these stars have very little systematic rotation. Additionally, the Galaxy has a non-rotating halo of stars. Although the density of stars in the halo is much less than the density of stars in the disk, even in the solar neighbourhood we see a handful of stars with high random velocities which presumably belong to the halo (we shall discuss this in more detail in §7.7). There are several distinct differences between the physical characteristics of these two subsystems. The stars in the non-rotating subsystem consisting mainly of the globular clusters and the spheroidal component are mainly very old stars. The bright O and B stars, which are short-lived, are not found in this subsystem, where formation of new stars does not take place. While discussing HR diagrams of globular clusters in §3.6.2, we already pointed out that these are very old systems in which the main sequence does not go all the way to very luminous stars. On the other hand, star formation out of interstellar matter continuously takes place in the rotating subsystem comprising stars and interstellar matter in the galactic disk. We see O and B stars in this subsystem. Finally, the stars in the non-rotating subsystem are deficient in 'metals' (i.e. elements heavier than He such as C, N and O, which are called 'metals' by astronomers) compared to stars in the galactic disk belonging to the rotating subsystem. We pointed out in §4.3 and §4.7 that the heavier elements are produced inside stars and get strewn in the interstellar matter when massive stars undergo supernova explosions. With more and more supernova explosions, the interstellar matter of the Galaxy is presumably getting more enriched with these heavier elements. The old stars of the non-rotating subsystem must have formed from a primordial interstellar matter which was not yet rich in metals. The stars in the other rotating subsystem are younger and formed out of interstellar matter after it became enriched with metals.

Based on these considerations, Baade (1944) introduced the idea of two stellar populations. The Population I stellar systems are relatively metal-rich, contain interstellar matter and very bright O/B stars, and revolve around the galactic centre to balance the pull of gravity. The Population II stellar systems are comparatively metal-poor, contain no interstellar matter or O/B stars, and counteract the gravitational field of the Galaxy by having random motions. The galactic disk is the prime example of a Population I system, whereas the globular clusters along with the spheroidal bulge and the halo belong to Population II. Since we believe that all stars form out of interstellar matter, even Population II systems must have contained interstellar matter at some early stage out of which they formed, even though now they do not have much interstellar matter any more. Presumably, all the interstellar matter has been used up in forming stars. Even though a classification into two distinct stellar populations may be an over-simplification of a complex situation, the concept of stellar populations has proved extremely useful and is still widely used by astronomers.

6.5 In search of the interstellar gas

We have seen in §6.1.3 that the existence of interstellar matter was established from the extinction and reddening of starlight produced by the interstellar dust. There was evidence that the interstellar space contained much more matter in the form of gas rather than in the form of dust. For example, evidence for the gas came from narrow absorption lines observed in the spectra of some stars. A spectral line gets broadened due to the random thermal motions of the atoms in the material which produces the spectral line (this is known as *thermal broadening*). An absorption line produced in a stellar atmosphere is expected to have a broadening appropriate for the temperature of the atmosphere. A narrow absorption line in a stellar spectrum indicates that the line must be produced by some considerably cooler gas, possibly distributed along the line of sight between the star and us. However, much of this gas (outside some limited regions as discussed in §6.6) emits no visible light. We shall discuss in §7.6.1 an important limit on the density of interstellar matter known as the *Oort limit* (Oort, 1932). It appeared that the mass in the interstellar matter may be comparable to the mass contained in the stars in the solar neighbourhood. Astronomers faced a peculiar problem in the 1930s and 1940s: even though they became aware of the existence of a considerable amount of gas in the interstellar space, they did not know how to study it systematically because they were unable to detect any radiation coming from the gas.

After the advent of radio astronomy, van de Hulst (1945) finally suggested a way out of the impasse when he predicted that the interstellar hydrogen gas would emit radiation at the radio wavelength of 21 cm. The proton and the electron in the hydrogen atom can have their spins either parallel or antiparallel. The state with parallel spins has slightly higher energy than the state with antiparallel spins. When transition from the higher state to the lower state takes place, radiation with wavelength 21 cm is expected to be emitted. This is, however, a 'forbidden' atomic line and it is not easy to see this line in laboratory experiments. Since interstellar space has a huge amount of hydrogen with very low density such that an atom in the higher state is unlikely to de-excite due to collisions, van de Hulst (1945) suggested that it should be possible to receive emission from interstellar hydrogen at this spectral line. Within a few years of this remarkable prediction, emission from interstellar gas at this wavelength was detected independently by Ewen and Purcell (1951) and by Muller and Oort (1951). The 21-cm line soon proved to be a very powerful tool for studying the distribution of the interstellar gas.

If the emitting gas has any radial velocity along the line of sight, that would cause the wavelength to shift from 21 cm. Since the intrinsic width of the 21-cm line from a cold gas would be narrow, it is ideally suited to measure the wavelength shift which gives the radial velocity. Suppose in the direction of galactic coordinates (l, b) we find the intensity $I(l, b, \lambda)$ as a function of wavelength. Since the wavelength shift gives the radial velocity v_R of the emitting gas, we can write the intensity as $I(l, b, v_R)$. Of particular interest is the intensity in various directions of the galactic plane for which $b = 0°$. Figure 6.8 shows $I(l, b = 0°, v_R)$ plotted in the l–v_R plane. The distribution of the interstellar gas has to be found out from plots like this.

We consider a line of sight in the galactic plane as shown in Figure 6.9. We assume the interstellar gas to revolve around the galactic centre exactly in circular orbits. Then the radial velocity v_R at different points along the line of sight is given by (6.13). It is clear that $|v_R|$ should be maximum when $|\omega - \omega_0|$ is maximum. Suppose ω increases as we go closer to the galactic centre (which is true except in a small region near the centre). Then $|v_R|$ should be maximum at point 1 where the line of sight is tangent to the innermost circular orbit touched by the path of light. We see in Figure 6.8 that, for a given l, the intensity drops to zero beyond a certain value of $|v_R|$. This maximum $|v_R|$ should correspond

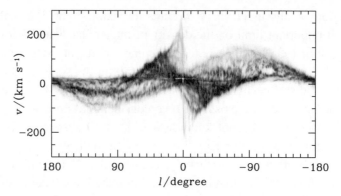

Fig. 6.8 The intensity $I(l, b = 0°, v_R)$ of the 21-cm line shown by grey scale in the l–v_R plane. As given by Binney and Merrifield (1998), based on the data provided by D. Hartmann. See Hartmann and Burton (1997) for details. Courtesy: D. Hartmann and M. Merrifield.

Fig. 6.9 A schematic line of sight through the Galaxy, along which we receive 21-cm emissions from interstellar clouds.

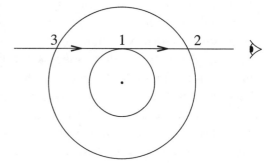

to that point along the light path where it is tangential to an orbit (like point 1 in Figure 6.9). On applying (6.13), we then find out ω at a distance $r = R_0 \sin l$ from the galactic centre, since it is the circular orbit at this distance to which the line of sight is a tangent. It is possible to find ω as a function of r till the solar orbit at $r = R_0$. This method does not apply for determining ω beyond R_0, for which we require other methods.

The interstellar gas is found to be quite clumpy. The clumps of interstellar gas are referred to as *clouds* by astronomers. The clumpiness of the interstellar gas can be easily inferred from the non-smooth distribution of the intensity $I(l, b = 0°, v_R)$ seen in Figure 6.8. Wherever there is a local peak of intensity in the l–v_R plane, we conclude that there must be a cloud in the l direction moving with radial velocity v_R. Clouds located at points 2 and 3 in Figure 6.9 should have the same v_R, which follows from (6.13). Hence, if we see a peak in the intensity at this v_R, we infer the existence of a cloud at 2 or 3. To determine whether the cloud is at 2 or 3, we can look at the angular size of the cloud perpendicular to the galactic plane. If this size is large, then we expect the cloud to be located at the nearer point. Proceeding in this way, we can

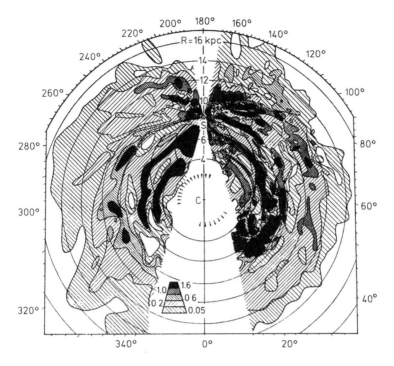

Fig. 6.10 The distribution of neutral hydrogen in the galactic plane, as found by Oort, Kerr and Westerhout (1958) from 21-cm observations. (©Royal Astronomical Society. Reproduced with permission from *Monthly Notices of Royal Astronomical Society.*)

reconstruct the gas distribution in the galactic plane from Figure 6.8. We show in Figure 6.10 the famous reconstruction due to Oort, Kerr and Westerhout (1958). It is clear that the distribution of neutral hydrogen gas in the Galaxy is highly non-homogeneous. We also notice that this distribution traces out the spiral arms of the Galaxy.

During the last few decades, it has been established that the interstellar medium (usually abbreviated as ISM) is an extremely complex system containing several distinct phases. We now discuss the phases of the interstellar medium and point out how we get information about them.

6.6   Phases of the ISM and the diagnostic tools

In §2.7 we discussed how one can analyse a spectral line which is formed by the passage of radiation through an absorbing medium. However, when we analyse radiation that has been emitted by the ISM or that has passed through the ISM, we need to keep in mind that the ISM is far from thermodynamic equilibrium. Hence usually the radiation present in the interstellar space would not be in equilibrium with matter, and Kirchhoff's law (2.26) may not hold. While the results derived in §2.7 should hold for visible light passing through the ISM, it is often necessary to study the radiative transfer through the ISM from a more microscopic point of view than what we adopted in Chapter 2.

Let us consider two energy levels of some atom. The transitions between these levels are accompanied by emission or absorption of photons with energy $h\nu_0$ equal to the energy difference between the levels. We shall use the subscripts u and l to denote the upper and lower levels. Let the atomic number densities in the upper and lower levels be n_u and n_l. It is useful to develop our discussion from the well-known Einstein coefficients of radiative transition treated in many textbooks (see, for example, Richtmeyer, Kennard and Cooper, 1969, §13.12). Let A_{ul} be the coefficient of spontaneous transition, whereas B_{ul} and B_{lu} are the coefficients of induced transition. The number of spontaneous transitions per unit volume per unit time is $n_u A_{ul}$ and the energy emitted in these transitions is $h\nu_0 n_u A_{ul}$. The energy emitted per unit volume per unit time per unit solid angle is given by dividing this by 4π. This should be equal to the emission coefficient j_ν integrated over the spectral line, i.e.

$$\int j_\nu d\nu = \frac{h\nu_0 n_u A_{ul}}{4\pi}.$$

Let $\phi(\Delta\nu)$ be the normalized line profile where $\Delta\nu$ is the departure of the frequency from the line centre at ν_0 and $\int \phi(\Delta\nu)d\nu = 1$. Then j_ν should be of the form

$$j_\nu = \frac{h\nu_0 n_u A_{ul}}{4\pi}\phi(\Delta\nu). \tag{6.51}$$

In the presence of a radiation field with energy density U_ν given by (2.5), the number of induced upward transitions per unit volume per unit time is $n_l B_{lu} U_\nu$, whereas the corresponding number of downward transitions is $n_u B_{ul} U_\nu$. The net energy absorbed per unit volume per unit time must be

$$\mathcal{E}_{abs} = \frac{h\nu_0}{c}(n_l B_{lu} - n_u B_{ul}) \int I_\nu \, d\Omega \tag{6.52}$$

on making use of (2.5). The energy absorbed from the beam I_ν in unit volume in unit time is $\alpha_\nu I_\nu$. The energy absorbed from radiation coming from all directions is obtained by integrating this over all solid angles. So another expression of $\mathcal{E}_{abs}$ is given by again integrating this over the absorption line (presumably α_ν is non-zero only for frequencies at which absorption takes place), i.e.

$$\mathcal{E}_{abs} = \int d\nu \int \alpha_\nu I_\nu \, d\Omega. \tag{6.53}$$

Comparing the above two expressions of $\mathcal{E}_{abs}$ and assuming for simplicity that the absorption coefficient α_ν also has the same profile $\phi(\Delta\nu)$, we conclude

$$\alpha_\nu = \frac{h\nu_0}{c}(n_l B_{lu} - n_u B_{ul})\phi(\Delta\nu). \tag{6.54}$$

From (6.51) and (6.54), we see that the source function is given by

$$S_\nu = \frac{j_\nu}{\alpha_\nu} = \frac{c}{4\pi} \frac{n_u A_{ul}}{n_l B_{lu} - n_u B_{ul}}. \tag{6.55}$$

The Einstein transition coefficients satisfy the following important relations amongst themselves

$$A_{ul} = \frac{8\pi h \nu^3}{c^3} B_{ul}, \quad g_u B_{ul} = g_l B_{lu}, \tag{6.56}$$

where g_u and g_l are the statistical weights of the upper and lower states. Since the relations (6.56) follow from the fundamental transitions of the atom, they should not depend on whether there is thermodynamic equilibrium around or not. We, therefore, expect (6.56) to be valid even when our system is not in thermodynamic equilibrium. However, only if the system is in thermodynamic equilibrium, should we have the Boltzmann relation

$$\frac{n_u}{n_l} = \frac{g_u}{g_l} \exp\left(-\frac{h\nu_0}{\kappa_B T}\right). \tag{6.57}$$

On using (6.56) and (6.57), it easily follows from (6.55) that the source function S_ν should be equal to the Planck function $B_\nu(T)$ as given by (2.6). This is the case only when the system is in thermodynamic equilibrium. If the system is out of thermodynamic equilibrium, then (6.57) may not hold and consequently the source function (6.55) may not be equal to $B_\nu(T)$.

For a system not in thermodynamic equilibrium, we have to determine the population n_i for a level i by solving microscopic rate equations. If R_{ij} is the transition probability from level i to level j, then $n_i \sum_j R_{ij}$ gives the rate of transitions out of level i. In the steady state, this has to equal the rate of transitions into level i from all other levels j given by $\sum_j n_j R_{ji}$. We thus have

$$n_i \sum_j R_{ij} - \sum_j n_j R_{ji} = 0. \tag{6.58}$$

We have one such equation for each atomic level i. If we can figure out the transition rates R_{ij} between various levels from fundamental physics, then we can solve these simultaneous equations to determine the populations in the various levels.

Let us consider the simplest case of two levels u and l as an illustration. In addition to the spontaneous emission and induced emission already discussed, there can be a transition from the upper level to the lower level by inelastic collisions with electrons present in the system. We expect the transition rate due to collisional de-excitation to be proportional to both the electron number density n_e and the number density n_u of atoms in the upper level. So we can write this transition rate as $\gamma_{ul} n_u n_e$. Similarly, there would be collisional excitation of atoms from the lower to the upper level, in addition to transitions induced by radiation. Since the transition rates for $u \to l$ and $l \to u$ have to balance in the steady state, we have

$$n_u(A_{ul} + B_{ul}U_\nu + \gamma_{ul}n_e) = n_l(B_{lu}U_\nu + \gamma_{lu}n_e). \tag{6.59}$$

The collisional transition rates should be independent of whether the radiation field is in equilibrium with matter. Hence, if we can derive any relation amongst them under the assumption of full thermodynamic equilibrium, then that relation should hold even when we do not have radiation in equilibrium with matter. The relations amongst the Einstein transition coefficients are always given by (6.56). Under the condition of thermodynamic equilibrium, U_ν follows from (2.6) and the Boltzmann relation (6.57) holds. On making use of them, we find that (6.59) holds only if

$$g_l\gamma_{lu} = g_u\gamma_{ul} \exp\left(-\frac{h\nu_0}{\kappa_B T}\right). \tag{6.60}$$

This should be valid even when no radiation field is present.

In the interstellar medium, we often have atoms excited to a higher level collisionally. Then the excited atoms return to the lower level either through collisions or through the spontaneous emission of photons. If the energy density of radiation is negligible, then we can put $U_\nu = 0$ in (6.59) and obtain

$$\frac{n_u}{n_l} = \frac{\gamma_{lu}n_e}{A_{ul} + \gamma_{ul}n_e}.$$

On making use of (6.60), this becomes

$$\frac{n_u}{n_l} = \frac{g_u}{g_l} \exp\left(-\frac{h\nu_0}{\kappa_B T}\right) \cdot \frac{1}{1 + (A_{ul}/\gamma_{ul}n_e)}. \tag{6.61}$$

It is clear that we would get back the Boltzmann distribution (6.57) if spontaneous emission is absent (i.e. if $A_{ul} = 0$). The spontaneous emission makes some atoms de-excite from the upper level and thereby decreases the population of the upper level compared to what we have got from the Boltzmann distribution (6.57).

In this simple example of an atom with two levels, we find that the upper level gets de-populated. However, this is not a generic result. If there is at least one more level of the atom to which transitions from the lower level are preferred, then it is possible for the upper level to be over-populated (see Exercise 6.6).

We now list the different phases of the interstellar medium and point out how we get information about them.

6.6.1 HI clouds

The neutral atomic hydrogen in the interstellar space is often referred to as HI. We see in Figure 6.10 that the distribution of HI inside the Galaxy is highly non-uniform. The clouds of HI concentration typically have densities of order 10^6–10^8 particles m^{-3} and temperatures of order 80 K. Although they may occupy only 5% of the volume of the interstellar space, they contribute nearly 40% of the mass of the interstellar matter. In the direction perpendicular to the galactic plane in the solar neighbourhood, HI clouds are found mostly within a distance of about 100 pc from the mid-plane. The 21-cm line is the most important diagnostic tool for studying HI clouds. However, they can also be studied by analysing the narrow absorption lines in the visible and UV parts of the stellar spectra, caused by the absorption of starlight by the interstellar gas at certain definite wavelengths. The low temperature of the gas ensures that the thermal broadening of the lines is much less than what would have been the case if they were formed in the stellar atmosphere. As discussed in §2.7, the composition of the interstellar gas can be determined from an analysis of these narrow absorption lines. It is found that elements like carbon, oxygen and nitrogen are much less abundant in the interstellar gas clouds compared to what we believe to be the *cosmic composition*. The usual reason given for this is that these elements are locked up in the dust grains (discussed in §6.1.3) which are present within the clouds. We still do not have a very good idea about the composition of the dust grains. However, as we pointed out in §6.1.3, a perusal of the extinction curve by dust grains suggests that carbon in the form of graphite should be an important component.

Much of the information about the neutral hydrogen gas in interstellar space comes from the 21-cm line. As we pointed out in §6.5, along the line of sight, there would be clouds moving with different radial velocities due to the differential rotation of the Galaxy. These clouds would emit at slightly different wavelengths. To focus on the basic physics, let us consider the simple situation of one optically thin cloud in the line of sight. It will produce a narrow emission line at a wavelength close to 21 cm. If there is a background radio source with a continuum spectrum, we also expect to see an absorption line at this wavelength. Let us now consider how we can extract important information about interstellar hydrogen from the emission and absorption lines.

In the upper level of the 21-cm transition, the spin of the electron and the proton are parallel, giving a combined spin of 1. It is a standard result of quantum mechanics that this level should have the statistical weight $g_u = 3$, whereas the lower state with antiparallel spins should have the statistical weight $g_l = 1$. For $T = 80$ K, the difference of energy $h\nu_0$ between these levels is small compared to $\kappa_B T$ and the exponential factor in (6.57) is close to 1, such that $n_u/n_l = 3$. If n_H is the number density of hydrogen atoms, then

$$n_u = \frac{3}{4}n_H, \quad n_l = \frac{1}{4}n_H \tag{6.62}$$

are the number densities of hydrogen atoms in the upper and lower levels. For an optically thin source, it easily follows from the radiative transfer equation (2.12) that the specific intensity is given by $\int j_\nu \, ds$, and the total intensity of emission in the spectral line is

$$I = \int ds \int j_\nu \, d\nu.$$

On substituting from (6.51) and (6.62), this becomes

$$I = \frac{3}{16\pi} h\nu_0 A_{ul} \int n_H \, ds. \tag{6.63}$$

It is known that $A_{ul} = 2.85 \times 10^{-15} \text{ s}^{-1}$ for the 21-cm transition, which means that an atom in the upper level is expected to make a downward transition once in 10^7 yr. It is no wonder that the 21-cm emission is difficult to produce in a laboratory setup, since a collisional downward transition would be much more likely. It follows from (6.63) that a measurement of I gives us the value of $\int n_H \, ds$, since all the other things are known. If we have an idea of the path length through the cloud, we get an estimate of n_H.

We now consider the absorption line. For an optically thin obstacle, it would follow from (2.20) that the intensity after passing through the obstacle would be

$$I_\nu(\tau_\nu) = I_\nu(0)e^{-\tau_\nu}, \tag{6.64}$$

where $I_\nu(0)$ is the intensity of the background source. Hence the depth of the spectral line depends on the optical depth $\tau_\nu = \int \alpha_\nu \, ds$, which we now estimate. On making use of (6.56) and (6.57), the expression (6.54) for the absorption coefficient becomes

$$\alpha_\nu = \frac{h\nu_0}{c} n_l B_{lu} \left[1 - \exp\left(-\frac{h\nu_0}{\kappa_B T} \right) \right] \phi(\Delta\nu). \tag{6.65}$$

Since $h\nu_0 \ll \kappa_B T$ for the 21-cm line, we get

$$\alpha_\nu = \frac{h\nu_0}{c} n_l B_{lu} \frac{h\nu_0}{\kappa_B T} \phi(\Delta\nu).$$

Making use of (6.56) and (6.62), this becomes

$$\alpha_\nu = \frac{3}{32\pi} n_H A_{ul} \frac{hc^2}{\nu_0 \kappa_B T} \phi(\Delta\nu).$$

Hence the optical depth is given by

$$\tau_\nu = \frac{3}{32\pi} A_{ul} \frac{hc^2}{\nu_0 \kappa_B} \phi(\Delta\nu) \int \frac{n_H}{T} \, ds. \tag{6.66}$$

It follows from (6.64) that the 21-cm absorption line gives us the integral

$$\int \frac{n_H}{T} \, ds$$

along the line of sight, whereas the 21-cm emission line gives the integral $\int n_H ds$, as we saw in (6.63). By combining the information obtained from the emission and the absorption lines, we can estimate the temperature of the neutral hydrogen gas.

It may be noted that we used the opposite limit $h\nu_0 \gg \kappa_B T$ in §2.7, where we considered visible light passing through HI regions. In that limit, the exponential term in (6.65) would be negligible so that we had a different expression (2.87) for the optical depth. We had used the oscillator strength f in our discussion in §2.7, whereas we are now using the Einstein coefficients. These are obviously related quantities (Exercise 6.5).

6.6.2 Warm intercloud medium

Let us now consider what happens to the emission and the absorption lines at 21 cm if the hydrogen gas is warmer. Certainly the line profile $\phi(\Delta\nu)$ should get broader due to thermal broadening. However, the total intensity at the emission line, as given by (6.63) should not change. On the other hand, the absorption line should become much weaker because the optical depth, as given by (6.66), is inversely proportional to temperature. Suppose, along our line of sight, we have both a cold cloud and some warm gas. What kinds of emission and absorption lines should we get? Since the warm gas would not absorb much because of its higher temperature, the absorption line will be a narrow line due to the absorption by the cloud. On the other hand, there will be both narrow-line emission from the cloud and broad-line emission from the warm gas. Thus the emission line should look like a narrow line above a broad shoulder. Hence, if neutral hydrogen gas in two phases with differing temperatures is present along the line of sight, it is in principle possible to isolate the two phases from a careful study of the emission and absorption lines at 21 cm.

Figure 6.11 shows the 21-cm emission line from the ISM close to a background radio source (top) as well as the 21-cm absorption line produced by the ISM in the spectrum of the background radio source (bottom). A careful look makes it clear that the emission line has a broader shoulder at the base, whereas the absorption line is narrow. Thus it appears that the 21-cm emission and absorption lines actually support the view that interstellar space contains neutral hydrogen in two distinct phases. We have already discussed the cloud phase. The space between clouds (as much as 40% of the interstellar space) appears filled with much warmer neutral hydrogen gas, with a temperature of about 8000 K and density in the range of 10^5–10^6 particles m^{-3}. This is the second important phase of the ISM with a much lower density compared to the clouds.

6.6.3 Molecular clouds

The ISM is known to contain varieties of molecules including some reasonably complex organic molecules. Usually these molecules are found in the cool

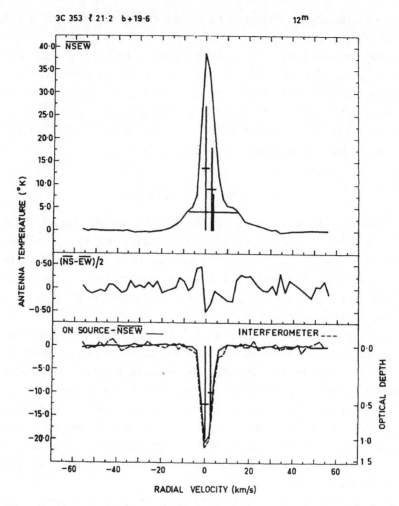

Fig. 6.11 The 21-cm emission line from the ISM close to the background radio source 3C 353 (top panel) and the 21-cm absorption line produced by the ISM in the spectrum of 3C 353 (bottom panel). From Radhakrishnan *et al.* (1972). (©American Astronomical Society. Reproduced with permission from *Astrophysical Journal.*)

dense regions of the ISM. These molecular clouds have densities more than 10^9 particles m^{-3} and temperatures in the range 10–30 K. Even though these clouds occupy less than 1% of interstellar space, they may contribute significantly to the mass of the ISM (as much as 40%). One important question is how the complex molecules form in these clouds. The subject of interstellar chemistry is still in its infancy. Many molecules are supposed to have been synthesized on the surfaces of dust grains.

Most molecules in the ISM are studied through the molecular radio lines. The hydrogen molecule H_2 is believed to be the most abundant molecule. Since this molecule does not have any radio lines, its presence is inferred from the absorption lines in the UV spectra of background sources. Perhaps the most extensively studied interstellar molecule is carbon monoxide CO, since it has very convenient radio lines arising from transitions between various rotational levels. A standard result of molecular physics is that frequencies in the rotational spectra should be equally spaced and be multiples of a fundamental frequency (see Exercise 6.7). The fundamental frequency for CO is 115 GHz, corresponding to a wavelength of 2.6 mm. The next higher frequencies are at 230 GHz, 345 GHz, and so on. The distribution of CO in the Galaxy has been studied quite extensively and is found to be somewhat different from the distribution of neutral hydrogen HI. Not much CO is found beyond 10 kpc from the galactic centre, whereas HI can be found at much greater distances.

A very big surprise was to find that the intensity of some sources in specific molecular lines (such as OH lines) was abnormally high. If the sources were assumed to be optically thick in the spectral lines and the specific intensity was equated to the Planck function $B_\nu(T)$, then temperatures as high as 10^9 K were inferred! The favoured explanation is that this high intensity is not caused by abnormally high temperatures, but by maser action. In our discussion of the two-level atom leading to (6.61), we saw an example in which the upper level is de-populated compared to what we expect in thermodynamic equilibrium. In more complex situations involving more levels, the upper level can become over-populated (see Exercise 6.6). If $n_u/n_l > g_u/g_l$, then it is easy to see from (6.54) that the absorption coefficient α_ν should be negative. In such a situation, a beam of radiation keeps getting stronger while passing through the material rather than being attenuated.

Molecular clouds, which are often of gigantic size, are of great interest to astrophysicists as birthplaces of stars. Many molecular clouds are believed to be contracting slowly under self-gravity and stars would eventually form in the central regions. Figure 6.12 shows such a molecular cloud from which stellar 'eggs' seem to be emerging. We shall discuss more about star formation in §6.8 and in §8.3.

6.6.4   HII regions

A UV photon with wavelength shorter than 912 Å can ionize a hydrogen atom by knocking off the electron from the ground level $n = 1$. The O and B stars, which have high surface temperatures, emit copious amounts of UV photons. Since these stars are short-lived (see §3.4), they are found in regions where star formation has recently taken place. We pointed out above that the cores of molecular clouds collapse to produce stars. Once the stars have been formed, the UV photons from the O and B stars ionize the ISM around them. Such regions of ionized hydrogen are called *HII regions*. The typical temperatures of such regions are of order 6000 K.

Fig. 6.12 The molecular cloud in M16, within which new stars are being born. Photographed with the Hubble Space Telescope. Courtesy: NASA, ESA, Space Telescope Science Institute, J. Hester and P. Scowen.

The H$_{II}$ regions are often found to be approximately spherical in shape and are known as *Strömgren spheres*. It is not difficult to estimate the Strömgren radius R_S of such a sphere (Strömgren, 1939). In a steady state, the number of ionizations in a unit volume within the Strömgren sphere has to balance the number of recombinations. Since the recombination rate should be proportional to the number of protons n_p and the number of electrons n_e, we can write the number of recombinations in unit volume in unit time as $\alpha n_p n_e$, where α is the recombination coefficient. This has to equal the number of ionizations in unit volume in unit time. Hence the total number of ionizations within the Strömgren sphere must be $(4\pi/3)R_S^3 \alpha n_p n_e$. If the central star emits N_γ UV photons per unit time with wavelength shorter than 912 Å, then we must have

$$N_\gamma = \frac{4}{3}\pi R_S^3 \alpha n_p n_e. \tag{6.67}$$

This is the relation which determines the sizes of Strömgren spheres.

If the recombination process involves a free electron jumping to the ground level $n = 1$, then a UV photon would be emitted. However, if the free electron is first captured in the $n = 2$ level and then it only makes a transition to the $n = 1$ level, then we would get two photons, one of which will be within the visible range. Very often an electron cascades through several energy levels, thereby emitting many photons. The HII region is one phase of the ISM which can be studied by the visible light emitted by it. When an electron makes a transition between two relatively high levels (say from $n = 100$ to $n = 99$), a radio photon is emitted, and even such radio emissions from HII regions have been detected. Additionally, the hot gas in the HII regions also emits bremsstrahlung (see §8.12) with a continuous spectrum in the radio range.

Apart from hydrogen emission lines, the HII regions radiate in emission lines from partially ionized atoms of elements like carbon, nitrogen and oxygen. Many of these emission lines, lying often in the visible part of the spectrum, correspond to 'forbidden' transitions with very slow transition rates. These lines are difficult to observe under laboratory conditions where the excited atoms are more likely to de-excite collisionally rather than by emitting a photon. Under the low-density conditions of interstellar space where collisions are much rarer, the excited atoms get a chance to de-excite by emitting a photon, even though this corresponds to a very slow transition. For any particular spectral line, there is a critical density beyond which the collisional de-excitation takes over and the emission line is quenched.

6.6.5   Hot coronal gas

A supernova explosion spews out hot gas in the interstellar space. Many supernova remnants are observationally known. It is thought that hot gases from very old supernovae ultimately fill up the interstellar spaces not occupied by the other phases (McKee and Ostriker, 1976). The coronal gas may have temperatures of the order of 10^6 K, but very low densities of about only 10^3 particles m^{-3}. This gas may occupy as much as 50% of the interstellar space, even though it contributes very little to the mass of the ISM. We shall see in §8.12 that hot gases emit radiation by the process of bremsstrahlung. The hot coronal gas in the Galaxy emits soft X-rays, which is the chief diagnostic tool for studying this phase.

6.7   The galactic magnetic field and cosmic rays

There is a large-scale magnetic field in the interstellar space of our Galaxy. This was first inferred when Hiltner (1954) was measuring the polarization of starlight and found that the light from most stars is slightly polarized. It is believed that interstellar grains are generally non-spherical and can be aligned by the galactic magnetic field, making the ISM act like a polarizing medium in the presence of a magnetic field. It should be mentioned that the alignment of grains involves some subtle physics and is not exactly analogous to the alignment of a compass needle by a magnetic field. Here we shall not get into the physics of grain alignment and light polarization, which was investigated by Davis and Greenstein (1951). A reader interested in knowing more about this subject should consult Chapter 8 of Spitzer (1978).

The nature of starlight polarization depends on the direction in which the star is seen. Figure 6.13 shows light polarizations of stars located in different galactic coordinates, the lengths of the small line segments indicating the magnitudes of polarization and their inclinations indicating the directions of polarization. In the direction of the galactic magnetic field, we do not expect to see any systematic polarization. This happens approximately in the directions $l \approx 60°$ and $l \approx 240°$ in Figure 6.13. These longitudes roughly correspond to the spiral arm in the solar neighbourhood. When we look at right angles with respect to these directions (i.e. with respect to the magnetic field), we see the maximum polarization, as expected from common sense. The polarization of starlight thus establishes that our Galaxy has a magnetic field running along the spiral arm. However, to estimate the amplitude of the magnetic field, we need a theory of grain alignment, which involves many uncertainties.

Signals from pulsars (introduced in §5.5) provide a method for estimating the galactic magnetic field. Not only are pulsars interesting objects by themselves, the signal from a pulsar gives us important information about the interstellar medium lying between the pulsar and us. We know that electromagnetic waves travelling in empty space are non-dispersive. However, we shall see in §8.13 that the speed of an electromagnetic wave passing through a plasma varies with the frequency of the wave. Since there are some free electrons in the interstellar space, the ISM can act like a plasma. Radio waves of lower frequency travel more slowly through the interstellar plasma. The effect on much higher-frequency visible light is practically negligible. When we analyse the pulse received from a pulsar, we find that higher-frequency waves arrive slightly before lower-frequency waves. The usual interpretation is that the pulsar emitted waves at all the frequencies simultaneously, but the lower-frequency waves got delayed while passing through the ISM. The reader is asked at the end of Chapter 8 (Exercise 8.8) to show that the variation of arrival time with frequency is given by

$$\frac{dT_a}{d\omega} = -\frac{e^2}{\epsilon_0 m_e c\, \omega^3} \int n_e\, ds, \tag{6.68}$$

where n_e is the electron number density in interstellar space and the integral is over the path from the pulsar to us. Since n_e in the solar neighbourhood is

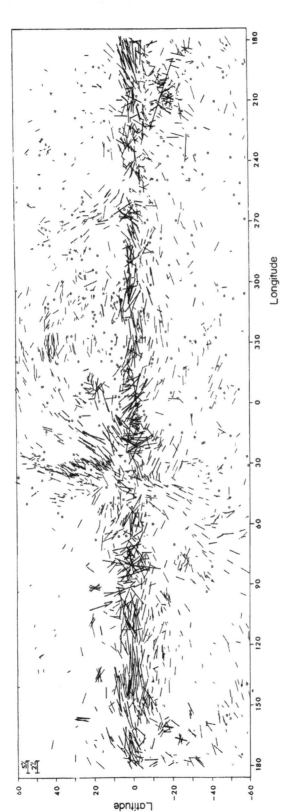

Fig. 6.13 The polarization of starlight measured for stars in different galactic coordinates. The length of the line segment indicates the amplitude of polarization for a star, whereas its direction indicates the polarization plane. From Mathewson and Ford (1970). (©Royal Astronomical Society. Reproduced with permission from *Memoirs of Royal Astronomical Society*.)

around 10^5 m^{-3}, a knowledge of $\int n_e \, ds$ from the time dispersion measurement can give an estimate of the distance of the pulsar. The signals from pulsars are also polarized. It can be shown that the magnetic field present in the plasma can make the plane of polarization rotate, the rotation being more for lower frequencies. A theoretical derivation of this phenomenon, known as *Faraday rotation*, can be found in §12.5 and §12.6 of Choudhuri (1998). The variation of the plane of polarization with frequency can be shown to be

$$\frac{d\theta}{d\omega} = -\frac{e^3}{\epsilon_0 m_e^2 c \, \omega^3} \int n_e B_\parallel \, ds, \qquad (6.69)$$

where $B_\parallel$ is the component of the magnetic field parallel to the line of sight. Thus the time dispersion gives the integral $\int n_e \, ds$, whereas the angular dispersion gives the integral $\int n_e B_\parallel \, ds$. An estimate of the magnetic field can be found from the ratio

$$\frac{\int n_e B_\parallel \, ds}{\int n_e \, ds}$$

between these two integrals. From the measurements of time dispersions and angular dispersions of many pulsars, the galactic magnetic field is estimated to have the value $(2$–$3) \times 10^{-10}$ T. As we already pointed out, the mean magnetic field is believed to run along the spiral arm of the Galaxy, although the fluctuations around the mean are probably as large as the mean.

Associated with the galactic magnetic field, there are highly energetic charged particles spiralling around the field lines. It was discovered by Hess (1912) that the Earth is continuously bombarded by *cosmic rays* coming from above the Earth's atmosphere. We shall discuss in §8.10 that the energetic charged particles of the cosmic rays are believed to be accelerated in supernova blast waves. Then they spiral around the galactic magnetic field and fill up the Galaxy. It will be shown in §8.11 that relativistically moving charged particles spiralling around a magnetic field give out a kind of radiation known as *synchrotron radiation*. For cosmic rays spiralling around the galactic magnetic field, the synchrotron spectrum lies mainly in the radio regime. Radio telescopes have detected synchrotron radiation not only from our Galaxy but also from other similar galaxies, making it clear that other similar galaxies also have magnetic fields and cosmic rays.

The energy density $B^2/2\mu_0$ associated with the galactic magnetic field is of order 10^{-14} J m^{-3}. It is found that HI clouds have typical random turbulent velocities of order 10 km s^{-1} (just like stars as pointed out in §6.3). One can easily check that the kinetic energy density $\rho v^2/2$ associated with interstellar turbulence is of the same order as the energy density of the magnetic field. The energy density associated with cosmic ray particles is also estimated to be comparable. There thus appears to be a remarkable *equipartition* of energy amongst the gas, the magnetic field and the cosmic rays. Where does the galactic magnetic field come from and why do we have this remarkable equipartition? In §8.7 we shall give an introduction to the *dynamo theory* – a theory which explains how turbulent motions in a plasma can generate magnetic fields under certain circumstances. Since the turbulent motions of the interstellar gas are responsible for the galactic magnetic field and the magnetic field is then responsible for the acceleration and confinement of cosmic rays within the Galaxy, perhaps an equipartition of the kind we find is not totally surprising. However, it is difficult to go beyond such hand-waving arguments and give very rigorous justifications.

6.8   Thermal and dynamical considerations

It should be clear to any reader by now that the ISM is an immensely complex system. Why does the ISM have several different phases in different regions of space instead of being a more uniformly spread gas? We, of course, know that the ISM is continuously disturbed by external sources. Supernova explosions keep adding more hot coronal gas to the ISM. Newly born O and B stars keep ionizing portions of the ISM to create HII regions. However, these external disturbances alone cannot explain why the neutral gas is found in such distinct phases as HI clouds and the warm intercloud medium. Why don't we see a more continuous distribution of densities and temperatures of the neutral gas?

Any phase of the ISM is giving out energy in the form of radiation. In §6.6 we have discussed the kinds of radiation emitted by the different phases. Let Λ be the rate at which energy is lost from unit volume in unit time. Often Λ is referred to as the *cooling function*. If the system is to remain in a steady state, then an equal amount of energy has to be supplied. The energy gained by the ISM per unit volume per unit time is called the *heating function* and is denoted by Γ. The energy dumped by the supernova explosions into the ISM and the UV photons from very hot stars absorbed in the ISM certainly make contributions to the heating function Γ. Another important contribution comes from cosmic rays. An energetic charged particle passing through matter can ionize some atoms and can lose some energy to the surrounding medium in this process. One important contribution to Γ comes from the ionization losses of cosmic ray particles. Certainly

$$\mathcal{L} = \Lambda - \Gamma = 0 \tag{6.70}$$

is a necessary condition for the thermal equilibrium of any phase of the ISM. However, it is not a sufficient condition. It was argued by Field (1965) that such a thermal equilibrium can exist only if $(\partial \mathcal{L}/\partial T) > 0$. Even without a detailed analysis, we can see how such a condition may arise. Let us think what would happen if the opposite $(\partial \mathcal{L}/\partial T) < 0$ were to hold. Suppose a system is in equilibrium with $\mathcal{L} = 0$. Some disturbance causes its temperature to decrease

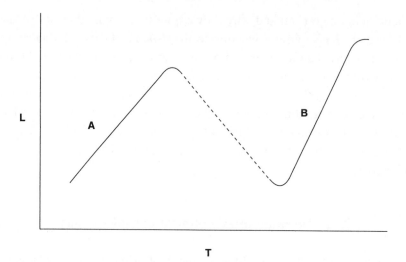

Fig. 6.14 A schematic sketch of $\mathcal{L}$ as a function of T for a system which has two possible stable equilibrium configurations.

slightly. This will increase $\mathcal{L}$ if $(\partial\mathcal{L}/\partial T) < 0$ holds. Since $\mathcal{L}$ is the net energy loss rate, an increase in $\mathcal{L}$ would mean that the system would start losing energy at a faster rate, leading to a further decrease in temperature. This causes a runaway situation and the system goes on becoming colder. Thus, once the equilibrium is disturbed, the system moves further away from equilibrium.

Suppose $\mathcal{L}$ as a function of T looks as sketched in Figure 6.14. Then A and B are the two possible regions of stable equilibrium where $(\partial\mathcal{L}/\partial T) > 0$. Thus the system can be in stable equilibrium corresponding to the temperatures at A and B, with the intermediate temperatures ruled out. Field, Goldsmith and Habing (1969) proposed that the HI clouds and the warm intercloud medium correspond to the two distinct thermal equilibrium states of the neutral gas.

Apart from the thermal balance, there has to be a force balance also in the ISM to ensure that no large-scale motions are driven by unbalanced forces. As we shall see in §8.2, a fluid system like the ISM basically can have two important kinds of forces, arising from pressure gradients and gravitational fields respectively. Let us first consider the pressure gradient forces. A large variation of pressure within the ISM would lead to gas flows from regions of high pressure to regions of low pressure. In spite of the very different physical conditions within the various phases of ISM, their pressures are comparable. This is seen in Figure 6.15, which plots the temperature T against the particle number density n. The different phases of the ISM correspond to different regions in this figure. The pressure nk_BT would be constant along the straight lines. The HI clouds and the intercloud medium are clearly in pressure equilibrium, with the other phases also having pressures not differing widely.

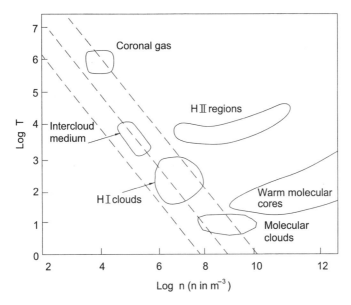

Fig. 6.15 Different phases of the ISM indicated in a temperature T versus number density n plot. The pressure would be constant along the dashed straight lines in this figure.

We at last come to the balance of gravitational forces. The ISM moves in circular orbits around the galactic centre such that the centrifugal force balances the dominant radial component of gravity. In the direction perpendicular to the galactic plane, the gravitational field is directed towards the mid-plane of the Galaxy. The hydrostatic equilibrium has to be maintained in this direction. The pressure of the neutral gas, in conjunction with the pressure of the magnetic field and cosmic rays, ensures that the ISM occupies a layer of thickness about 200 pc around the mid-plane of the Galaxy.

For a complicated system like the ISM, considering the force balance alone is not sufficient to determine the equilibrium. One needs to check if the equilibrium is stable. We shall show in §8.3 that, if a sufficiently large region of the ISM becomes over-dense, then the enhanced gravity of that region may cause the gas of that region to collapse further. This is the celebrated *Jeans instability* (Jeans, 1902), which triggers the process of star formation. We believe that the cores of some molecular clouds are such collapsing regions where stars are ultimately born. However, the Jeans instability allows us to understand only how the collapse is initiated. The subsequent progress of the collapse is an extremely complex process and is still poorly understood. For example, a simple estimate shows that the mass of a collapsing over-dense region has to be much more than the masses of individual stars (see §8.3). Obviously the collapsing cloud has to fragment into individual stars at some stage of the collapse. The distribution of masses amongst the newly forming stars was studied by Salpeter (1955) from observational data. If $\xi(M)\, dM$ is the number of stars born with masses between M and $M + dM$, then the *Salpeter initial mass function* is

$$\xi(M)\, dM \propto M^{-2.35}\, dM. \tag{6.71}$$

We still do not have a proper theoretical explanation of this initial mass function.

Star formation is one of the important theoretically ill-understood problems of modern astrophysics. Observationally also this is a process impossible to study directly, since the dust particles present in a molecular cloud make the core region inaccessible in visible light. However, as the cloud collapses and becomes hotter, the dust particles emit in the infrared. The infrared observations have revealed a very complicated scenario. The cores of the clouds do not collapse in a spherically symmetric fashion. The angular momentum present in a typical gas cloud (due to the rotation of the Galaxy) makes the collapsing core take the shape of a disk. One of the most striking discoveries of infrared astronomy is that there are often *bipolar outflows* from the polar regions of these collapsing disks. The Doppler shifts of molecular lines give the speeds of such outflows, which can sometimes be as large as $100\,\mathrm{km\ s^{-1}}$.

Exercises

6.1 We have presented a very elementary discussion of star count analysis in §6.1.1 by assuming that all stars have the same absolute magnitude M and there is no absorption in interstellar space. Now assume that a fraction of stars $\Phi(M)\, dM$ have absolute magnitudes between M and $M + dM$, whereas $a(r)$ is the change in magnitude of a star at a distance r due to absorption. Suppose $A(m)\, dm$ is the number of stars within a solid angle ω having apparent magnitude between m and $m + dm$. If $D(r)$ is the number density of stars at a distance r, show that

$$A(m) = \omega \int_0^\infty \Phi[m + 5 - 5\log_{10} r - a(r)]\, D(r)\, r^2 dr.$$

Show that this expression reduces to the form (6.3) if all stars have the same absolute magnitude with no absorption and are uniformly distributed.

6.2 The interstellar medium in the galactic disk diminishes the luminosity of stars by about 1.5 magnitude (i.e. increases the magnitude by 1.5) per kpc. Show that this implies that the brightnesses of stars fall off with distance r in the galactic disk as

$$\frac{e^{-\alpha r}}{r^2}.$$

Find the value of α.

6.3 Suppose the gravitational field is falling as r^{-2} in a region of the Galaxy. Find the A and B constants, as defined by (6.21) and (6.22), for such a region. Show that the frequency of the epicyclic motion is going to be equal to the angular velocity. What is the physical significance of this?

6.4 Make a simplified model of the galactic disk by assuming it to be an infinite sheet of constant thickness with constant density inside. Show that a star displaced from the mid-plane of the Galaxy in the vertical direction undergoes simple harmonic oscillations around the mid-plane (assuming that the star always remains within the region of constant density). Taking the density in the mid-plane to correspond to about 5×10^6 hydrogen atoms m^{-3}, estimate the period of oscillation. How does it compare with the period of revolution of a star in the solar neighbourhood around the galactic centre?

6.5 Let B_{lu} be the Einstein coefficient for transition from the lower level l of an atom to its upper level u, separated by energy $h\nu_0$. The oscillator strength f for this transition as introduced in §2.7 satisfies

$$\frac{e^2}{4\epsilon_0 m_e} f = h\nu_0 B_{lu}.$$

Show that this relation makes the discussions in §2.7 and §6.6 consistent with each other.

6.6 Consider an atom with three levels denoted by 1, 2 and 3 in order of increasing energy. Suppose no transitions take place between the upper two levels 2 and 3. Writing balance equations of the type (6.59) and assuming that the radiation present is not strong enough to make radiative transitions important, show that

$$\frac{n_3}{n_2} = \frac{g_3(1 + A_{21}/n_e\gamma_{21})}{g_2(1 + A_{31}/n_e\gamma_{31})} e^{-E_{23}/\kappa_B T}.$$

Here all the symbols have obvious meanings. It is clear that we shall have the Boltzmann distribution law when n_e is large. Discuss the conditions which would lead to population inversion. If there is no transition between the upper two levels, then this population inversion may not give rise to maser action. But this simple example of a three-level system should give some idea of how population inversions can arise.

6.7 We have pointed out in §6.6.3 that CO molecules in molecular clouds emit at frequencies which are integral multiples of 115 GHz. If I is the moment of inertia of the molecule around an axis perpendicular to the axis of the molecule, then show that the energy levels of the molecule are given by

$$E_J = \frac{\hbar^2}{2I} J(J+1),$$

where J can have integral values. If the selection rule $\Delta J = -1$ has to be obeyed for emission, then show that the emission spectrum should be as seen. Make an estimate of the distance between the carbon and the oxygen atoms in the molecule.

7

Elements of stellar dynamics

7.1 Introduction

Since gravity is a long-range attractive force, any star in a galaxy attracts all the other stars in the galaxy all the time. For simplicity, we can regard the stars as point particles. Then a galaxy or a star cluster can be regarded as a collection of particles in which all the particles are attracting each other through an inverse square law of force. The aim of *stellar dynamics* is to study the dynamics of such a system of self-gravitating particles. We, of course, know that there is also gas between the stars in a galaxy, which can add extra complications. However, it is generally believed that stellar dynamics holds the key to understanding the structure of galaxies or star clusters.

We have discussed our Galaxy in Chapter 6 and shall discuss external galaxies in Chapter 9. Although some galaxies are irregular in appearance, we shall see in §9.2 that most galaxies have very regular shapes. The fundamental question of stellar dynamics is: why do collections of self-gravitating mass particles tend to take certain particular configurations in preference to many other possible configurations? A fully satisfactory answer to this question is still not known. Hence the subject of galactic structure is on a much less firm footing compared to the subject of stellar structure. We know that the gravitational attraction of the stars has to be balanced by their motions, to ensure that the stars do not all fall towards the centre of the stellar system together due to their mutual gravitational attraction. The considerations involved here are analogous to the considerations of why the gas particles in an atmosphere do not all settle at the bottom, even though they are pulled by gravity downwards. It is basically the random motions of the gas particles which prevent this from happening. We shall see in §7.2 that the motions (they need not necessarily be random) of stars can 'balance' the gravity in such a way that the system remains in a steady state. We shall obtain a relation connecting the total kinetic energy with the total gravitational energy. However, going beyond this to calculate the detailed structure of a galaxy or a star cluster is not easy.

If the mass and the chemical composition of a star are given, we saw in Chapter 3 how the structure of the star can be theoretically calculated. We do not have to know the details of the initial conditions, such as the nature of the gas cloud from which the star formed. In the case of a galaxy, is it even in principle possible to calculate its structure from a knowledge of, say, the mass distribution of the stars which make up the galaxy and the total kinetic energy? Or do the details of the initial conditions in the formation stage of the galaxy determine what the galaxy is going to be like?

Let us consider the particles of a gas in a container. We know that the velocity distribution of the particles would obey a universal law: the Maxwellian velocity distribution given in (2.27). It can be shown from the principles of statistical mechanics that this distribution corresponds to a configuration of maximum probability. That is why the gas is likely to be found in this config-uration. Suppose we do something to make the velocity distribution of the gas very different from a Maxwellian. If the gas is again left to itself, the velocity distribution will relax to a Maxwellian after a few collisions. In a typical system of stars, the probability of an actual physical collision between two stars is very low. A star usually moves in a smooth gravitational field produced by all the stars around it. However, when two stars come sufficiently close to each other, their trajectories are deflected by mutual gravitational interaction. In a stellar system, such encounters between stars play the role of collisions and tend to relax the velocity distribution of the stars. We shall discuss in §7.3 how the collisional relaxation time in a stellar system can be estimated. Simple estimates show that the relaxation time of a typical galaxy is much longer than the age of the Universe and galaxies must be unrelaxed system. On the other hand, the relaxation time of globular clusters is less and they are expected to be systems in which collisional relaxation is important. The subject of stellar dynamics is usually divided into two parts. *Collisional stellar dynamics* deals with stellar systems in which collisional relaxation has been important, whereas *collisionless stellar dynamics* deals with stellar systems in which we can ignore collisions.

One may naively expect that collisional relaxation would lead to an equilib-rium configuration with the stars obeying the Maxwellian distribution. We shall show in §7.4 that this naive expectation leads to inconsistencies, since self-gravity is not compatible with thermodynamic equilibrium. Hence collisional stellar dynamics is a much more complex subject than what one may expect. Other than showing that a simple thermodynamic equilibrium is not possible, we shall not be able to go into the details of this subject. An elementary introduction to collisionless stellar dynamics will be presented in §7.5 and §7.6. Although we cannot calculate the detailed structures of collisionless stellar systems from first principles, we shall see that various aspects of stellar motions in galaxies are inter-related and can be understood from a stellar dynamical analysis.

It should be emphasized that the aim of this chapter is only to give the readers a feeling of what the subject of stellar dynamics is like. For a full treatment of stellar dynamics, which is beyond the scope of this elementary book, the readers should consult books like Binney and Tremaine (1987). Even topics of considerable astrophysical interest are left out because their stellar dynamical analysis involves rather advanced theoretical techniques. An example of such a topic is the theory of spiral structures in galaxies. We shall see in §9.2 that many galaxies have spiral structures. The most successful effort in explaining the spiral structures theoretically is the density wave theory of Lin and Shu (1964), which we shall not be able to discuss because of its complexity. For an excellent non-technical account of the subject, the readers are urged to look up Shu (1982, pp. 275–281).

7.2   Virial theorem in stellar dynamics

Since the inward pull of gravity inside a star is balanced by thermal energy, we saw in §3.2.2 that there is a relation between the gravitational potential energy and the total thermal energy, as given by (3.10), which is known as the virial theorem. We now consider a collection of particles attracting each other through gravity. If this collection is in a steady state (i.e. if its overall size is neither increasing nor decreasing), then it is the motions of the particles in the collection which must balance the inward gravitational pull. We thus expect a relation between the total gravitational potential energy and the total kinetic energy of the system, which must be analogous to (3.10). We now derive such a relation, which is valid for both collisionally relaxed and unrelaxed systems.

Let the position and the velocity of the i-th particle at an instant of time be $\mathbf{x}_i$ and $\mathbf{v}_i$ respectively. The momentum of the particle is $\mathbf{p}_i = m_i \mathbf{v}_i$. We have

$$\frac{d}{dt}(\mathbf{p}_i.\mathbf{x}_i) = \frac{d\mathbf{p}_i}{dt}.\mathbf{x}_i + \mathbf{p}_i.\frac{d\mathbf{x}_i}{dt} = \mathbf{F}_i.\mathbf{x}_i + 2T_i, \tag{7.1}$$

where $\mathbf{F}_i$ is the force acting on the i-th particle and T_i is its kinetic energy. We now integrate (7.1) over a sufficiently long time τ and divide all the terms by τ. This gives

$$\frac{1}{\tau}\delta(\mathbf{p}_i.\mathbf{x}_i) = \overline{\mathbf{F}_i.\mathbf{x}_i} + 2\overline{T_i}, \tag{7.2}$$

where the overline indicates averaging over this time interval τ, while $\delta(\mathbf{p}_i.\mathbf{x}_i)$ is the difference between the values of $\mathbf{p}_i.\mathbf{x}_i$ at the beginning and the end of the interval. We can write an equation like (7.2) for each of the particles in the collection. On summing them up, we have

$$\frac{1}{\tau}\delta\left(\sum_i \mathbf{p}_i.\mathbf{x}_i\right) = \sum_i \overline{\mathbf{F}_i.\mathbf{x}_i} + 2\overline{T}, \tag{7.3}$$

where $T = \sum_i T_i$ is the total kinetic energy of the system. For a system with size not changing in time, we do not expect the value of $\sum_i \mathbf{p}_i . \mathbf{x}_i$ to change with time. Hence the left-hand side of (7.3) must be zero, leading to

$$\sum_i \mathbf{F}_i . \mathbf{x}_i + 2\overline{T} = 0. \tag{7.4}$$

Now the force on the i-th particle due to all the other particles is

$$\mathbf{F}_i = \sum_{j \neq i} Gm_i \frac{m_j}{|\mathbf{x}_j - \mathbf{x}_i|^3} (\mathbf{x}_j - \mathbf{x}_i)$$

so that

$$\sum_i \mathbf{F}_i . \mathbf{x}_i = \sum_i \sum_{j \neq i} \frac{Gm_i m_j}{|\mathbf{x}_j - \mathbf{x}_i|^3} (\mathbf{x}_j - \mathbf{x}_i).\mathbf{x}_i. \tag{7.5}$$

It is to be noted that the double summation on the right-hand side implies a summation over all possible pairs of particles. For a particular pair of particles i and j, it is obvious that the summation will have two terms

$$\frac{Gm_i m_j}{|\mathbf{x}_j - \mathbf{x}_i|^3}[(\mathbf{x}_j - \mathbf{x}_i).\mathbf{x}_i + (\mathbf{x}_i - \mathbf{x}_j).\mathbf{x}_j] = -\frac{Gm_i m_j}{|\mathbf{x}_j - \mathbf{x}_i|}.$$

Then, from (7.4) and (7.5), we have

$$2\overline{T} - \sum_{\text{all pairs}} \frac{\overline{Gm_i m_j}}{|\mathbf{x}_j - \mathbf{x}_i|} = 0. \tag{7.6}$$

We can write this equation as

$$2\overline{T} + \overline{V} = 0, \tag{7.7}$$

where

$$\overline{V} = -\sum_{\text{all pairs}} \frac{\overline{Gm_i m_j}}{|\mathbf{x}_j - \mathbf{x}_i|} \tag{7.8}$$

is the total gravitational potential energy. The virial theorem for stellar dynamics, as given by (7.7), has the same form as the virial theorem (3.10) for stellar structure. We have the thermal energy instead of the kinetic energy in the virial theorem for stellar structure. However, since the thermal energy is nothing but the kinetic energy of the gas particles in the stellar interior, the basic physics is the same in both the cases.

Suppose we have a cluster of N stars each having mass m. There are $N(N-1)/2 \approx N^2/2$ pairs in the system and the gravitational potential energy of a typical pair is

$$\frac{Gm^2}{\langle R \rangle},$$

where $\langle R \rangle$ is the average distance between the stars in the pair, which must be of the same order as the radius of the star cluster. Noting that $2\overline{T} = Nm\langle v^2 \rangle$, we have from (7.6) that

$$Nm\langle v^2 \rangle \approx \frac{N^2}{2} \frac{Gm^2}{\langle R \rangle},$$

from which

$$\langle v^2 \rangle \approx \frac{GM}{\langle R \rangle}, \tag{7.9}$$

where $M = Nm$ is the mass of the star cluster. Astronomers use (7.9) quite regularly to estimate masses of star clusters. One can estimate the velocity dispersion $\langle v^2 \rangle^{1/2}$ from Doppler measurements of spectral lines of the stars in the cluster. If the distance to the cluster is known, then we can get $\langle R \rangle$ from the apparent size of the cluster. The only remaining quantity in (7.9) is the cluster mass M, which can then be calculated.

Suppose the velocity dispersion in a star cluster of mass M and radius R is less than what would be expected from (7.9). Then gravity cannot be balanced by the motions of stars and the cluster has to shrink in size. In this process, the gravitational potential energy will decrease. This gravitational potential energy has to go into the kinetic energy of the stars, making the velocity dispersion larger. Eventually, if the velocity dispersion becomes large enough to satisfy (7.9), the cluster will stop shrinking in size any further. While applying the virial theorem (7.9), one should ensure that the system is gravitationally bound and is *virialized*. Otherwise, the application of the virial theorem may lead to erroneous results.

7.3 Collisional relaxation

After establishing the virial theorem which should be valid for any gravitationally bound stellar system in steady state (i.e. which is not growing or shrinking in size), irrespective of whether the system is collisionally relaxed or not, we now come to the important question of estimating the collisional relaxation time of a stellar system.

Let us consider a galaxy or a cluster with stars of mass m. Suppose a star is moving with speed v. If no other star is very close, then this star will move in a smooth gravitational field collectively produced by all the stars in the system. On the other hand, if another star happens to be close by, then the trajectory of this star may get deflected by the gravitational attraction of the other star, and we would refer to this as a *collision*. This statement may appear vague. How close do the two stars have to be in order for their interaction to be called a collision? We give a working definition. If the deflection of the trajectory of the

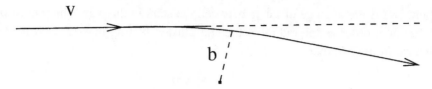

Fig. 7.1 A sketch of a collision between two stars.

star involves a change in momentum at least as large as the original momentum
of the star, then we would regard it as a collision. Using this working definition,
we now determine the distance b from the trajectory of the star within which
another star has to be in order for their interaction to qualify as a collision.
Figure 7.1 shows a star initially moving in a straight line with speed v. Another
star is at a distance b from the trajectory. If this is a limiting collision, then
the change in momentum of the moving star should be equal to its original
momentum mv. Now, when the two stars are close, the force of gravity between
them is of order Gm^2/b^2. The two stars are close to each other for an interval of
time of order b/v during which this gravitational force acts. Hence the change
of momentum of the moving star (which should be in a direction perpendicular
to its original momentum) is of order

$$\Delta p \approx \frac{Gm^2}{b^2}\frac{b}{v}.$$

Equating this to mv, the limiting distance b for a collision is given by

$$b \approx \frac{Gm}{v^2}. \tag{7.10}$$

In unit time the moving star sweeps out a volume $\pi b^2 v$ within which another
star has to lie for a collision to take place. If n is the number density of stars,
then the number of collisions per unit time is given by $\pi b^2 vn$. The typical time
between collisions is the inverse of this. Since this collision time is the time in
which the memory of any initial velocity distribution is effectively erased, we
call it the relaxation time T_{rel}. This is given by

$$T_{rel} \approx (\pi b^2 vn)^{-1} \approx \frac{v^3}{\pi nG^2m^2} \tag{7.11}$$

on substituting for b from (7.10). If v is in km s^{-1} and n is the number per pc^3,
then this becomes

$$T_{rel} \approx 10^{10}\frac{v^3}{n} \text{ yr} \tag{7.12}$$

if we take $m \approx M_\odot$. A more rigorous treatment of collisional relaxation would
involve an integration over the effects of stars at different distances. See pp.
187–190 of Binney and Tremaine (1987) for a rigorous treatment. The simple
estimate given above gives the correct order of magnitude of the relaxation
time T_{rel}.

Table 7.1 Relaxation times for different stellar systems.

	v (in km s^{-1})	n (in pc^{-3})	T_{rel} (in yr)
Galaxy	100	0.1	10^{17}
Open cluster	0.5	1	10^{9}
Globular cluster	10	10^{3}	10^{10}

Table 7.1 gives the typical stellar velocity v, stellar number density n and collisional relaxation time T_{rel} (calculated by using (7.12)) for different stellar systems. The age of the Universe is of order 10^{10} yr. It is clear that a galaxy would not have sufficient time for collisional relaxation. On the other hand, a cluster of stars may at least be partially relaxed.

Since the collisional relaxation time in a galaxy is so enormous, one may tend to think that the stellar velocity distribution in a galaxy would be completely unrelaxed and would have the signature of some initial primordial velocity distribution. This is not entirely correct. We saw in §6.3.3 that there is an ellipsoidal distribution of velocity amongst stars in our neighbourhood. If a galaxy forms by contracting from a larger volume, then the gravitational field at a point inside the galaxy will keep changing drastically during the contraction time. It can be shown that a rapidly changing gravitational field has some effects analogous to the effects of collision (Lynden-Bell, 1967). This is called *violent relaxation*.

We end this discussion of collisional relaxation by pointing out an interesting relation. A star moving with speed v inside a stellar system of size R takes time of order R/v to cross the system. Hence

$$\frac{T_{rel}}{T_{cross}} \approx \frac{v^4}{\pi n G^2 m^2 R} \tag{7.13}$$

on substituting from (7.11). If the system is in virial equilibrium, then v^2 should be equal to GNm/R by (7.9), where N is the total number of stars in the system. Then from (7.13) we have

$$\frac{T_{rel}}{T_{cross}} \approx \frac{(GNm/R)^2}{\pi n G^2 m^2 R} \approx \frac{N^2}{\pi n R^3} \approx N, \tag{7.14}$$

since $N \approx \pi n R^3$. Hence, if a stellar system of N stars is in virial equilibrium, then the collisional relaxation time is N times the crossing time for a typical star in the system.

7.4 Incompatibility of thermodynamic equilibrium and self-gravity

If a stellar system has lasted for enough time for collisional relaxation to take place, what should it relax to? This is a question much more difficult to answer than what would appear at the first sight. We know the answer to the similar question for a gas in a container. No matter what initial velocity distribution we create for the gas particles, collisions would make the velocity distribution relax to a Maxwellian, which is the distribution appropriate for thermodynamic equilibrium of the system. We may naively expect a similar thermodynamic equilibrium to be established in the relaxed stellar system. Suppose we consider a stellar system made of stars of the same mass, which we take as the unit of mass. The energy of a star at position $\mathbf{x}$ moving with velocity $\mathbf{v}$ is given by

$$E(\mathbf{x}, \mathbf{v}) = \frac{1}{2}v^2 + \Phi(\mathbf{x}), \tag{7.15}$$

where $\Phi(\mathbf{x})$ is the gravitational potential at the point $\mathbf{x}$. In thermodynamic equilibrium, we would expect the distribution function for the stars to be

$$f(\mathbf{x}, \mathbf{v}) = Ae^{-\beta E(\mathbf{x},\mathbf{v})} = Ae^{-\beta[\frac{1}{2}v^2 + \Phi(\mathbf{x})]}, \tag{7.16}$$

where A is a normalization constant and we use the standard convention that $f(\mathbf{x}, \mathbf{v}) \, d^3x \, d^3v$ is the number of stars within volume d^3x having the ends of their velocity vectors lying within the volume d^3v in the velocity space. If we consider a region of the Earth's atmosphere over which temperature does not vary much, then we would find air molecules to obey a distribution function like (7.16), where $\Phi(\mathbf{x})$ would be the potential due to the Earth's gravitational field. In a stellar system, however, we have an additional requirement of self-consistency that the gravitational potential has to be due to stars in the stellar system itself. We shall now show that, if we demand self-consistency, then the distribution function (7.16) would lead to absurd conclusions.

Let us begin by explaining the meaning of self-consistency a little bit more. Suppose we have a distribution function like (7.16) depending on the gravitational potential $\Phi(\mathbf{x})$ and we are somehow able to guess or know the gravitational potential $\Phi(\mathbf{x})$. Then we know the spatial dependence of the distribution function from (7.16) and can find out the density of stars in various regions of space. On using Poisson's equation for gravitation, a knowledge of density would lead to a determination of gravitational potential $\Phi(\mathbf{x})$. If this $\Phi(\mathbf{x})$ turns out to be the same $\Phi(\mathbf{x})$ we began with and used in the distribution function while calculating density, then our solution is self-consistent.

We now try to impose this condition of self-consistency mathematically on the distribution function (7.16). The density at the position $\mathbf{x}$ is given by

$$\rho(\mathbf{x}) = \int f(\mathbf{x}, \mathbf{v}) \, d^3v = \frac{C}{4\pi G}e^{-\beta\Phi(\mathbf{x})} \tag{7.17}$$

on using (7.16) and writing

$$\int_0^\infty A e^{-\beta v^2/2} 4\pi v^2 \, dv = \frac{C}{4\pi G}.$$

On substituting (7.17) in Poisson's equation

$$\nabla^2 \Phi = 4\pi G \rho,$$

we get

$$\nabla^2 \Phi = C e^{-\beta \Phi}. \tag{7.18}$$

Self-consistency essentially requires that $\Phi(\mathbf{x})$ should satisfy (7.18). If we are able to solve (7.18) to obtain $\Phi(\mathbf{x})$ and use that in the distribution function (7.16), then everything will turn out to be consistent.

For simplicity, let us assume the stellar system to be spherically symmetric and try to solve (7.18) in that situation. Then $\Phi(\mathbf{x})$ becomes a function of r only and we need to keep only the radial derivatives in the expression of the Laplacian. In this situation, (7.18) reduces to

$$\frac{1}{r^2} \frac{d}{dr} \left(r^2 \frac{d\Phi}{dr} \right) = C e^{-\beta \Phi(r)}. \tag{7.19}$$

It follows from (7.17) that

$$\Phi(r) = -\frac{1}{\beta} \ln \frac{4\pi G \rho(r)}{C}.$$

Substituting this in (7.19), we get an equation for $\rho(r)$ as follows

$$\frac{1}{r^2} \frac{d}{dr} \left(r^2 \frac{d}{dr} \ln \rho \right) = -4\pi G \beta \rho. \tag{7.20}$$

This equation has to be solved with the boundary condition that there should be no cusp at the origin, i.e.

$$\frac{d\rho}{dr} = 0 \text{ at } r = 0.$$

Instead of trying to find the full solution of (7.20) (which is not difficult to do numerically), let us figure out the asymptotic form of the solution for large r. Let us see if a solution of the form

$$\rho(r \to \infty) = \frac{\rho_0}{r^b}$$

works at large r. On substituting this in (7.19), we get

$$-\frac{b}{r^2} = -4\pi G \beta \frac{\rho_0}{r^b}.$$

This can be satisfied only if $b = 2$, leading to the conclusion that density has to fall off as

$$\rho(r \to \infty) \propto \frac{1}{r^2}$$

to meet the requirement of self-consistency. It is trivial to show that the total mass would diverge to infinity for such a density distribution. Thus, if we begin with the thermodynamic distribution (7.16), the requirement of self-consistency forces us to the absurd conclusion that the total mass of the system has to be infinite! A stellar system of finite mass (i.e. finite number of stars) cannot relax to the thermodynamic distribution given by (7.16).

It is possible to construct well-behaved self-consistent solutions of spherically symmetric stellar systems if the distribution function $f(\mathbf{x}, \mathbf{v})$ is assumed to depend on energy $E(\mathbf{x}, \mathbf{v})$ in certain particular ways, i.e. different from the exponential dependence assumed in (7.16). The reader may look up §4.4 of Binney and Tremaine (1987) for some such solutions. Such solutions are of great mathematical interest as examples of self-consistent solutions for stellar systems. However, there is no physical reason why the distribution function should have the form necessary for obtaining these self-consistent solutions. The fact that the distribution function corresponding to thermodynamic equilibrium leads to unphysical results merely shows that thermodynamic equilibrium is not possible for a self-gravitating system (i.e. a system in which the gravitational attraction of one part on another is important). If the stellar system does not have an end state of thermodynamic equilibrium, what then is the outcome of collisional relaxation in a stellar system? Presumably such systems keep on evolving, usually leading to the formation of black holes in the central regions. Many galaxies and star clusters are indeed believed to have black holes in their centres. Chapter 8 of Binney and Tremaine (1987) gives an introduction to the collisional evolution of stellar systems.

Although a detailed discussion of the collisional evolution of stellar systems is beyond the scope of the present book, we point out one important effect. For stars to fall into the deep potential well at the centre of the stellar system, it is necessary for them to lose some kinetic energy through a frictional process. Since stars do not physically collide with each other, it may seem at the first sight that there is no friction in the system. However, Chandrasekhar (1943) derived the famous result that a star moving through a stellar system should encounter a drag opposing its motion, giving rise to a frictional term in the evolution equation. Let us qualitatively explain why this should be so. Suppose a star has moved from point P to point Q as shown in Figure 7.2. While passing from P to Q, the star attracted the surrounding stars towards itself. Hence we expect the number density of stars around PQ to be slightly larger than that ahead of Q. The star at Q, therefore, experiences a net gravitational attraction

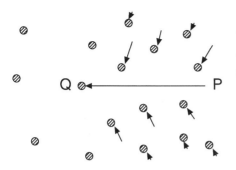

Fig. 7.2 The illustration of dynamical friction. A star has moved from P to Q creating a region of enhanced density of surrounding stars behind it.

in the backward direction (i.e. in the direction of QP). This important effect is known as *dynamical friction*.

7.5   Boltzmann equation for collisionless systems

After discussing the difficulty of obtaining realistic solutions of collisional stellar systems, let us now look at collisionless stellar systems. Since different initial conditions may produce different types of collisionless stellar systems, we would not expect to obtain unique models of such systems from basic principles alone. Since the probability of two stars coming sufficiently close to each other (such that their trajectories are appreciably deflected) would be fairly low in a collisionless system, a typical star would move in the smooth gravitational field produced by all the other stars in the system collectively. As all stars move in this way, the distribution function of stars $f(\mathbf{x}, \mathbf{v})$ may change with time. We now derive the equation which describes how the distribution function would change with time. This equation, known as the *collisionless Boltzmann equation*, is a special form of an equation first derived by Boltzmann (1872) while studying the dynamics of gas particles and is the fundamental equation in collisionless stellar dynamics.

Let us consider the six-dimensional phase space made up of three components of the position vector $\mathbf{x}$ and three components of the velocity vector $\mathbf{v}$. All stars are assumed to be identical particles with the same mass. A star at position $\mathbf{x}$ with velocity $\mathbf{v}$ is represented by a point in this phase space. Hence a stellar system of N stars would correspond to N points in this phase space. The distribution function $f(\mathbf{x}, \mathbf{v})$ is nothing but the density of points at $(\mathbf{x}, \mathbf{v})$ in this phase space. As the position and the velocity of a star change with time, the point in the phase space corresponding to this star will trace out a trajectory in the phase space. Since all the points in the phase space keep moving, the density of points $f(\mathbf{x}, \mathbf{v})$ may in general be expected to change in time. One can prove a very important result known as *Liouville's theorem* if the particles in the system obey Hamiltonian dynamics, i.e. if the dynamics of the particles can be obtained from a Hamiltonian of the form $H(\mathbf{x}, \mathbf{v})$. Liouville's theorem is a fundamental result in statistical mechanics and is proved in many standard textbooks (see, for example, Landau and Lifshitz, 1980, §3; Pathria, 1996, §2.2; Choudhuri, 1998, §1.4). We therefore quote it without proof. Let us consider the trajectory of a point in the phase space. If we keep on considering the distribution function $f(\mathbf{x}, \mathbf{v})$ along the trajectory as the trajectory is being traced out in time, then we would find that the distribution function does not change (provided, of course, the dynamics is Hamiltonian). Mathematically it can be represented as

$$\frac{df}{dt} = 0, \tag{7.21}$$

where d/dt represents the time derivative as we move along the trajectory.

A point in the phase space located at $(\mathbf{x}, \mathbf{v})$ at time t will get shifted to $(\mathbf{x} + \dot{\mathbf{x}}\,\delta t, \mathbf{v} + \dot{\mathbf{v}}\,\delta t)$ at time $t + \delta t$, as the point moves along the trajectory. Hence

$$\frac{df}{dt} = \lim_{\delta t \to 0} \frac{f(\mathbf{x} + \dot{\mathbf{x}}\,\delta t, \mathbf{v} + \dot{\mathbf{v}}\,\delta t, t + \delta t) - f(\mathbf{x}, \mathbf{v}, t)}{\delta t}. \tag{7.22}$$

Expansion in a Taylor series to linear terms in δt gives

$$f(\mathbf{x} + \dot{\mathbf{x}}\,\delta t, \mathbf{v} + \dot{\mathbf{v}}\,\delta t, t + \delta t) = f(\mathbf{x}, \mathbf{v}, t)$$
$$+ \sum_i \delta t\, \dot{x}_i \frac{\partial f}{\partial x_i} + \sum_i \delta t\, \dot{v}_i \frac{\partial f}{\partial v_i} + \delta t \frac{\partial f}{\partial t}.$$

Substituting the above expression in (7.22), we have

$$\frac{df}{dt} = \frac{\partial f}{\partial t} + \sum_i \dot{x}_i \frac{\partial f}{\partial x_i} + \sum_i \dot{v}_i \frac{\partial f}{\partial v_i}.$$

It thus follows from (7.21) that

$$\frac{\partial f}{\partial t} + \sum_i \dot{x}_i \frac{\partial f}{\partial x_i} + \sum_i \dot{v}_i \frac{\partial f}{\partial v_i} = 0, \tag{7.23}$$

which is the *collisionless Boltzmann equation*. As we already pointed out, this equation holds only if the dynamics in the phase space can be obtained from a Hamiltonian of the form $H(\mathbf{x}, \mathbf{v})$. This is the case if all the stars move in a smooth gravitational field. However, if two stars *collide*, then the gravitational potential will have to be a function of the positions of both the stars and a Hamiltonian of the form $H(\mathbf{x}, \mathbf{v})$ will not be able to describe the collision. Thus (7.23) is valid only in the absence of collisions. When collisions are important, the right-hand side of (7.23) ceases to be zero (see, for example, Choudhuri, 1998, §2.2).

While considering a stellar system, it is often convenient to use cylindrical coordinates (r, θ, z) instead of Cartesian coordinates. Writing the components of velocity in cylindrical coordinates as $\Pi = \dot{r}, \Theta = r\dot{\theta}, Z = \dot{Z}$, it is easily seen that the collisionless Boltzmann equation takes the form

$$\frac{\partial f}{\partial t} + \Pi\frac{\partial f}{\partial r} + \frac{\Theta}{r}\frac{\partial f}{\partial \theta} + Z\frac{\partial f}{\partial z} + \dot{\Pi}\frac{\partial f}{\partial \Pi} + \dot{\Theta}\frac{\partial f}{\partial \Theta} + \dot{Z}\frac{\partial f}{\partial Z} = 0. \qquad (7.24)$$

If we take the mass of particles to be equal to unity, then the Lagrangian of a particle (or a star) is given by

$$L = \frac{1}{2}(\dot{r}^2 + r^2\dot{\theta}^2 + \dot{z}^2) - \Phi(r, \theta, z).$$

On substituting this in Lagrange's equation (see, for example, Goldstein, 1980), the three components of the equation of motion are found to be

$$\ddot{r} - r\dot{\theta}^2 = -\frac{\partial \Phi}{\partial r},$$

$$\frac{d}{dt}(r^2\dot{\theta}) = -\frac{\partial \Phi}{\partial \theta},$$

$$\ddot{z} = -\frac{\partial \Phi}{\partial z}.$$

These three equations can be written as

$$\dot{\Pi} = \frac{\Theta^2}{r} - \frac{\partial \Phi}{\partial r},$$

$$\Pi\Theta + r\dot{\Theta} = -\frac{\partial \Phi}{\partial \theta},$$

$$\dot{Z} = -\frac{\partial \Phi}{\partial z}.$$

On making use of these equations, (7.24) can be written as

$$\frac{\partial f}{\partial t} + \Pi\frac{\partial f}{\partial r} + \frac{\Theta}{r}\frac{\partial f}{\partial \theta} + Z\frac{\partial f}{\partial z} + \left(\frac{\Theta^2}{r} - \frac{\partial \Phi}{\partial r}\right)\frac{\partial f}{\partial \Pi}$$

$$- \left(\frac{\Pi\Theta}{r} + \frac{1}{r}\frac{\partial \Phi}{\partial \theta}\right)\frac{\partial f}{\partial \Theta} - \frac{\partial \Phi}{\partial z}\frac{\partial f}{\partial Z} = 0, \qquad (7.25)$$

which is the form of the collisionless Boltzmann equation used extensively in stellar dynamics.

If we are somehow able to determine the complete distribution function $f(r, \theta, z, \Pi, \Theta, Z, t)$, then we would have full information about the dynamics of the stellar system. As we would not expect to determine the dynamics of an unrelaxed system without a knowledge of the initial conditions, it should be clear that we cannot obtain a full solution of the distribution function on the basis of (7.25) alone. What information does (7.25) provide then? It is the same equation as (7.21), which tells us that the distribution function should not change along a trajectory in phase space. Suppose I_1, I_2, ... are some constants of motion which do change along a trajectory. If the distribution function is a function of these constants of motion alone, i.e. if we can write it as $f(I_1, I_2, \ldots)$, then it can be easily shown that this distribution function should satisfy (7.21). For an axisymmetric stellar system, the total energy E and the angular momentum component L_z should be constants of motion. Hence a distribution function of the form $f(E, L_z)$ should satisfy (7.25). The early years of stellar dynamics research were marked by a search for a third integral of motion. If the gravitational potential Φ is provided by the stars themselves, then we have to impose the condition of self-consistency as we saw in §7.4. Any arbitrary function of E and L_z that satisfies the self-consistency requirement is an admissible solution of the collisionless Boltzmann equation (7.25). This equation, therefore, does not give us a unique solution in a particular situation.

7.6 Jeans equations and their applications

Although the collisionless Boltzmann equation (7.25) alone does not provide a complete solution to the dynamics of a stellar system, we shall now show that this equation can be used to derive several important conclusions regarding stellar motions in our Galaxy.

We consider a galaxy which is axisymmetric and is in a steady state. Then the derivatives with respect to θ and t can be set to zero, so that (7.25) becomes

$$\Pi\frac{\partial f}{\partial r} + Z\frac{\partial f}{\partial z} + \left(\frac{\Theta^2}{r} - \frac{\partial\Phi}{\partial r}\right)\frac{\partial f}{\partial \Pi} - \frac{\Pi\Theta}{r}\frac{\partial f}{\partial \Theta} - \frac{\partial\Phi}{\partial z}\frac{\partial f}{\partial Z} = 0. \qquad (7.26)$$

Let us consider some dynamical variable $q(r, \theta, z, \Pi, \Theta, Z)$, which would have a particular value at each point of the phase space. The energy as given by (7.15) is an example of such a dynamical variable. Now think of all stars in a unit volume of physical space. These stars would have different velocities and would in general have different values of q. The average value of q for stars in this unit volume, indicated by $\langle q \rangle$, would be given by

$$n\langle q \rangle = \iiint qf\, d\Pi\, d\Theta\, dZ, \qquad (7.27)$$

where we carry out the integration over all possible velocities and n is the number density given by

$$n = \iiint f\, d\Pi\, d\Theta\, dZ. \qquad (7.28)$$

We now derive a very useful equation by multiplying (7.26) by Π and then integrating over all velocities. Since integration over velocities commutes with differentiation with respect to r or z, the first two terms give

$$\frac{\partial}{\partial r}(n\langle\Pi^2\rangle) + \frac{\partial}{\partial z}(n\langle\Pi Z\rangle),$$

where the averages $\langle \Pi^2 \rangle$ and $\langle \Pi Z \rangle$ are defined through (7.27). The third term gives

$$\iint \left(\frac{\Theta^2}{r} - \frac{\partial \Phi}{\partial r} \right) d\Theta \, dZ \int \Pi \frac{\partial f}{\partial \Pi} d\Pi = -\frac{n}{r} \langle \Theta^2 \rangle + n \frac{\partial \Phi}{\partial r}$$

on noting that

$$\int \Pi \frac{\partial f}{\partial \Pi} d\Pi = \Pi f |_{-\infty}^{+\infty} - \int f \, d\Pi = -\int f \, d\Pi.$$

The next (fourth) term gives

$$-\iint \frac{\Pi^2}{r} d\Pi \, dZ \int \Theta \frac{\partial f}{\partial \Theta} d\Theta = n \frac{\langle \Pi^2 \rangle}{r}$$

in exactly the same way. Combining all these together and noting that the contribution of the last term in (7.26) would be zero, we obtain

$$\frac{\partial}{\partial r}(n \langle \Pi^2 \rangle) + \frac{\partial}{\partial z}(n \langle \Pi Z \rangle) - \frac{n}{r} \left[\langle \Theta^2 \rangle - \langle \Pi^2 \rangle \right] = -n \frac{\partial \Phi}{\partial r}. \tag{7.29}$$

Let us now multiply (7.26) by Z and integrate over all velocities. By proceeding in exactly the same way, we obtain the following equation

$$\frac{\partial}{\partial r}(n \langle \Pi Z \rangle) + \frac{\partial}{\partial z}(n \langle Z^2 \rangle) + \frac{n \langle \Pi Z \rangle}{r} = -n \frac{\partial \Phi}{\partial z}. \tag{7.30}$$

As we shall show below, equations (7.29) and (7.30) are of great help in analysing stellar motions near the solar neighbourhood. These equations are known as the *Jeans equations*, after Jeans (1922) who first obtained them.

The quantity $\langle \Pi Z \rangle$ appears in both the Jeans equations (7.29) and (7.30). Let us discuss how one can evaluate it. We pointed out in §6.3.3 that velocities of stars in the solar neighbourhood have an ellipsoidal distribution. Let us consider a point P away from the central plane of the Galaxy, as shown in Figure 7.3. We expect the velocity ellipsoid at this point to have its major axis elongated towards the galactic centre. Then the velocity distribution should be elliptical in Π' and Z', the components of velocity along the major and minor axes of the ellipsoid at P, i.e.

$$f(\Pi', Z') = C \exp \left(-\frac{\Pi'^2}{\sigma_{\Pi'}^2} - \frac{Z'^2}{\sigma_{Z'}^2} \right). \tag{7.31}$$

If the velocity ellipsoid at P is inclined to the galactic plane by an angle α, then

$$\Pi = \Pi' \cos \alpha - Z' \sin \alpha$$

and

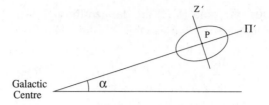

Fig. 7.3 A sketch showing the velocity ellipsoid at a point P away from the mid-plane of our Galaxy.

$$Z = Z' \cos \alpha + \Pi' \sin \alpha$$

so that

$$\Pi Z = (\Pi'^2 - Z'^2) \sin \alpha \cos \alpha + \Pi' Z' (\cos^2 \alpha - \sin^2 \alpha). \tag{7.32}$$

We now have to take the average of this as defined by (7.27). It is trivial to see that $\langle \Pi' Z' \rangle = 0$ if the distribution function is given by (7.31). Keeping in mind that $\sin \alpha \approx z/r$ and $\cos \alpha \approx 1$ for small α, (7.32) gives

$$\langle \Pi Z \rangle \approx \frac{z}{r} \left[\langle \Pi'^2 \rangle - \langle Z'^2 \rangle \right].$$

If α is small, then $\langle \Pi'^2 \rangle \approx \langle \Pi^2 \rangle$ and $\langle Z'^2 \rangle \approx \langle Z^2 \rangle$ so that

$$\langle \Pi Z \rangle \approx \frac{z}{r} \left[\langle \Pi^2 \rangle - \langle Z^2 \rangle \right]. \tag{7.33}$$

We now consider some applications of the Jeans equations to make sense of the observational data of stellar motions in the solar neighbourhood.

7.6.1   Oort limit

The distance of the solar neighbourhood from the galactic centre is much larger than the thickness of the Galaxy, so that a vertical gradient in the solar neighbourhood should be much stronger than a radial gradient. It should be easy to see that the vertical gradient term on the left-hand side of (7.30) should be the dominant term so that (7.30) would reduce to

$$\frac{d}{dz} \left(n \langle Z^2 \rangle \right) = n g_z, \tag{7.34}$$

where $g_z = -\partial \Phi / \partial z$ is the vertical gravitational field.

Oort (1932) used (7.34) to find the average matter density near the solar neighbourhood of our Galaxy. Even if there is some matter in the solar neighbourhood which does not emit light and is not detected in direct observations, it will produce a gravitational field and hence will affect the motions of visible stars. Therefore, by analysing the motions of visible stars, it is possible to estimate the total amount of matter in the solar neighbourhood. If the number density n and vertical velocity dispersion $\langle Z^2 \rangle$ for a particular type of stars are known at different distances from the galactic plane, then it is possible to calculate g_z with the help of (7.34). K giants are very bright stars which can be observed to sufficiently large distances from the galactic plane and for which sufficiently good data existed in Oort's time about their number density and line-of-sight velocity at different heights from the galactic plane. Oort (1932) used the statistics of K giant stars to obtain the gravitational field at different heights from the galactic plane. Once g_z is obtained as a function of z, one can calculate the matter density producing this gravitational field from the Poisson equation for gravity, $\nabla.\mathbf{g} = -4\pi G \rho_{\text{matter}}$, which here becomes

$$\frac{dg_z}{dz} = -4\pi G \rho_{\text{matter}}. \tag{7.35}$$

When the total matter density in the solar neighbourhood is estimated in this fashion, it turns out to be around

$$\rho_{\text{matter}} \approx 10 \times 10^{-21} \text{ kg m}^{-3}. \tag{7.36}$$

On the other hand, if we calculate the density by estimating the amount of matter in the visible stars, then we find

$$\rho_{\text{star}} \approx 4 \times 10^{-21} \text{ kg m}^{-3}. \tag{7.37}$$

Thus there must be unseen matter present in the solar neighbourhood in addition to the visible stars. This was a very important conclusion in 1932 when not much was known about the interstellar matter. This analysis also provides an upper limit for the amount of interstellar matter, since its density cannot exceed $(\rho_{\text{matter}} - \rho_{\text{star}})$. This is known as the *Oort limit*.

7.6.2 Asymmetric drift

Let us consider a group of stars in the solar neighbourhood with an average value of Θ given by $\langle \Theta \rangle$. For an individual star, the value of Θ would differ from this average by an amount ϑ, i.e.

$$\Theta = \langle \Theta \rangle + \vartheta. \tag{7.38}$$

On squaring and averaging this (keeping in mind $\langle \vartheta \rangle = 0$), it follows that

$$\langle \Theta^2 \rangle = \langle \Theta \rangle^2 + \langle \vartheta^2 \rangle. \tag{7.39}$$

Using the notation of §6.3, we know that, if all the stars in our group had $\Theta = \Theta_{\text{circ}}$, then they would all move in exactly circular orbits. Our aim now is to find out the physical effects which may make $\langle \Theta \rangle$ different from Θ_{circ}. We have

$$\langle \Theta \rangle^2 - \Theta_{\text{circ}}^2 = 2\Theta_{\text{circ}}(\langle \Theta \rangle - \Theta_{\text{circ}}) \tag{7.40}$$

on writing $\langle\Theta\rangle + \Theta_{circ} = 2\Theta_{circ}$ because $\langle\Theta\rangle$ is expected to be close to Θ_{circ}. Since $\Theta_{circ} = (A - B)R_0$ where A and B are the Oort constants introduced through (6.21) and (6.22), we have from (7.39) and (7.40) that

$$2(A - B)R_0(\langle\Theta\rangle - \Theta_{circ}) = \langle\Theta^2\rangle - \Theta_{circ}^2 - \langle\vartheta^2\rangle. \tag{7.41}$$

We now need to adapt (7.29) for our group of stars in the solar neighbourhood. Using the fact that

$$-\frac{\partial\Phi}{\partial r} = -\frac{\Theta_{circ}^2}{r}$$

and substituting for $\langle\Pi Z\rangle$ from (7.33), we can rewrite (7.29) in the form

$$\langle\Theta^2\rangle - \Theta_{circ}^2 = \frac{r}{n}\frac{\partial}{\partial r}(n\langle\Pi^2\rangle) + \frac{r}{n}\frac{\partial}{\partial z}\left[n\frac{z}{r}(\langle\Pi^2\rangle - \langle Z^2\rangle)\right]$$
$$+ \langle\Pi^2\rangle. \tag{7.42}$$

In the term involving differentiation with respect to z, the main contribution would come from the variation in z so that we can write that term as

$$\frac{r}{n}\frac{\partial}{\partial z}\left[n\frac{z}{r}(\langle\Pi^2\rangle - \langle Z^2\rangle)\right] \approx \langle\Pi^2\rangle - \langle Z^2\rangle.$$

From (7.41) and (7.42), we then have

$$\langle\Theta\rangle - \Theta_{circ} = \frac{\langle\Pi^2\rangle}{2R_0(A - B)}\left[\frac{\partial\ln n}{\partial\ln r} + \frac{\partial\ln\langle\Pi^2\rangle}{\partial\ln r}\right.$$
$$\left. + \left(1 - \frac{\langle\vartheta^2\rangle}{\langle\Pi^2\rangle}\right) + \left(1 - \frac{\langle Z^2\rangle}{\langle\Pi^2\rangle}\right)\right]. \tag{7.43}$$

This is an extremely important equation which tells us what would make $\langle\Theta\rangle$ for a group of stars to be different from Θ_{circ}.

Let us try to understand the physical significance of (7.43). First of all, if there were no random motions in the radial direction, i.e. if $\langle\Pi^2\rangle = 0$, then the right-hand side of (7.43) has to be zero and $\langle\Theta\rangle$ has to equal Θ_{circ}. In other words, in the absence of random motions, stars have to move in circular orbits with speed Θ_{circ} in order to be in a steady state. Only when some amount of random motion is present in a group of stars, is it possible for the group to go around the galactic centre with an average speed $\langle\Theta\rangle$ different from Θ_{circ}. Now, amongst the terms within the square bracket in the right-hand side of (7.43), the term $\partial\ln n/\partial\ln r$ typically turns out to be the dominant term. Since stellar density n decreases with radius, this term is negative. This implies that $\langle\Theta\rangle$ has to be less than Θ_{circ}, i.e. a typical group of stars in the solar neighbourhood would lag behind the LSR (defined in §6.3). If the other terms on the right-hand side of (7.43) are unimportant compared to $\partial\ln n/\partial\ln r$, it should be evident from (7.43) that whether a group of stars would lag behind the LSR or not will be determined by whether n decreases with radius or not. Why should this be the case? We have seen in §6.3.1 that stars moving with respect to the LSR do not actually move 'randomly'. Rather they follow epicycles. In the solar neighbourhood, there would be stars with their centres of epicycle lying both slightly inward (i.e. at $r < R_0$) and slightly outward (i.e. at $r > R_0$). It follows from conservation of angular momentum that stars having centres of epicycle inward would have Θ less than Θ_{circ} when they are at R_0, whereas stars having centres of epicycle outward would have Θ more than Θ_{circ}. Since n decreases with r, the number of stars coming from the inward side is larger. Since these stars lag behind the LSR when they are at R_0, it is expected that $\langle \Theta \rangle$ averaged over all stars would be less than Θ_{circ}.

On the basis of (7.43), we expect an approximate relation

$$(\Theta_{circ} - \langle \Theta \rangle) = \alpha \langle \Pi^2 \rangle, \tag{7.44}$$

where α is a constant of proportionality. To determine whether such a relation actually exists, we need to study the kinematics of stars in the solar neighbourhood belonging to different spectral classes. It is found that stars with larger $B - V$ (i.e. stars which are more reddish in colour) have larger dispersions $\langle \Pi^2 \rangle$. Their average velocities $\langle v \rangle$ in the negative θ direction with respect to the Sun (i.e. $\Theta_\odot - \langle \Theta \rangle$) are also observationally found to increase with $B - V$. We expect from (7.44) that

$$\langle v \rangle = \Theta_\odot - \langle \Theta \rangle = \Theta_\odot - \Theta_{circ} + \alpha \langle \Pi^2 \rangle. \tag{7.45}$$

This relation was first found empirically by Strömberg (1924). Figure 7.4 is a modern plot of $\langle v \rangle$ against $\langle \Pi^2 \rangle$ for stars of different spectral types, $B - V$ increasing towards the right side of the figure. It is clearly seen that a linear relation between $\langle v \rangle$ and $\langle \Pi^2 \rangle$, as expressed by (7.45), is a reasonable fit to the observational data. From the point where the straight line cuts the vertical axis, we conclude that $\Theta_\odot - \Theta_{circ}$ has to be close to about 5.2 km s^{-1}. This result was quoted in §6.3.2 without explaining there how it was obtained.

Stars with higher $B - V$, which are more reddish and are found to have larger velocity dispersions, have longer lifetimes (as pointed out in Chapter 3) and would statistically be older than lower $B - V$ stars. Why do older stars have larger velocity dispersions? A theoretical explanation was provided by Spitzer and Schwarzschild (1951). As we have seen, our Galaxy is a collisionless system and close interactions between stars can be neglected. However, the gravitational attraction of interstellar gas clouds tends to perturb stellar orbits. As a star grows older, it is expected to have more interactions with gas clouds, of which the effects accumulate. This explains why older stars have more velocity dispersions, as can be seen in Figure 6.7.

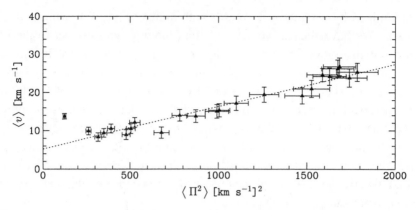

Fig. 7.4 A plot of $\langle v \rangle$ (i.e. the average velocity in the negative θ direction with respect to the Sun) against the velocity dispersion $\langle \Pi^2 \rangle$, for stars of different spectral types in the solar neighbourhood. From Dehnen and Binney (1998). (©Royal Astronomical Society. Reproduced with permission from *Monthly Notices of Royal Astronomical Society*.)

7.7 Stars in the solar neighbourhood belonging to two subsystems

As we pointed out in §6.2 and §6.4, our Galaxy has two subsystems. Objects in the first subsystem revolve around the galactic centre in nearly circular orbits, whereas objects in the second subsystem have very low general rotation and are principally balanced against gravity by random motions. Most of the stars in the solar neighbourhood belong to the first subsystem. However, we expect a few stars belonging to the second subsystem also to be present in the solar neighbourhood. Oort (1928) carried out a beautiful analysis to establish this from stellar kinematics. Figure 7.5 is a famous figure taken from Oort (1928), plotting Π and Θ for stars found in the solar neighbourhood. The dashed large circle corresponds to $\sqrt{\Pi^2 + \Theta^2} = 365\,\mathrm{km\ s^{-1}}$, which is presumably the escape velocity from the Galaxy so that stars with larger velocities are not found. The Sun is represented by the dot at the centre of a small circle of which the radius corresponds to $20\,\mathrm{km\ s^{-1}}$. Stars within this small circle are not plotted because of uncertainties in selection effects. One clearly sees that many stars make up an ellipsoidal distribution near the Sun, with the major axis of the ellipse in the direction of Π. This is the Schwarzschild velocity ellipsoid introduced in §6.3.3. The stars making up this ellipsoid certainly belong to the first subsystem of the Galaxy. These stars move in nearly circular orbits, with the small departures from circular orbits giving rise to epicyclic motions responsible for the ellipsoidal velocity dispersion (see §6.3.3). The majority of these stars lie within a circle shown in Figure 7.5 of which the radius corresponds to a velocity of $65\,\mathrm{km\ s^{-1}}$ with respect to the LSR. Some of the stars having velocities much

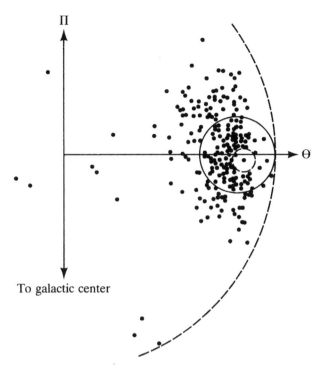

Fig. 7.5 A plot showing the values of Π and Θ for stars in the solar neighbourhood. From Oort (1928).

larger than $65\,\text{km s}^{-1}$ clearly do not belong to the ellipsoid and should be members of the second subsystem of our Galaxy, which does not have much systematic rotation.

Exercises

7.1 Suppose $K(\mathbf{x}, \mathbf{v})$ is a constant of motion as a star moves around within a stellar system (it can be energy or angular momentum). Show that a distribution function of the form $f(K(\mathbf{x}, \mathbf{v}))$ will give a time-independent solution of the collisionless Boltzmann equation. This result is known as the *Jeans theorem*.

7.2 Write down the collisionless Boltzmann equation in the cylindrical coordinates. Without any assumptions (i.e. without assuming axisymmetry or steady state), integrate over the velocity space and show that

$$\frac{\partial n}{\partial t} + \frac{\partial}{\partial r}(n\langle\Pi\rangle) + \frac{1}{r}\frac{\partial}{\partial\theta}(n\langle\Theta\rangle) + \frac{\partial}{\partial z}(n\langle Z\rangle) + \frac{n\langle\Pi\rangle}{r} = 0.$$

7.3 Consider the collisionless Boltzmann equation in the Cartesian co-ordinates

$$\frac{\partial f}{\partial t} + v_i \frac{\partial f}{\partial x_i} + \frac{F_i}{m} \frac{\partial f}{\partial v_i} = 0, \tag{1}$$

where F_i is the force acting on a particle of mass m at the point x_i. (Note that we are using the summation convention that an index like i repeated twice in a term implies summation over i.) Integrating over the velocity space, show that

$$\frac{\partial}{\partial t}(nm) + \frac{\partial}{\partial x_i}(nm\langle v_i \rangle) = 0, \tag{2}$$

where n is the number density. Now multiply (1) by mv_j and integrate over the velocity space to obtain

$$\frac{\partial}{\partial t}(nm\langle v_j \rangle) + \frac{\partial}{\partial x_i}(nm\langle v_i v_j \rangle) - nF_j = 0. \tag{3}$$

Define the pressure tensor

$$P_{ij} = nm\langle (v_i - \langle v_i \rangle)(v_j - \langle v_j \rangle)\rangle \tag{4}$$

and show that (3) can be put in the form

$$nm\left(\frac{\partial}{\partial t}\langle v_j \rangle + \langle v_i \rangle \frac{\partial}{\partial x_i}\langle v_j \rangle\right) = -\frac{\partial P_{ji}}{\partial x_i} + nF_j. \tag{5}$$

Do the equations (2) and (5) resemble the basic fluid equations which we shall discuss in detail in the next chapter?

7.4 Suppose a collection of self-gravitating particles has a distribution function somewhat different from (7.16), given by

$$f(\mathbf{x}, \mathbf{v}) = \begin{cases} A[e^{-\beta E(\mathbf{x},\mathbf{v})} - 1] & \text{if } E(\mathbf{x}, \mathbf{v}) < 0, \\ 0 & \text{if } E(\mathbf{x}, \mathbf{v}) > 0, \end{cases}$$

where $E(\mathbf{x}, \mathbf{v})$ is given by (7.15), with the gravitational potential $\Phi(\mathbf{x})$ defined in such a way that it tends to zero at infinity. Find the expression of the density $\rho(r)$ (note that the expression will involve the error function). Then write down the equation you will have in the place of (7.19). Solve that equation numerically for the values $-\beta \Phi(r = 0) = 12, 9, 6, 3$. Plot the density $\rho(r)$ to show that it falls to zero at finite radius, indicating that the total mass is finite, unlike what happens for the distribution function (7.16). This stellar dynamical model is known as the *King model* (King, 1966).

8

Elements of plasma astrophysics

8.1 Introduction

A plasma is a gas in which at least some atoms have been broken into positively charged ions and negatively charged electrons. Most of the matter in the Universe exists in the plasma state. The gases inside stars are ionized because of the high temperature, as can be shown easily with the help of the Saha equation (2.29). We have seen in §6.6.4 that HII regions in the interstellar medium are fully ionized due to energetic photons from very hot stars. Even the HI regions are partially ionized, with some free electrons present in them. Our aim in this chapter is to give an introduction to some dynamical principles as well as some radiation processes involving plasmas, which are of great relevance to astrophysics.

The reader may wonder why this introductory chapter on plasma astrophysics is put exactly in this place of the book. We could, of course, introduce the subject much earlier. However, since we shall illustrate the dynamical principles by applications to stars and the interstellar medium, I felt that a prior acquaintance with these systems will put you in a better position to appreciate the relevance of plasma processes in astrophysics. There is also some justification for introducing this subject before a discussion of extragalactic astronomy. In Chapter 9 we shall discuss some extragalactic systems such as active galaxies in which plasma processes are extremely important. So it will be helpful to have some knowledge of plasma astrophysics before we launch into a study of extragalactic astronomy.

Because of the electrical attraction between opposite charges, the positively and the negatively charged particles in a plasma remain well mixed. In other words, if you consider a small volume element of a plasma which has sufficiently large number of charged particles, the positive and negative charges in that volume would very nearly balance each other. So the volume element would be nearly charge-neutral. Does it then follow that all the physical properties of this volume element will be identical with those of a volume element of an ordinary neutral gas? Certainly not! A neutral gas like the air is generally a poor conductor of electricity. On the other hand, if you put an electric field in a plasma, the positively charged ions would move in the direction of the field and the electrons would move in the opposite direction, giving rise to an electrical current. In other words, a plasma is an extremely good conductor of electricity. Currents in a plasma naturally give rise to magnetic fields, and there are lots of intriguing phenomena connected with magnetic fields. For example, we have discussed in §6.7 that our Galaxy is filled with cosmic ray particles, which are charged particles accelerated to very high energies. We shall see in §8.10 that magnetic fields play the key role in the acceleration of cosmic ray particles. Such accelerated particles spiralling around magnetic field lines give rise to a kind of radiation called synchrotron radiation. Since many astrophysical systems have accelerated charged particles in them and emit synchrotron radiation, the emission of such radiation (to be discussed in §8.11) is a very important radiation process in astrophysics. The detection and analysis of synchrotron radiation is crucial in understanding the nature of many astrophysical systems.

Although water is made up of molecules, we can study the flow of water at a macroscopic level by considering water as a continuum governed by a set of macroscopic equations. In exactly the same way, many (but not all!) phenomena involving plasmas can be studied by treating the plasma as a continuous fluid which is a good conductor of electricity. The branch of plasma physics in which the plasma is treated as a continuum is known as *magnetohydrodynamics*, abbreviated as MHD. The first few sections of this chapter will develop the continuum model. Only when electromagnetic phenomena are present, will a plasma behave differently from an ordinary neutral fluid. In the absence of electromagnetic phenomena, the plasma behaves exactly like a neutral fluid, which is governed by equations simpler than the equations of MHD. We first develop fluid mechanics appropriate for neutral fluids in §8.2 and §8.3. Then we discuss MHD in §8.4–8.9. We shall consider several important astrophysical topics while developing fluid mechanics and plasma physics. Then §8.10 and §8.11 are devoted to particle acceleration and synchrotron radiation respectively, which require a more microscopic treatment of the plasma. Finally, the last two sections deal with some other radiation processes important in astrophysics.

8.2 Basic equations of fluid mechanics

Our aim is to develop a dynamical theory of fluids, with which we can study how a fluid configuration evolves with time. Any dynamical theory has two requirements. Firstly, we need some means by which we can mathematically prescribe the state of the system at any particular instant of time. Secondly, we need some equations which will tell us how the state changes with time. Let us begin with a discussion of how the state of a fluid, treated as a continuum, can be prescribed mathematically at an instant of time. We know that the thermodynamic state of a gas in a cylinder can be prescribed with the help of two thermodynamic parameters, such as density and temperature. Inside a fluid, the density and temperature would in general vary from point to point. However, if we consider a small volume of the fluid within which we can neglect the variations of physical parameters, then the thermodynamic state of that small volume is given by the density $\rho(\mathbf{x}, t)$ and temperature $T(\mathbf{x}, t)$ within that volume. Additionally, if there are motions inside the fluid, then we need to know the velocity $\mathbf{v}(\mathbf{x}, t)$ of the volume with respect to some inertial frame. The state of a neutral fluid at a particular time t is completely prescribed by the values of $\rho(\mathbf{x}, t)$, $T(\mathbf{x}, t)$ and $\mathbf{v}(\mathbf{x}, t)$ at all points inside the fluid at that time t. For a plasma which is a good conductor of electricity, we need something more to prescribe the state, as we shall see in §8.4.

To develop a dynamical theory, we have to derive equations which will describe how the dynamical variables $\rho(\mathbf{x}, t)$, $T(\mathbf{x}, t)$ and $\mathbf{v}(\mathbf{x}, t)$ evolve with time. Let us begin by drawing attention to the two different kinds of time derivatives: *Eulerian* and *Lagrangian*. The *Eulerian* derivative denoted by $\partial/\partial t$ implies differentiation with respect to time at a fixed point. On the other hand, one can think of moving with a fluid element with the fluid velocity $\mathbf{v}$ and time-differentiating some quantity associated with this moving fluid element. This type of time derivative is called *Lagrangian* and is denoted by d/dt. If $\mathbf{x}$ and $\mathbf{x} + \mathbf{v}\,\delta t$ are the positions of a fluid element at times t and $t + \delta t$, then the Lagrangian time derivative of some quantity $Q(\mathbf{x}, t)$ is given by

$$\frac{dQ}{dt} = \lim_{\delta t \to 0} \frac{Q(\mathbf{x} + \mathbf{v}\,\delta t, t + \delta t) - Q(\mathbf{x}, t)}{\delta t}. \tag{8.1}$$

Keeping the first-order terms in the Taylor expansion, we have

$$Q(\mathbf{x} + \mathbf{v}\,\delta t, t + \delta t) = Q(\mathbf{x}, t) + \delta t \frac{\partial Q}{\partial t} + \delta t\, \mathbf{v}.\nabla Q.$$

Putting this in (8.1), we have the very useful relation between the Lagrangian and the Eulerian derivatives:

$$\frac{dQ}{dt} = \frac{\partial Q}{\partial t} + \mathbf{v}.\nabla Q. \tag{8.2}$$

We now derive the first fluid dynamical equation giving the time derivative of $\rho(\mathbf{x}, t)$. The mass $\int \rho\, dV$ inside a volume can change only due to the motion of matter across the surface bounding this volume. Since the mass flux across an element of surface $d\mathbf{S}$ is $\rho \mathbf{v}.d\mathbf{S}$, we must have

$$\frac{\partial}{\partial t} \int \rho\, dV = - \oint \rho \mathbf{v}.d\mathbf{S},$$

where the minus sign implies that a mass flux out of the volume reduces the mass inside the volume. Transforming the right-hand side of the above equation by Gauss's theorem, we have

$$\int \left[\frac{\partial \rho}{\partial t} + \nabla.(\rho \mathbf{v}) \right] dV = 0.$$

Since this equation must be valid for any arbitrary volume dV, we must have

$$\frac{\partial \rho}{\partial t} + \nabla.(\rho \mathbf{v}) = 0. \tag{8.3}$$

This is known as the *equation of continuity*.

To find the equation of motion for the fluid velocity, we consider a fluid element of volume δV. The mass of this fluid element is $\rho \, \delta V$ and its acceleration is given by the Lagrangian derivative $(d\mathbf{v}/dt)$. Hence it follows from Newton's second law of motion that

$$\rho \, \delta V \frac{d\mathbf{v}}{dt} = \delta \mathbf{F}_{\text{body}} + \delta \mathbf{F}_{\text{surface}}, \tag{8.4}$$

where we have split the force acting on the fluid element into two parts: the body force $\delta \mathbf{F}_{\text{body}}$ and the surface force $\delta \mathbf{F}_{\text{surface}}$. A body force is something which acts at all points within the body of a fluid. Gravity is an example of such a force. It is customary to denote the body force per unit mass as $\mathbf{F}$ so that

$$\delta \mathbf{F}_{\text{body}} = \rho \, \delta V \mathbf{F}. \tag{8.5}$$

The surface force on a fluid element is the force acting on it across the surface bounding the fluid element. Let $d\mathbf{S}$ be an element of area on the bounding surface. If the fluid is at rest, then we know that the force across this element of area is normal to it and is given by

$$d\mathbf{F}_{\text{surface}} = -P \, d\mathbf{S}, \tag{8.6}$$

where P is the pressure and we put the minus sign because we want to consider the force acting on the fluid element inside the bounding surface. We shall assume that (8.6), which is strictly valid for a fluid at rest, holds even when the fluid is moving. This is known as the *ideal fluid approximation*. In reality, however, when layers of fluid on the two sides of a surface move differently, there is a tangential stress across the surface. This stress tries to damp out the differential motion on the two sides of the surface and gives rise to the phenomenon of *viscosity*. In our elementary treatment, we shall neglect viscosity and treat the fluid as ideal. The total surface force acting across the whole bounding surface is then given by a surface integral

$$\mathbf{F}_{\text{surface}} = -\oint P \, d\mathbf{S}.$$

The right-hand side can be transformed into the volume integral $-\int \nabla P \, dV$. For the small volume δV, we can write

$$\delta \mathbf{F}_{\text{surface}} = -\nabla P \, \delta V. \qquad (8.7)$$

Substituting (8.5) and (8.7) into (8.4), we finally have

$$\rho \frac{d\mathbf{v}}{dt} = \rho \mathbf{F} - \nabla P. \qquad (8.8)$$

If we use (8.2) to change from the Lagrangian derivative to the Eulerian derivative, then we get

$$\frac{\partial \mathbf{v}}{\partial t} + (\mathbf{v}.\nabla)\mathbf{v} = -\frac{1}{\rho}\nabla P + \mathbf{F}. \qquad (8.9)$$

This is known as the *Euler equation* (Euler, 1755, 1759). If viscosity is included, then, in the place of the Euler equation, we have a more complicated equation known as the *Navier–Stokes equation*, which will not be discussed in this book.

To complete our discussion of basic equations, we need an energy equation, which may tell us how the temperature evolves with time. Instead of getting into a general discussion, we shall consider here only the case of a perfect gas under adiabatic conditions, i.e. we shall neglect heat conduction between an element of the gas and its surroundings. If an element of the gas moves under adiabatic conditions, a well-known perfect gas relation implies that P/ρ^{γ} will remain invariant for this element, where γ is the adiabatic index. Mathematically this can be expressed as

$$\frac{d}{dt}\left(\frac{P}{\rho^{\gamma}}\right) = 0. \qquad (8.10)$$

For a perfect gas, it may be more convenient to treat the pressure as the primary dynamical variable rather than the temperature. If we know the force $\mathbf{F}$ acting on the system, then the equations (8.3), (8.9) and (8,10) together provide a complete dynamical theory of a perfect gas, describing how the state of the gas, given by $\rho(\mathbf{x}, t)$, $P(\mathbf{x}, t)$ and $\mathbf{v}(\mathbf{x}, t)$, evolves with time. We shall now consider a very important astrophysical application to illustrate how these equations are used.

8.3 Jeans instability

We believe that stars form out of the interstellar medium. Star formation is an extremely complex and still ill-understood phenomenon. It is initiated by a fluid dynamical process known as the *Jeans instability*, which breaks the initially uniform interstellar medium into clumps.

Suppose we initially have a uniformly distributed gas and some disturbance has compressed it in a certain region. The excess pressure in this compressed region would give rise to acoustic waves which spread out the compression in surrounding regions so that the gas can again come back to its initial uniform state. The compressed region, however, has also enhanced gravitational attraction and this tends to pull more gas into the compressed region. How the system will evolve depends on whether the acoustic waves or the enhanced gravity will win over. If the region of compression is small, then it can be shown that the enhanced gravity is not so important and the acoustic waves take over. On the other hand, if the region of compression is larger than a critical size, then the enhanced gravity in the region of compression may overpower the acoustic waves, pulling more material into the region and triggering an instability. Since Jeans (1902) was the first person to demonstrate the existence of this instability, it is called the *Jeans instability* in his honour.

To analyse a fluid dynamical instability mathematically, we have to consider some perturbations around an equilibrium configuration. If these perturbations grow in time, then we expect that disturbances present in the system would make it move away from the equilibrium. On the other hand, if the perturbations die out or oscillate with time, then the system is *stable*. Let us consider the gas to be in an initial static equilibrium configuration with density ρ_0 and pressure P_0. We assume that some perturbations have caused the density and the pressure to be $\rho_0 + \rho_1$ and $P_0 + P_1$ respectively. The subscript 0 should refer to the unperturbed equilibrium configuration and 1 to perturbations. Since there can be motions induced in the perturbed gas, we also have to consider the velocity, which can be written as $\mathbf{v}_1$ because it has no unperturbed part. Apart from these fluid dynamical variables, we also introduce the gravitational potential $\Phi = \Phi_0 + \Phi_1$ broken into unperturbed and perturbed parts. The force $\mathbf{F}$ in (8.9) should then be given by

$$\mathbf{F} = -\nabla \Phi. \tag{8.11}$$

To consider perturbations around an equilibrium configuration, we first have to make sure that the unperturbed variables ρ_0, P_0 and Φ_0 satisfy the requirements of static equilibrium. Out of the three basic fluid dynamics equations (8.3), (8.9) and (8.10), it is easily seen that (8.3) and (8.10) trivially have all terms zero in a static equilibrium situation. The only non-trivial equation (8.9) gives us

$$\nabla P_0 = -\rho_0 \, \nabla \Phi_0 \tag{8.12}$$

on making use of (8.11). In addition to the fluid dynamical equations, we also need to satisfy the Poisson equation for gravity, which gives

$$\nabla^2 \Phi_0 = 4\pi G \rho_0. \tag{8.13}$$

It is trivial to show that a uniform infinite gas does not satisfy the two equations (8.12) and (8.13). From (8.12), a constant P_0 would imply a constant Φ_0. When a constant Φ_0 is substituted in (8.13), we are driven to the conclusion that the unperturbed density ρ_0 has to be zero everywhere! For a proper stability analysis, one should first find a proper equilibrium solution and then consider perturbations around that solution. Jeans (1902), however, proceeded to perform a perturbation analysis on the uniform infinite gas as if the unperturbed configuration satisfied the equilibrium equations (8.12) and (8.13)! Hence this approach is often referred to as the *Jeans swindle*. We reproduce here the analysis based on the Jeans swindle because of its historical importance and simplicity. It is possible to carry out proper stability analyses for realistic density distributions without recourse to the Jeans swindle. For example, if we consider a slab of gas in static equilibrium under its own gravity, then we can carry out a proper stability analysis. See Spitzer (1978, pp. 283–285) for a discussion of this problem. As it happens, the correct (and much more complicated!) analysis yields results which are qualitatively similar to those we get from the perturbation analysis of the uniform infinite gas with the help of the Jeans swindle.

We shall now use the fluid dynamical equations along with the Poisson equation for gravity to find out how the perturbations $\rho_1(\mathbf{x}, t)$, $P_1(\mathbf{x}, t)$, $\mathbf{v}_1(\mathbf{x}, t)$ and $\Phi_1(\mathbf{x}, t)$ will evolve with time. We shall assume that the perturbed quantities are small (i.e. $\rho_1 \ll \rho_0$, $P_1 \ll P_0$, $|\Phi_1| \ll |\Phi_0|$) and the quadratic terms of these quantities will be neglected. The technique of keeping only the linear terms in perturbed quantities and neglecting the higher terms is called the *linearization of the perturbation equations*. From (8.10), it follows that

$$\frac{P_0 + P_1}{P_0} = \left(\frac{\rho_0 + \rho_1}{\rho_0}\right)^\gamma.$$

Neglecting terms higher than linear in ρ_1, we get

$$P_1 = c_s^2 \, \rho_1, \tag{8.14}$$

where

$$c_s = \sqrt{\frac{\gamma \, P_0}{\rho_0}}. \tag{8.15}$$

The perturbed quantities substituted in the equation of continuity (8.3) give

$$\frac{\partial \rho_1}{\partial t} + \nabla.[(\rho_0 + \rho_1)\mathbf{v}_1] = 0.$$

To linearize this perturbation equation, we neglect the term involving $\rho_1 \mathbf{v}_1$ which is quadratic in small quantities, so that we are left with

$$\frac{\partial \rho_1}{\partial t} + \rho_0 \, \nabla.\mathbf{v}_1 = 0. \tag{8.16}$$

We now have to linearize the Euler equation (8.9), which becomes

$$(\rho_0 + \rho_1)\left[\frac{\partial \mathbf{v}_1}{\partial t} + (\mathbf{v}_1.\nabla)\mathbf{v}_1\right] = -\nabla(P_0 + P_1) - (\rho_0 + \rho_1)\nabla(\Phi_0 + \Phi_1).$$

Using (8.12) to cancel two terms on the right-hand side and keeping only the linear terms in perturbed quantities, we get

$$\rho_0 \frac{\partial \mathbf{v}_1}{\partial t} = -\nabla P_1 - \rho_0 \nabla \Phi_1.$$

Using (8.14) to substitute for P_1, we have

$$\rho_0 \frac{\partial \mathbf{v}_1}{\partial t} = -c_s^2 \nabla \rho_1 - \rho_0 \nabla \Phi_1. \tag{8.17}$$

Finally, subtracting (8.13) from the full equation $\nabla^2 \Phi = 4\pi G \rho$, we get

$$\nabla^2 \Phi_1 = 4\pi G \rho_1. \tag{8.18}$$

We now have three equations (8.16)–(8.18) satisfied by the three perturbation variables ρ_1, $\mathbf{v}_1$ and Φ_1. These have to be solved to find out how the perturbations will evolve in time.

Before proceeding to solve the full equations, let us consider the special case in which the enhanced gravity is negligible. For example, in the case of ordinary sound waves in the atmosphere, the enhanced gravity in the regions of compression is utterly insignificant. In such a situation, the last term in (8.17) can be omitted. Then we can take the divergence of (8.17) and use (8.16) to substitute for $\nabla.\mathbf{v}$. This gives

$$\left(\frac{\partial^2}{\partial t^2} - c_s^2 \nabla^2 \right) \rho_1 = 0, \tag{8.19}$$

which is the equation for acoustic waves, and c_s as given by (8.15) is the sound speed.

To solve the equations (8.16)–(8.18), we note that any arbitrary perturbation may be represented as a superposition of Fourier components and that each Fourier component will evolve independently of the others because these equations are linear. For a particular Fourier component, let us take all our variables to vary as $\exp[i(\mathbf{k}.\mathbf{x} - \omega t)]$. Then (8.16)–(8.18) give

$$-\omega \rho_1 + \rho_0 \mathbf{k}.\mathbf{v}_1 = 0,$$

$$-\rho_0 \omega \mathbf{v}_1 = -c_s^2 \mathbf{k} \rho_1 - \rho_0 \mathbf{k} \Phi_1,$$

$$-k^2 \Phi_1 = 4\pi G \rho_1.$$

Combining these three equations, we readily find that

$$\omega^2 = c_s^2 (k^2 - k_J^2), \tag{8.20}$$

where

$$k_J^2 = \frac{4\pi G \rho_0}{c_s^2}. \tag{8.21}$$

When $k < k_J$, we see from (8.20) that ω has to be imaginary and can be written as

$$\omega = \pm i\alpha, \tag{8.22}$$

where α is a real positive quantity given by

$$\alpha = +c_s\sqrt{k_J^2 - k^2}.$$

Since all Fourier components grow as $\exp(-i\omega t)$, it follows from (8.22) that one mode should grow as $\exp(+\alpha t)$. Thus, any perturbation in which such a mode is present should lead to a runaway situation enhancing the perturbation and leading to an instability. If $k > k_J$, it should be easy to check that the perturbation will be oscillatory and will not grow in a runaway fashion.

We thus come to the conclusion that a perturbation would be unstable if its wavenumber k is less than k_J as given by (8.21). In other words, if the size of the perturbation is larger than some critical wavelength of order $\lambda_J = 2\pi/k_J$, then the enhanced self-gravity can overpower the acoustic waves so that the perturbation grows. The corresponding critical mass

$$M_J = \frac{4}{3}\pi\lambda_J^3\rho_0$$

is often referred to as the *Jeans mass*. Substituting from (8.21) and using (8.15) for c_s with γ taken as 1 for large-wavelength slowly evolving perturbations (which can be regarded as isothermal), we get

$$M_J = \frac{4}{3}\pi^{5/2}\left(\frac{\kappa_B T}{Gm}\right)^{3/2}\frac{1}{\rho_0^{1/2}}, \tag{8.23}$$

where m is the mass of the gas particles. If a perturbation in a uniform gas involves a mass larger than the Jeans mass, then we expect the gas in the perturbed region to keep contracting due to the enhanced gravity. Thus an initially uniform distribution of gas may eventually fragment into pieces due to the Jeans instability.

Jeans instability is the basic reason why the matter in the Universe is not spread uniformly. Stars and galaxies are believed to be the end-products of perturbations which initially started growing due to the Jeans instability. We can estimate the Jeans mass for the interstellar matter by assuming it to have 10^6 hydrogen atoms per m^3 at temperature 100 K. Then (8.23) gives a Jeans mass of about 8×10^{35} kg. This is several orders of magnitude larger than the typical mass of a star (about 10^{30} kg). Presumably the interstellar matter first breaks into large chunks with masses corresponding to clusters of stars rather than individual stars. Then somehow these contracting chunks of gas have to break further to produce stars. The presence of angular momentum or magnetic fields can make the process quite complicated. See Spitzer (1978, §13.3) for an introduction to the complex subject of star formation.

8.4 Basic equations of MHD

After familiarizing ourselves with the basic equations of fluid mechanics, let us now consider how these equations have to be generalized to MHD, which essentially treats fluids which are good conductors of electricity. We pointed out at the beginning of §8.2 that the state of a neutral fluid can be prescribed by two thermodynamic variables plus the velocity field $\mathbf{v}(\mathbf{x}, t)$. Since a plasma or an electrically conducting fluid responds to electromagnetic interactions, it may at first seem that we have to additionally specify the electric and magnetic fields $\mathbf{E}(\mathbf{x}, t)$ and $\mathbf{B}(\mathbf{x}, t)$ to complete the prescription of a state of the system. However, actually only the magnetic field $\mathbf{B}(\mathbf{x}, t)$ is needed for the prescription of the state, since the positive and negative charges in the plasma remain well mixed, as we pointed out in §8.1, and the electric field cannot be too large. We shall soon show that even the weak electric field can be found from a knowledge of $\mathbf{v}(\mathbf{x}, t)$ and $\mathbf{B}(\mathbf{x}, t)$. The electric field is, therefore, not an additional dynamical variable. It may be noted that, when dealing with a plasma, it does not make sense to distinguish between $\mathbf{E}$ and $\mathbf{D}$ or $\mathbf{B}$ and $\mathbf{H}$. These distinctions are useful only when we can distinguish between charges and currents in the conductors versus charges and currents induced in the surrounding medium.

According to Ohm's law, the current density $\mathbf{j}$ in the plasma should be given by

$$\mathbf{j} = \sigma \mathbf{E},$$

where σ is the electrical conductivity. However, if the plasma moves with velocity $\mathbf{v}$ in a magnetic field $\mathbf{B}$, then the forces on charged particles in the plasma are given by $q(\mathbf{E} + \mathbf{v} \times \mathbf{B})$ rather than $q\mathbf{E}$. Hence, Ohm's law also should be modified to

$$\mathbf{j} = \sigma(\mathbf{E} + \mathbf{v} \times \mathbf{B}). \tag{8.24}$$

The currents in the plasma give rise to magnetic fields. We know that this is described by one of Maxwell's equations:

$$\nabla \times \mathbf{B} = \mu_0 \mathbf{j} + \epsilon_0 \mu_0 \frac{\partial \mathbf{E}}{\partial t},$$

where the last term is the displacement current discovered by Maxwell (1865) himself. As we know, this is the crucial term in deriving the equation of electromagnetic waves. However, when we consider plasma motions at speeds small compared to c, this term is unimportant for studying the dynamics of the plasma. Hence we take

$$\nabla \times \mathbf{B} = \mu_0 \mathbf{j}. \tag{8.25}$$

By combining (8.24) and (8.25), we can write the electric field as

$$\mathbf{E} = \frac{\nabla \times \mathbf{B}}{\mu_0 \sigma} - \mathbf{v} \times \mathbf{B}. \tag{8.26}$$

It should be clear from this that $\mathbf{E}$ is not an independent dynamical variable in MHD, since it can be found from $\mathbf{v}$ and $\mathbf{B}$.

Since the magnetic field $\mathbf{B}$ is the important additional dynamical variable in MHD, we would need an equation for the time evolution of $\mathbf{B}$ to complete our dynamical theory. For this, we turn to one of Maxwell's equations:

$$\frac{\partial \mathbf{B}}{\partial t} = -\nabla \times \mathbf{E},$$

which is the mathematical expression of Faraday's law of electromagnetic induction. On substituting for $\mathbf{E}$ from (8.26) into this equation, we get

$$\frac{\partial \mathbf{B}}{\partial t} = \nabla \times (\mathbf{v} \times \mathbf{B}) + \eta \nabla^2 \mathbf{B}, \tag{8.27}$$

where

$$\eta = \frac{1}{\mu_0 \sigma} \tag{8.28}$$

and we have assumed that σ does not vary with position. The equation (8.27) is known as the *induction equation*.

The induction equation is the central equation of MHD. In order to have a complete dynamical theory, we also need time derivative equations for the other dynamical variables – two thermodynamic quantities and $\mathbf{v}$. For a neutral gas, these are given by (8.3), (8.9) and (8.10). We now need to figure out if these equations get modified in MHD. Since the equation of continuity (8.3) follows simply from mass conservation, it has to remain unchanged. We shall not discuss here how the presence of the magnetic field modifies the energy equation. Let us only consider how the Euler equation (8.9) has to be modified. When there is a magnetic field in the plasma, there can be a magnetic force in addition to the other forces. We know that the magnetic force per unit volume is given by $\mathbf{j} \times \mathbf{B}$ (see, for example, Panofsky and Phillips, 1962, §7–6), and the magnetic force per unit mass is obtained by dividing this by ρ. We add this extra term on the right-hand side of (8.9) and use (8.25) to eliminate $\mathbf{j}$, which gives

$$\frac{\partial \mathbf{v}}{\partial t} + (\mathbf{v}.\nabla)\mathbf{v} = \mathbf{F} - \frac{1}{\rho}\nabla P + \frac{1}{\mu_0 \rho}(\nabla \times \mathbf{B}) \times \mathbf{B}. \tag{8.29}$$

Using the vector identity

$$(\nabla \times \mathbf{B}) \times \mathbf{B} = (\mathbf{B}.\nabla)\mathbf{B} - \nabla\left(\frac{B^2}{2}\right),$$

we can write (8.29) as

$$\frac{\partial \mathbf{v}}{\partial t} + (\mathbf{v}.\nabla)\mathbf{v} = \mathbf{F} - \frac{1}{\rho}\nabla\left(P + \frac{B^2}{2\mu_0}\right) + \frac{(\mathbf{B}.\nabla)\mathbf{B}}{\mu_0\rho}. \tag{8.30}$$

It is clear from this that the magnetic field introduces a pressure $B^2/2\mu_0$. The other magnetic term $(\mathbf{B}.\nabla)\mathbf{B}/\mu_0$ is of the nature of a tension force along magnetic field lines.

We thus see that the MHD equations have two main complications with respect to the fluid dynamical equations. Firstly, the Euler equation gets modified by the addition of magnetic pressure and magnetic tension, as we see in (8.30). Secondly, we have an additional equation (8.27) to describe the evolution of the magnetic field – the induction equation. We now discuss a very important consequence of the induction equation.

8.5 Alfvén's theorem of flux freezing

Suppose the magnetic field inside the plasma has the typical value B and the velocity field has the typical value V, whereas L is the typical length scale over which the magnetic or velocity fields vary significantly. Then the term $\nabla \times (\mathbf{v} \times \mathbf{B})$ in the induction equation (8.27) should be of order VB/L, while the other term $\eta\nabla^2\mathbf{B}$ in (8.27) should be of order $\eta B/L^2$. The ratio of these two terms is a dimensionless number known as the *magnetic Reynolds number* and is given by

$$\mathcal{R}_{\mathrm{M}} \approx \frac{VB/L}{\eta B/L^2} \approx \frac{VL}{\eta}. \tag{8.31}$$

The important point to note here is that $\mathcal{R}_{\mathrm{M}}$ goes as L, which is much larger for an astrophysical system than what it is for a laboratory plasma. In fact, it turns out that $\mathcal{R}_{\mathrm{M}}$ is usually much smaller than 1 for laboratory plasmas and much larger than 1 for astrophysical systems. This means that $\eta\nabla^2\mathbf{B}$ is the dominant term on the right-hand side of (8.27) when we are dealing with laboratory plasmas and $\nabla \times (\mathbf{v} \times \mathbf{B})$ is the dominant term when we are dealing with astrophysical plasmas. For laboratory plasmas, we can often write

$$\text{Laboratory:} \qquad \frac{\partial \mathbf{B}}{\partial t} \approx \eta\nabla^2\mathbf{B}. \tag{8.32}$$

This equation is not difficult to interpret. We see from (8.28) that η is essentially the inverse of conductivity σ, which means that η goes as the resistivity of the plasma. We know that the resistivity of a system makes currents in the system decay and thereby magnetic fields produced by those currents also decay. The significance of (8.32) is that the magnetic field in the plasma diffuses away with time due to the resistivity, with the resistivity η appearing as the diffusion coefficient. On the other hand, magnetic fields in astrophysical plasmas often evolve primarily due to the other term in (8.27), i.e. we can write

$$\text{Astrophysics:} \qquad \frac{\partial \mathbf{B}}{\partial t} \approx \nabla \times (\mathbf{v} \times \mathbf{B}).$$

We now discuss the significance of this equation.

If the magnetic Reynolds number $\mathcal{R}_M$ of an astrophysical system is extremely large, then it is often justified to replace the approximation sign in the last equation by an equality sign, i.e.

$$\frac{\partial \mathbf{B}}{\partial t} = \nabla \times (\mathbf{v} \times \mathbf{B}). \tag{8.33}$$

When the magnetic field in the plasma evolves according to this equation, we can prove a very remarkable theorem called *Alfvén's theorem of flux freezing* (Alfvén, 1942a). A very similar theorem involving vorticity $\omega = \nabla \times \mathbf{v}$ was, however, known to fluid dynamicists for a long time (see, for example, Choudhuri, 1998, §4.6). We first state the theorem of flux freezing before proving it.

Consider a surface S_1 inside a plasma at time t_1. The flux of magnetic field linked with this surface is $\int_{S_1} \mathbf{B}.d\mathbf{S}$. At some future time t_2, the parcels of plasma which made up the surface S_1 at time t_1 will move away and will make up a different surface S_2. The magnetic flux linked with this surface S_2 at time t_2 will be $\int_{S_2} \mathbf{B}.d\mathbf{S}$. The theorem of flux freezing states that

$$\int_{S_1} \mathbf{B}.d\mathbf{S} = \int_{S_2} \mathbf{B}.d\mathbf{S}$$

if $\mathbf{B}$ evolves according to (8.33). We write this more compactly in the form

$$\frac{d}{dt} \int_S \mathbf{B}.d\mathbf{S} = 0, \tag{8.34}$$

where the Lagrangian derivative d/dt implies that we are considering the variation of the magnetic flux $\int_S \mathbf{B}.d\mathbf{S}$ linked with the surface S as we follow the surface S with the motion of the plasma parcels constituting it.

To proceed with the proof now, we note that the flux $\int_S \mathbf{B}.d\mathbf{S}$ linked with the surface S can change with time due to two reasons: (i) intrinsic variation in $\mathbf{B}$, and (ii) motion of the surface S. Mathematically we write

$$\frac{d}{dt} \int_S \mathbf{B}.d\mathbf{S} = \int_S \frac{\partial \mathbf{B}}{\partial t}.d\mathbf{S} + \int_S \mathbf{B}.\frac{d}{dt}(d\mathbf{S}). \tag{8.35}$$

Figure 8.1 shows an element of area which has changed from $d\mathbf{S}$ at time t to $d\mathbf{S}'$ at time $t' = t + \delta t$. We see that $d\mathbf{S}$ and $d\mathbf{S}'$ make up the two ends of a cylinder. The vector area of a side strip of this cylinder is $-\delta t \, \mathbf{v} \times \delta\mathbf{l}$, where $\delta\mathbf{l}$ is a length element from the curve encircling the surface $d\mathbf{S}$ as shown in Figure 8.1. Since the vector area $\oint d\mathbf{S}$ for a closed surface is zero, the surfaces of this cylinder

Fig. 8.1 Displacement of a surface element due to motions in the plasma.

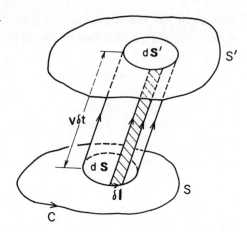

satisfy the equation

$$dS' - dS - \delta t \oint \mathbf{v} \times \delta \mathbf{l} = 0,$$

where the line integral is taken around the surface element dS. It then follows that

$$\frac{d}{dt}(dS) = \lim_{\delta t \to 0} \frac{dS' - dS}{\delta t} = \oint \mathbf{v} \times \delta \mathbf{l}.$$

The last term of (8.35) now becomes

$$\int_S \mathbf{B}.\frac{d}{dt}(dS) = \int \oint \mathbf{B}.(\mathbf{v} \times \delta \mathbf{l}) = \int \oint (\mathbf{B} \times \mathbf{v}).\delta \mathbf{l}.$$

Here the double integral $\int \oint$ means that we first take a line integral around surface elements like dS and then sum up such line integrals for the many surface elements which would make up the surface S. It is easy to see that this ultimately gives a line integral along the curve C encircling the whole surface S, because the contributions from the line integrals in the interior cancel out when we sum over all surface elements. Hence we have

$$\int_S \mathbf{B}.\frac{d}{dt}(dS) = \oint_C (\mathbf{B} \times \mathbf{v}).\delta \mathbf{l} = \int_S [\nabla \times (\mathbf{B} \times \mathbf{v})].dS$$

by Stokes's theorem. We then have from (8.35) that

$$\frac{d}{dt}\int_S \mathbf{B}.dS = \int_S dS.\left[\frac{\partial \mathbf{B}}{\partial t} - \nabla \times (\mathbf{v} \times \mathbf{B})\right]. \tag{8.36}$$

We now see that (8.34) follows from (8.33) and (8.36). This completes our proof.

In astrophysical systems with high $\mathcal{R}_M$, we can imagine the magnetic flux to be frozen in the plasma and to move with the plasma flows. Suppose we have straight magnetic field lines going through a plasma column as shown in

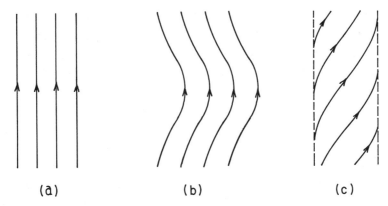

Fig. 8.2 Illustration of flux freezing. (a) A straight column of magnetic field. (b) Magnetic configuration after bending the column. (c) Magnetic configuration after twisting the column.

Figure 8.2(a). If the plasma column is bent, then, in the high $\mathcal{R}_M$ limit, the magnetic field lines are also bent with it as shown in Figure 8.2(b). On the other hand, if one end of the plasma column is twisted, then the magnetic field lines are also twisted as in Figure 8.2(c). As a result of the theorem of flux freezing, the magnetic field in an astrophysical system can almost be regarded as a plastic material which can be bent, twisted or distorted by making the plasma move appropriately. This view of a magnetic field is radically different from that which we encounter in laboratory situations, where the magnetic field appears as something rather passive which we can switch on or off by sending a current through a coil. In the astrophysical setting, the magnetic field appears to acquire a life of its own.

We thus see that magnetic fields behave very differently in laboratory and astrophysical settings, due to the fact that the magnetic field evolves respectively according to the two different equations (8.32) and (8.33) in these two situations. Alfvén coined the name *cosmical electrodynamics* to distinguish electrodynamics at cosmic scales from ordinary laboratory electrodynamics, although we start from the same Maxwell's equations and Ohm's law in both the cases. In the astrophysical setting, if we know the initial configuration of the magnetic field and the nature of plasma flows, we can almost guess on the basis of the flux-freezing theorem what the subsequent magnetic field configuration is going to be (as we saw in Figure 8.2). The human mind is more attuned to thinking geometrically rather than thinking analytically. We may be able to solve an equation describing a process, but only when we are able to make a mental picture of how the process proceeds, do we feel that we have understood the process. The beauty of cosmical electrodynamics is that the flux-freezing theorem allows us to make a mental picture of how the magnetic field evolves in an astrophysical plasma.

When an astronomical object shrinks due to gravitational attraction, its magnetic field is expected to become stronger. If a is the equatorial cross-section of the body through which a magnetic field of order B is passing, then the magnetic flux linked with the equatorial plane is of order Ba^2. If the magnetic field is perfectly frozen, then this flux should remain an invariant during the contraction of the object. Some neutron stars are believed to have magnetic fields as high as 10^8 T, as pointed out in §5.5.2. Let us see if we can explain this magnetic field by assuming that the neutron star formed due to the collapse of an ordinary star of which the magnetic field got compressed. A star like the Sun has a radius of order 10^9 m, and the magnetic field near its pole is about 10^{-3} T. Since the radius of a typical neutron star is about 10^4 m, the equatorial area would decrease by a factor of 10^{10} if an ordinary star were to collapse to become the neutron star. If the magnetic flux remained frozen during this collapse, then the initial field of 10^{-3} T would finally become 10^7 T, which is of the same order of magnitude as the magnetic fields of neutron stars.

8.6 Sunspots and magnetic buoyancy

We have summarized some properties of sunspots in §4.8. Now we shall discuss how these properties can be explained with the basic principles of MHD.

First of all, a sunspot is a region of concentrated magnetic field (of order 0.3 T) with very little magnetic field in the surrounding region. Why does the magnetic field get bundled up in a limited region which appears darker compared to its surroundings? We have pointed out in §4.4 that energy is transported by convection in the layers immediately below the solar surface. A sunspot is, therefore, a bundle of magnetic flux sitting in a region where convection is taking place. To understand the formation of sunspots, we need to know how convection is affected by the presence of a magnetic field. This subject is known as *magnetoconvection*, of which the foundations were laid down by Chandrasekhar (1952). We have seen in (8.30) that a magnetic field has a tension force associated with it, which would oppose gas motions connected with convection. If magnetic fields are present in a region of convection, they tend to get swept in confined regions within which convection is inhibited by magnetic tension, but the remaining regions are free from magnetic fields where convection can take place freely. This is clearly seen in numerical simulations of magnetoconvection (Weiss, 1981). Sunspots are then merely regions within which magnetic fields are kept bundled up by convection. Since magnetic tension inhibits convection within a sunspot, heat transport is less efficient within a sunspot, leading to a cooler surface temperature there. That is why a sunspot appears darker than the surroundings.

We also pointed out in §4.8 that often two sunspots appear side by side at nearly the same latitude with opposite polarities. The most obvious explanation

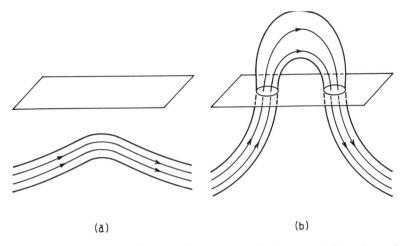

(a)   (b)

Fig. 8.3 Magnetic buoyancy of a flux tube. (a) A nearly horizontal flux tube under the solar surface. (b) The flux tube after its upper part has risen through the solar surface.

for this is that there must have been a strand of magnetic field underneath the solar surface aligned in the toroidal direction, of which a part has come out through the solar surface as shown in Figure 8.3(b). If the two sunspots are merely the two locations where this strand of magnetic field intersects the solar surface, then we readily see that one sunspot must have magnetic field lines coming out and the other must have field lines going in. We now address the question how such a magnetic configuration may come about. As we already discussed, magnetic fields in a region of convection may be expected to remain concentrated within localized regions. Let us consider a nearly horizontal cylindrical region within which some magnetic field is concentrated, as sketched in Figure 8.3(a). Such a region of concentrated magnetic field with very little magnetic field outside is often called a *magnetic flux tube*. Parker (1955a) pointed out that a horizontal flux tube may become buoyant. The argument is quite straightforward. Let P_i be the gas pressure inside the magnetic flux tube and P_e be the external pressure. We have seen in (8.30) that a magnetic field causes a pressure $B^2/2\mu_0$ wherever it exists. In order to have a pressure balance across the bounding surface of the flux tube, we must have

$$P_e = P_i + \frac{B^2}{2\mu_0}. \tag{8.37}$$

It readily follows that

$$P_i < P_e. \tag{8.38}$$

This usually, though not always, implies that the internal density ρ_i is also less than the external density ρ_e. In the particular case when the temperature inside and outside are both T, (8.37) leads to

$$R\rho_e T = R\rho_i T + \frac{B^2}{2\mu_0},$$

from which we obtain

$$\frac{\rho_e - \rho_i}{\rho_e} = \frac{B^2}{2\mu_0 P_e}. \tag{8.39}$$

We thus see that the fluid in the interior of the flux tube is lighter and must be buoyant. In the limit of high $\mathcal{R}_M$, the magnetic field is frozen in this lighter fluid. As a result, the flux tube as an entity becomes buoyant and rises against the gravitational field. This very important effect, discovered by Parker (1955a), is known as *magnetic buoyancy*. Since (8.38) does not always imply that the interior of a flux tube is lighter, it is possible that one part of a flux tube becomes buoyant and not the other parts. Here we shall not get into a discussion as to how this may come about. Suppose only the middle part of the flux tube shown in Figure 8.3(a) has become buoyant. Then this middle part is expected to rise, eventually piercing through the surface and creating the configuration of Figure 8.3(b). With the help of this idea of magnetic buoyancy, one can thus explain how a bipolar magnetic region arises. We have the photograph of a freshly emerged bipolar magnetic region in Figure 4.16. The granules of convection lying between the two large sunspots seem somewhat distorted and elongated. Looking at the photograph carefully, one almost gets the feeling that something has recently come up through the solar surface between the two large sunspots.

We have pointed out in §4.8 that the Sun does not rotate like a solid body. The regions near the equator rotate with a higher angular velocity compared to regions near the poles. Let us consider a magnetic field line passing through the solar interior as shown in Figure 8.4(a). Since the magnetic field line must be nearly frozen in the plasma due to the high $\mathcal{R}_M$, we expect that the varying angular velocity, which is called *differential rotation*, should stretch out this field line as shown in Figure 8.4(b). Thus the differential rotation has a tendency to produce strong magnetic fields in the toroidal direction (i.e. in the ϕ-direction in spherical coordinates), and the magnetic fields in the interior of the Sun are believed to be predominantly toroidal. Parts of the toroidal field, concentrated into flux tubes by interaction with convection, may then become buoyant and produce bipolar sunspots by piercing through the solar surface. It is straightforward to see from Figure 8.4(b) that the bipolar sunspots in the two hemispheres would have opposite polarity alignments, as we saw in Figure 4.17. Choudhuri (1989) was the first person to carry out a three-dimensional simulation to study the formation of bipolar sunspots by magnetic buoyancy.

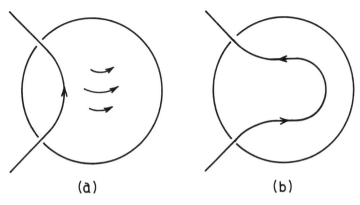

Fig. 8.4 The production of a strong toroidal magnetic field underneath the solar surface. (a) An initial poloidal field line. (b) A sketch of the field line after it has been stretched by the faster rotation near the equatorial region.

8.7 A qualitative introduction to dynamo theory

If there is any magnetic field in the solar interior, we saw in §8.6 that one can combine the ideas of flux freezing, magnetoconvection and magnetic buoyancy to explain the bipolar sunspots. But why should there be any magnetic field to begin with? Most stars are believed to have magnetic fields. We pointed out in §6.7 that our Galaxy has a magnetic field roughly running along the spiral arms. Magnetic fields are almost ubiquitous in the astrophysical Universe. Dynamo theory is the basic theory based on MHD which tries to explain how magnetic fields are generated in astrophysical systems. Magnetic fields of many astrophysical objects have complicated spatio-temporal variations, an example of which is the butterfly diagram for the Sun shown in Figure 4.18. An aim of solar dynamo models is to explain the butterfly diagram. Dynamo theory is a somewhat complicated subject and it is beyond the scope of this book to treat it fully. We merely present below some qualitative ideas of dynamo theory.

The component of the magnetic field in the toroidal direction with respect to the rotation axis of the astrophysical object (i.e. the ϕ-direction in spherical coordinates) is called the *toroidal* magnetic field. On the other hand, the part of the magnetic field lying in the poloidal plane (i.e. $B_r \hat{\mathbf{e}}_r + B_\theta \hat{\mathbf{e}}_\theta$ in spherical coordinates) is called the *poloidal* magnetic field. We have already shown in Figure 8.4 that it is possible to generate the toroidal field by the stretching of poloidal field lines due to differential rotation, in a body like the Sun of which the equator is rotating faster than the poles. However, if the poloidal field cannot be sustained, then it will eventually decay away and consequently the production of the toroidal field will also stop.

In a famous paper, Parker (1955b) gave the crucial idea of how the poloidal field can be generated. If there are turbulent convective motions inside the astronomical body, then the upward (or downward) moving plasma blobs stretch out the toroidal field in the upward (or downward) direction due to flux freezing.

Fig. 8.5 Different stages
of the dynamo process.
See the text for explana-
tions.

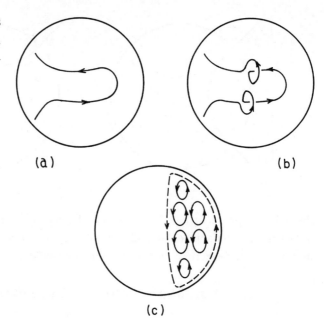

(a)   (b)

(c)

If the convection takes place in a rotating frame of reference, then the upward (or
downward) moving plasma blobs rotate like corkscrews as they rise (or fall). We
see the evidence of such helical motions in cyclones in the Earth's atmosphere.
Figure 8.5(b) shows that a toroidal field line has been twisted by such helical
turbulent motions in such a way that its projections in the meridional plane are
magnetic loops. Several such magnetic loops produced by the helical turbulent
motions are shown projected in the meridional plane in Figure 8.5(c). One can
show that the helical motions in the two hemispheres have opposite sense,
just as the cyclones in the Earth's two hemispheres rotate in opposite senses.
If we keep this in mind and also note that B_ϕ has opposite directions in the
two hemispheres, it then follows that the magnetic loops produced in the two
hemispheres have the same sense. This is indicated in Figure 8.5(b). Although
magnetic fields are partially frozen in the plasma, turbulence associated with
convection makes the magnetic fields mix and diffuse to some extent. As a
result, we eventually expect the magnetic fields of the loops in Figure 8.5(c)
to get smoothened out and give rise to a large-scale magnetic field. Since all
the loops in Figure 8.5(c) have the same sense, their diffusion gives rise to a
global field with the same sense as indicated by the broken field line. Thus we
ultimately end up with a poloidal field in the meridional plane starting from a
toroidal field.

　　Figure 8.6 summarizes the main points of the argument. The poloidal and
toroidal fields can sustain each other through a cyclic feedback process. The
poloidal field can be stretched by the differential rotation to generate the toroidal
field. The toroidal field, in its turn, can be twisted by the helical turbulence
(associated with convection in a rotating frame) to give back a field in the

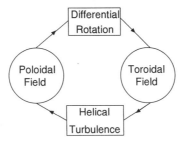

Fig. 8.6 Schematic representation of Parker's idea of the turbulent dynamo.

poloidal plane. Readers desirous of knowing how this central idea of dynamo theory is given a mathematical expression should consult Choudhuri (1998, Chapter 16). The magnetic fields of most astrophysical systems are believed to be produced by the process encapsulated in Figure 8.6.

8.8 Parker instability

The interstellar medium inside a galaxy is usually found to be distributed rather non-uniformly. Figure 8.7 shows how the interstellar medium is distributed in the galaxy M81. In parts of the spiral arms, the interstellar medium seems to form a succession of clumps like beads on a string. It was Parker (1966) who first pointed out that a uniform distribution of the interstellar medium would be unstable. This instability, known as the *Parker instability*, is related to magnetic buoyancy and is presumably the cause behind the interstellar medium fragmenting into clumps.

The magnetic field of the galaxy can be assumed to be frozen in the interstellar medium. Let us consider an initial configuration with the interstellar medium distributed uniformly in a layer having straight magnetic field lines passing through it. Now suppose the system has some small perturbations with parts of the magnetic field lines bulging upward, as sketched in Figure 8.8(a). From symmetry, the gravitational field is directed towards the central plane of the layer. So the gravitational field in the bulging region of magnetic field lines must be downward. If the magnetic field is frozen in the plasma, then the plasma can come down vertically in the bulge region only if the magnetic field lines are also brought down. It is, however, possible for the plasma to flow down along the magnetic field lines as indicated by the arrows in Figure 8.8(a). Alfvén's theorem of flux freezing allows such flows without bringing down the field lines in the bulge, and hence we expect that the downward gravitational field in the bulge region would make the plasma flow in this fashion. As a result of the plasma draining down from the top region of the bulge, this region

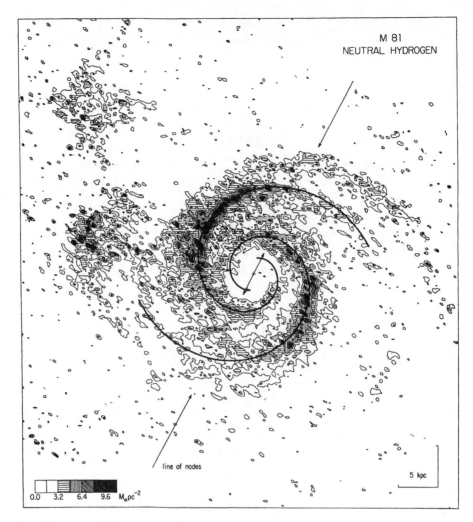

M 81
NEUTRAL HYDROGEN

line of nodes

5 kpc

0.0   3.2   6.4   9.6 $M_\odot pc^{-2}$

Fig. 8.7 The distribution of interstellar matter in the galaxy M81, as measured by radio emission from neutral hydrogen atoms. From Rots (1975). (©European Southern Observatory. Reproduced with permission from *Astronomy and Astrophysics*.)

becomes lighter and more buoyant. We therefore expect this region to rise up further. In other words, the initial bulge keeps on getting enhanced, leading to an instability (Parker, 1966). As the magnetic field lines become more bent, the magnetic tension gets stronger. Eventually the magnetic tension halts the rise of the upper part of the bulge. This was clearly seen in the detailed numerical simulations of Parker instability by Mouschovias (1974). Figure 8.8(b) sketches what the final configuration may look like. The magnetic field lines bulge out of the galactic plane, whereas the interstellar plasma collects in the valleys of the magnetic field lines. This is presumably the reason why the interstellar medium is intermittent and clumpy.

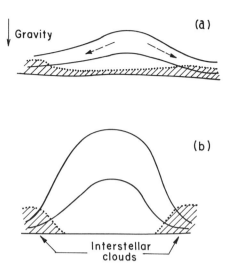

Fig. 8.8 Sketch of Parker instability. (a) Perturbed magnetic field lines bulging out of the galactic plane. (b) The final configuration.

8.9 Magnetic reconnection

We have pointed out in §8.5 that the magnetic Reynolds number is very high in most astrophysical systems and the diffusion term $\eta\nabla^2\mathbf{B}$ in (8.27) can be neglected, leading to the flux freezing condition. It would appear that the diffusion of magnetic fields would be a very unimportant and slow process in an astrophysical system. However, under certain circumstances, it is often found that a large amount of magnetic energy gets dissipated rather quickly. Within our solar system, solar flares provide an example of this. These are explosions taking place in the solar atmosphere above sunspots, where as much energy as 10^{26} J may get released in a few minutes. We have seen in §8.6 that magnetic fields in the Sun rise due to magnetic buoyancy and there would be magnetic loops in the solar atmosphere above sunspots. A solar flare is basically an event in which a large amount of magnetic energy in the solar atmosphere quickly gets converted into heat and other forms. If the magnetic Reynolds number is high, how is it possible to have such a quick dissipation of magnetic energy? We now turn to this question.

Even if the magnetic diffusion coefficient η is small, it is possible for the gradient of the magnetic field to be very large in a region so that the term $\eta\nabla^2\mathbf{B}$ cannot be neglected within that region. Figure 8.9 shows a region with oppositely directed magnetic fields above and below the line OP. Such a magnetic configuration implies that there must be a concentrated sheet of electric current between the oppositely directed magnetic fields. This is called a *current sheet*. Since the gradient of magnetic field would be large in the central region of Figure 8.9, the diffusion term $\eta\nabla^2\mathbf{B}$ may become significant there and hence the magnetic field would decay away in this central region. Since

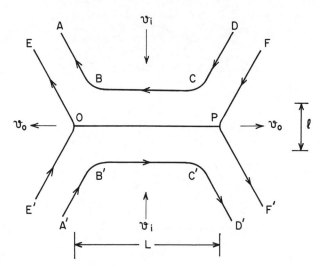

Fig. 8.9 Magnetic reconnection in a current sheet. See the text for explanations.

magnetic fields have the pressure $B^2/2\mu_0$ associated with them, a decrease in the magnetic field would cause a pressure decrease in the central region. In many astrophysical situations, the magnetic pressure may be comparable to or even greater than the gas pressure. If that is the situation, then the decay of the magnetic field in the central region of Figure 8.9 would cause an appreciable depletion of the total pressure there, and we expect that the plasma from above and below with fresh magnetic fields would be sucked into the central region. This fresh magnetic field would then decay and more plasma from above and below would be sucked in to compensate for the pressure decrease due to this decay. This process, known as *magnetic reconnection*, may go on as long as fresh magnetic fields are brought to the central region.

Let us look at Figure 8.9 more carefully to understand the physics of magnetic reconnection. The field lines $ABCD$ and $A'B'C'D'$ are moving with inward velocity v_i towards the central region. Eventually the central parts BC and $B'C'$ of these field lines decay away. The part AB is moved to EO and the part $A'B'$ to $E'O$. These parts originally belonging to different field lines now make up one field line EOE'. Similarly the parts CD and $C'D'$ eventually make up the field line FPF'. We thus see that the cutting and pasting of field lines take place in the central region. Since plasmas from the top and the bottom in Figure 8.9 push against the central region, the plasma in the central region is eventually squeezed out sideways through the points O and P. The resultant outward velocity v_o carries the reconnected field lines EOE' and FPF' away from the reconnection region. Carrying out a full mathematical analysis of magnetic reconnection is extremely difficult. Several scientists (Parker, 1957; Sweet, 1958; Petschek, 1964) attempted to calculate theoretically the rate at which magnetic reconnection may be expected to proceed. Since we cannot get into this complex subject here, let us end our discussion of magnetic reconnection with one comment. Even when the magnetic Reynolds number of an astrophysical system based on its overall length scale is large, it may be possible for magnetic reconnection to take place in localized regions, thereby converting magnetic energy into other forms much more rapidly than what one might expect.

As a result of magnetic reconnection, magnetic energy gets converted into other forms like heat. If the plasma has a low density and hence low heat capacity, then the heat produced by magnetic reconnection in the plasma can raise the temperature of the plasma significantly. One example is the corona of the Sun. We have pointed out in §4.6 that regions of corona can have temperatures of order of millions of degrees, although the solar surface has a temperature of only about 6000 K. Figure 8.10 is an X-ray image of the Sun obtained with the spacecraft Solar and Heliospheric Observatory (SOHO). Since the solar surface at 6000 K does not emit much X-rays, the surface appears dark. The X-rays mainly seem to come from loop-like regions of the corona. These loops are essentially magnetic loops above sunspots – like the loop shown in Figure 8.3(b). It is believed that magnetic reconnections taking place within these loops raise their temperatures to values of the order of a million degrees, leading to copious emission of X-rays. What causes magnetic reconnections to take place in these coronal loops is a complex question which is beyond the scope of this book.

8.10   Particle acceleration in astrophysics

In the previous few sections, we have considered several important astrophysical applications of MHD, treating the plasma as a continuum. There are some astrophysical plasma problems which require a more microscopic point of view and we have to go beyond MHD. One such problem is to understand why many astrophysical systems have a small number of charged particles accelerated to very high energies.

It was established by the balloon flight experiments of Hess (1912) that the Earth is exposed to some ionizing rays coming from above the Earth's atmosphere. It was later ascertained that these *cosmic rays* are actually not rays, but highly energetic charged particles – mostly electrons and light nuclei. A question of fundamental importance was to determine if cosmic rays are something local existing in the neighbourhood of the Earth and the solar system, or if they fill up the whole Galaxy or even the whole Universe. We pointed out in §6.7 the present astrophysical opinion that the cosmic rays are a galactic phenomenon. These charged particles are accelerated within our Galaxy and remain confined within it by the galactic magnetic field. There is evidence for particles with energies as high as 10^{20} eV. For comparison, remember that the

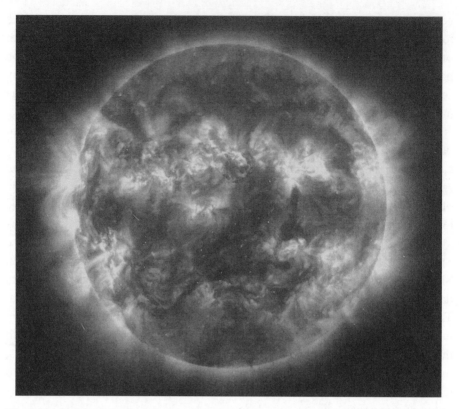

Fig. 8.10 An X-ray image of the Sun obtained by the spacecraft SOHO. This image was taken in 2000, around the time of sunspot maximum. Courtesy: SOHO (ESA and NASA).

rest mass energy of a nucleon is of the order of 10^9 eV, implying that most of these particles must be highly relativistic. We should mention that such energetic charged particles are believed to exist in other galaxies as well. We shall show in the next section that relativistic charged particles gyrating around magnetic fields emit a special kind of radiation known as *synchrotron radiation*, which is often (but not always) found in the radio band of the electromagnetic spectrum. Radio telescopes have discovered synchrotron radiation from many extragalactic systems, implying that the acceleration of charged particles to very high energies must be a fairly ubiquitous process in the astrophysical Universe.

Figure 8.11 shows the spectrum of cosmic ray electrons at the top of the Earth's atmosphere. From energies of the order of about 10^3 MeV to energies of the order of 10^6 MeV, the spectrum can be fitted quite well with a power law

$$N(E)dE \propto E^{-2.6}dE. \tag{8.40}$$

From the study of the synchrotron radiation coming from extragalactic sources, it can be inferred that the electrons in many other sources also have power-law distributions in energy with an index close to 2.5, in fair agreement with what

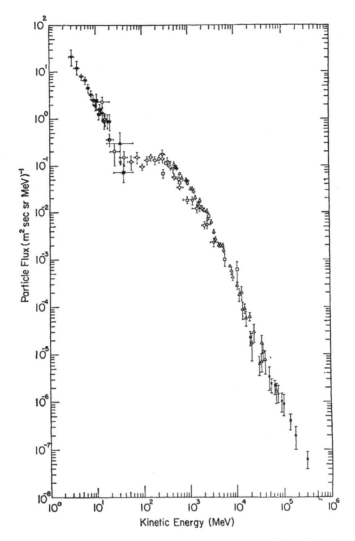

Fig. 8.11 The spectrum of cosmic ray electrons at the top of the Earth's atmosphere. From Meyer (1969). (©Annual Reviews Inc. Reproduced with permission from *Annual Reviews of Astronomy and Astrophysics.*)

is observed in cosmic ray measurements. The aim of any particle acceleration theory is to explain the origin of this power-law spectrum with the observed index. A very influential theory was proposed in a pioneering paper by Fermi (1949).

By studying orbits of charged particles in a non-uniform magnetic field, it can be shown that the particles are reflected from regions of concentrated magnetic field. We shall not derive this result here, but refer the reader to the standard literature (see, for example, Jackson, 2001, §12.5; Choudhuri, 1998, §10.3). Since interstellar clouds are known to carry magnetic fields, the surfaces of the clouds should act as magnetic mirrors and reflect charged particles. Just as a ball picks up energy on being hit by a bat, Fermi (1949) visualized that charged particles can be accelerated by being hit repeatedly by the moving magnetic clouds. Although a particle gains energy in a 'head-on' collision with a moving cloud, there can also be 'trailing' collisions in which energy is lost. Hence we have to show that the particle on average gains energy in collisions. To understand the basic physics, let us consider a simple model of only one-dimensional motions of clouds and charged particles, so all the collisions can neatly be divided into head-on and trailing collisions. We present a Newtonian treatment of the problem here. Since the energetic particles are relativistic, one should actually use a relativistic treatment. We refer the reader to Longair (1994, §21.3) for the relativistic treatment along with a clear discussion of several other aspects of the problem.

Let us consider the clouds to move with velocity U in one dimension, i.e. half of the clouds are moving in one direction and the other half moving in the opposite direction. Let a particle of initial velocity u undergo a head-on collision with a cloud. The initial velocity seen from the rest frame of the cloud is $u + U$. If the collision is elastic, it would appear from this frame that the particle also bounces back in the opposite direction with the same magnitude of velocity $u + U$. In the observer's frame, this reflected velocity appears to be $u + 2U$. Hence the energy gain according to the observer is

$$\Delta E_+ = \frac{1}{2}m(u + 2U)^2 - \frac{1}{2}mu^2 = 2mU(u + U). \qquad (8.41)$$

The energy loss in a trailing collision can similarly be shown to be

$$\Delta E_- = -2mU(u - U). \qquad (8.42)$$

The probability of head-on collisions is proportional to the relative velocity $u + U$, whereas the probability of trailing collisions is proportional to the relative velocity $u - U$. The average energy gain in a collision is therefore equal to

$$\Delta E_{\text{ave}} = \Delta E_+ \frac{u + U}{2u} + \Delta E_- \frac{u - U}{2u} = 4mU^2 \qquad (8.43)$$

on using (8.41) and (8.42). We now write down the corresponding expression for the average energy gain in a relativistic treatment derived in Longair (1994, §21.3). It is

$$\Delta E_{\text{ave}} = 4\left(\frac{U}{c}\right)^2 E. \qquad (8.44)$$

It is easy to see that (8.44) reduces to (8.43) in the non-relativistic limit on putting $E = mc^2$. Readers good at special relativity may try to derive (8.44) themselves.

The main point to note in (8.44) is that the average energy gain is proportional to the energy. Hence the energy of a particle suffering repeated collisions is expected to increase, obeying an equation of the form

$$\frac{dE}{dt} = \alpha E, \tag{8.45}$$

where α is a constant. The solution of (8.45) is

$$E(t) = E_0 \exp(\alpha t), \tag{8.46}$$

where E_0 is the initial energy. If all particles started with the same initial energy E_0, then a particle acquires energy E after remaining confined in the acceleration region for time

$$t = \frac{1}{\alpha} \ln \left(\frac{E}{E_0} \right). \tag{8.47}$$

We expect particles to be continuously lost from the acceleration region. If τ is the mean confinement time, then the probability that the confinement time of a particle is between t and $t + dt$ is

$$N(t)dt = \frac{\exp(-t/\tau)}{\tau} dt. \tag{8.48}$$

This is exactly like the kinetic theory result of finding the probability that the time between two collisions for a particle is in the range t to $t + dt$ and is discussed in any elementary textbook presenting kinetic theory (see, for example, Reif, 1965, §12.1; Saha and Srivastava, 1965, §3.30). Now a particle with confinement time between $t, t + dt$ would acquire the energy between E, $E + dE$. Substituting for t from (8.47) and for dt from (8.45), we are led from (8.48) to

$$N(E)dE = \frac{\exp\left[-\frac{1}{\alpha \tau} \ln \left(\frac{E}{E_0} \right) \right]}{\tau} \cdot \frac{dE}{\alpha E},$$

from which

$$N(E) \propto E^{-\left(1 + \frac{1}{\alpha \tau} \right)}. \tag{8.49}$$

We thus end up with a power-law spectrum.

This theory of Fermi (1949), although somewhat heuristic in nature and based on several ad hoc assumptions, gives us a clue as to how a power-law spectrum may arise. There are, however, very big gaps in the theory. Since it is not so straightforward to estimate α and τ, the index of the power law cannot be calculated easily. Furthermore, there is no indication in the theory why this index should be close to some universal value in different astrophysical systems. We also see from (8.44) that the average energy gain is proportional to $(U/c)^2$. Since the clouds are moving at non-relativistic speeds, this is a very small number and the acceleration process is quite inefficient. Because of this quadratic dependence on U, this process is referred to as the *second-order Fermi acceleration*.

If we could somehow arrange that only head-on collisions take place, then the acceleration process would be much more efficient. For $u \gg U$, it follows from (8.41) that the energy gain will depend linearly on U rather than quadratically. The acceleration resulting from such a situation is called the *first-order Fermi acceleration*. But is it possible for this to happen in Nature? It was pointed out by several authors in the late 1970s (Axford, Leer and Skadron, 1977; Krymsky, 1977; Bell, 1978; Blandford and Ostriker, 1978) that shock waves produced in supernova explosions may provide sites for the first-order Fermi acceleration. Magnetic irregularities are expected on both sides of the advancing shock wave. It is possible that a charged particle is trapped at the shock front and is reflected repeatedly from magnetic irregularities on both sides. Such collisions are always head-on and lead to much more efficient acceleration compared to Fermi's original proposal of acceleration by moving interstellar clouds. We again refer the reader to Longair (1994, §21.4) for a detailed discussion of this theory. Although many questions are still unanswered, acceleration in supernova shocks seems at present to be the most promising mechanism for producing cosmic rays.

8.11 Relativistic beaming and synchrotron radiation

A very famous result of classical electrodynamics is that accelerated charged particles emit electromagnetic radiation. Whenever the velocity of a charged particle in a plasma changes, we, therefore, expect radiation to come out. In this section and the next, we shall discuss two astrophysically important plasma radiation processes. When relativistic charged particles gyrate around magnetic fields, we get *synchrotron radiation*, which is discussed in this section. When charged particles undergo Coulomb collisions amongst each other, we get a type of radiation called *bremsstrahlung*, which will be discussed in §8.12.

To understand synchrotron radiation, we first have to derive a special relativistic effect known as *relativistic beaming*, which is important in many astrophysical problems. We consider an object moving in the x direction with velocity v. Let S and S' be the frames of reference attached with us and with the moving object respectively (both assumed inertial). Now the moving object ejects a projectile with velocity u' in its own frame S' making an angle θ' with the x direction. From our frame, it will appear that the projectile is moving with u making an angle θ. We want to relate θ and θ'.

Suppose it is seen from the moving frame S' that the projectile is at (x', y') and $(x' + dx', y' + dy')$ at times t' and $t' + dt'$ respectively. From our frame S, we would record these events at t and $t + dt$ with the projectile at (x, y)

and $(x + dx, y + dy)$ respectively. From the standard Lorentz transformation formulae,

$$dx = \gamma(dx' + v \, dt'),\tag{8.50}$$

$$dt = \gamma\left(dt' + \frac{v}{c^2}dx'\right),\tag{8.51}$$

$$dy = dy',\tag{8.52}$$

where γ is the usual Lorentz factor $1/\sqrt{1 - v^2/c^2}$. Keeping in mind that $u_x = dx/dt$, $u_y = dy/dt$, $u'_x = dx'/dt'$ and $u'_y = dy'/dt'$, we can divide (8.50) by (8.51) to obtain

$$u_x = \frac{u'_x + v}{1 + vu'_x/c^2},\tag{8.53}$$

whereas the division of (8.52) by (8.51) gives

$$u_y = \frac{u'_y}{\gamma(1 + vu'_x/c^2)}.\tag{8.54}$$

The angle θ which the projectile motion makes with respect to the x direction in our frame S is obviously given by

$$\tan\theta = \frac{u_y}{u_x} = \frac{u'_y}{\gamma(u'_x + v)}\tag{8.55}$$

from (8.53) and (8.54). Since $u'_x = u' \cos\theta'$ and $u'_y = u' \sin\theta'$, we finally have

$$\tan\theta = \frac{u' \sin\theta'}{\gamma(u' \cos\theta' + v)}\tag{8.56}$$

relating θ and θ'.

Let us consider the special case in which the projectile is a beam of light emitted by the moving object so that $u' = c$. Substituting this in (8.56), we obtain

$$\tan\theta = \frac{\sin\theta'}{\gamma(\cos\theta' + v/c)}.\tag{8.57}$$

It is not difficult to verify that θ will in general be smaller than θ'. We can consider the special case in which the moving object emits a light signal perpendicular to its direction of motion, i.e. $\theta' = \pi/2$. Then (8.57) gives

$$\tan\theta = \frac{c}{\gamma v}.\tag{8.58}$$

Suppose the object is moving highly relativistically. Then $v \sim c$ and $\gamma \gg 1$. It follows from (8.58) that θ will be a small angle of order $1/\gamma$. In other words, even if a relativistically moving object emits radiation in different directions in its own rest frame, it will appear to us that all the radiation is emitted in the forward direction of its motion within a cone of angle $1/\gamma$. This is the relativistic beaming effect, which is very important in many astrophysical situations.

We now have to find out what kind of radiation an observer will receive from a relativistic charged particle spiralling around a magnetic field. A rigorous treatment of synchrotron radiation is somewhat complicated. So we shall present a heuristic discussion which captures much of the essential physics. Since a charged particle moving in a spiral path must be having an acceleration directed towards the axis of the spiral, such a particle should obviously emit radiation. The rate of energy loss by a particle moving non-relativistically is given by an expression derived in any standard electrodynamics textbook (see, for example, Panofsky and Phillips, 1962, §20-2). For a relativistic particle spiralling in a magnetic field, we have to consider a relativistic generalization of this and have to average over charged particles moving at different angles with respect to the magnetic field. It can be shown that the average energy loss rate due to radiation for a highly relativistic charged particle of Lorentz factor γ moving in a uniform magnetic field is

$$P = \frac{4}{3}\sigma_T c \gamma^2 U_B, \tag{8.59}$$

where $U_B = B^2/2\mu_0$ is the magnetic energy density and σ_T is the Thomson cross-section given by (2.81). See Rybicki and Lightman (1979, §6.1) or Longair (1994, §18.1) for a derivation of (8.59). It is clear from (2.81) and (8.59) that P is inversely proportional to m^2. Hence electrons emit much more efficiently than heavier nuclei. Even if different accelerated charged particles are present in a system, it is the electrons which emit the synchrotron radiation.

If the electron is relativistic, then we do not have to bother about the direction in which the radiation will be emitted, because the relativistic beaming effect will make sure that we see the radiation coming out in the direction of motion within a cone of angle $1/\gamma$, no matter in which direction the radiation is emitted in the rest frame of the electron. Only if the observer lies within this cone of angle $1/\gamma$, will the observer see the radiation from the electron. Figure 8.12 shows an electron moving in a circular orbit. When the electron is at position A, the observer comes within the cone of radiation and starts receiving the radiation. When the electron reaches B, the observer goes out of the cone and ceases to receive any more radiation. We now need to find out the duration of time during which the observer receives the radiation.

Let L be the distance between A and B, which is also the arc length between them if θ is small. The electron moving with speed v takes time L/v to travel from A to B. This is the interval of time between the emissions of the earliest and latest radiations which are seen by the observer. Keeping in mind that the radiation from B takes time L/c less to travel to the observer compared to the

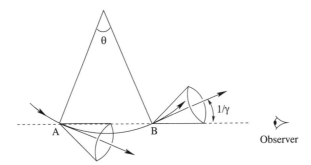

Fig. 8.12 A sketch illustrating how synchrotron radiation arises. The observer receives the radiation emitted by the charged particle only during its transit from A to B.

radiation from A, it should be clear that the time during which the observer receives the radiation is

$$\Delta t = \frac{L}{v} - \frac{L}{c} = \frac{L}{v}\left(1 - \frac{v}{c}\right). \tag{8.60}$$

The frequency of gyration of an electron of charge e and rest mass m_e gyrating in a magnetic field B is $Be/\gamma m_e$ (see, for example, Jackson, 2001, §12.2), which can be written as $\omega_{g,nr}/\gamma$, where

$$\omega_{g,nr} = \frac{Be}{m_e} \tag{8.61}$$

is the non-relativistic gyration frequency. Since $\theta \approx 2/\gamma$, we can write

$$\frac{L}{v} = \frac{\theta}{v/r} \approx \frac{2/\gamma}{\omega_{g,nr}/\gamma} \approx \frac{2}{\omega_{g,nr}}. \tag{8.62}$$

Also

$$1 - \frac{v}{c} = \frac{1 - v^2/c^2}{1 + v/c} \approx \frac{1}{2\gamma^2} \tag{8.63}$$

if $v \approx c$. On making use of (8.62) and (8.63), we get from (8.60) that

$$\Delta t \approx \frac{1}{\gamma^2 \omega_{g,nr}}. \tag{8.64}$$

Hence, as the electron gyrates around the magnetic field, the observer receives a radiation pulse of this duration once every gyration period. If we take a Fourier transform of this signal, the spectrum should peak at a frequency of about $\gamma^2 \omega_{g,nr}$.

Instead of considering a single electron, we now consider a collection of electrons having the energy distribution

$$N(E)\,dE \propto E^{-p}\,dE \tag{8.65}$$

all spiralling around a magnetic field. Since $\omega_{g,nr} = Be/m_e$ will be the same for all these electrons, the frequency at which an electron of energy E will predominantly emit should be proportional to γ^2 or E^2 (because of the special relativistic relation $E = m_e\gamma c^2$). We can write

$$\nu = CE^2, \tag{8.66}$$

where ν is the frequency at which electrons of energy E emit radiation. From this, we get

$$dE = \frac{d\nu}{2\sqrt{C\nu}}. \tag{8.67}$$

In other words, electrons having energy in the range E to $E + dE$ will emit radiation with frequencies in the range ν to $\nu + d\nu$, with dE and $d\nu$ related by (8.67). Since the number of such electrons is $E^{-p}dE$ according to (8.65) and the rate of emission is proportional to E^2 according to (8.59), the rate of radiation emitted by these electrons should be proportional to

$$E^2 E^{-p} dE.$$

On substituting $\sqrt{\nu/C}$ for E and using (8.67), we conclude that the spectrum of emitted radiation should be of the form

$$f(\nu)\, d\nu \propto \nu^{-s} d\nu, \tag{8.68}$$

where

$$s = \frac{p-1}{2}. \tag{8.69}$$

We thus arrive at an extremely important conclusion: if an astrophysical system has relativistic electrons obeying a power-law distribution with index p, then the emitted synchrotron spectrum also should obey a power-law with index s given by (8.69).

We pointed out in §8.10 that accelerated particles typically tend to have power-law indices around $p = 2.6$. According to (8.69), such electrons should emit synchrotron radiation with index $s = 0.8$. Many astrophysical systems indeed emit synchrotron radiation with power-law indices not very different from this. It can be shown that synchrotron radiation is polarized. Hence a power-law spectrum with some degree of polarization is a signature of synchrotron radiation. Whenever we detect synchrotron radiation from an astrophysical source, we can immediately conclude that the source must have magnetic fields and relativistic electrons.

8.12 Bremsstrahlung

Synchrotron radiation discussed in the previous section is an example of *non-thermal radiation*, i.e. a type of radiation arising from a cause other than temperature. The radiation emitted by a body just because of its heat is called *thermal radiation*. In §2.2 we have discussed the emission of radiation by matter in local thermodynamic equilibrium (LTE). We saw that an optically thick source emits like a blackbody. On the other hand, the spectrum of radiation coming out of an optically thin source is essentially given by the emission coefficient j_ν. An optically thin, moderately hot gas primarily emits at spectral lines. However, if the gas is a plasma with temperature of the order of millions of degrees, then all the atoms are broken up and the radiation is produced only when charged particles in the plasma are accelerated or decelerated due to mutual Coulomb interactions amongst themselves. Such radiation, called *bremsstrahlung*, is observed from many astrophysical systems such as the coronae of stars like the Sun or hot gas in clusters of galaxies (to be discussed in §9.5). The radiation from such extremely hot plasmas is often seen in the X-ray part of the spectrum.

Here we shall only quote the main results without the full derivation, which can be found in Rybicki and Lightman (1979, Ch. 5) or Longair (1992, §3.4). Since electrons are much lighter than ions, they are accelerated much more during Coulomb collisions with ions, and it is these electrons which are primarily responsible for bremsstrahlung. An approximate mathematical derivation is not difficult. For a Coulomb collision with impact parameter b, an approximate expression for acceleration can be written down by arguments similar to the arguments given in §7.3 for gravitational collisions. By taking a Fourier transform of the acceleration, one can find the acceleration associated with a frequency ω. Then standard results of electrodynamics give the rate of radiation emitted at that frequency. Finally, we have to allow for different values of the impact parameter b and average over different possible velocities of the electrons (assuming the Maxwellian distribution). The emissivity ϵ_ν (in W m^{-3} Hz^{-1}) is found to be given by

$$\epsilon_\nu = 6.8 \times 10^{-51} \frac{n_e n_i Z^2}{\sqrt{T}} e^{-h\nu/\kappa_B T} g(\nu, T), \qquad (8.70)$$

where T is the temperature, n_e is the number density of electrons (in m^{-3}), n_i is the number density of ions with charge Ze (in m^{-3}) and $g(\nu, T)$ is a dimensionless factor of order unity known as the Gaunt factor which depends on ν and T rather weakly. It should be easy to check that the emission coefficient j_ν introduced in §2.2.2 is simply obtained by dividing ϵ_ν by 4π. To get the total emissivity ϵ, we simply have to integrate ϵ_ν over all frequencies. This gives

$$\epsilon = 1.4 \times 10^{-40} \sqrt{T} n_e n_i Z^2 \overline{g} \qquad (8.71)$$

(in W m^{-3}), where $\overline{g}$ is the averaged Gaunt factor. The formulae (8.70) and (8.71) are regularly used in the astrophysical literature to calculate radiation from very hot plasmas.

8.13   Electromagnetic oscillations in cold plasmas

We end this chapter by pointing out how electromagnetic waves are affected by the presence of plasma. The electric field of the wave accelerates the electrons in the plasma, which then has an effect on the propagation. Because of the inertia of the electrons, very high frequency waves cannot move the electrons much. So we expect the plasma effects to be more important on low-frequency electromagnetic waves like radio waves, which get affected while propagating through the interstellar medium or the solar wind. We pointed out in §6.7 how important inferences can be made about the interstellar medium by analysing radio signals from pulsars.

   We pointed out in §8.4 that the MHD model of the plasma neglects the displacement current, which is crucial for studying electromagnetic waves. Hence we have to go beyond MHD and assume the plasma to be a collection of electrons and ions. Since the ions are much heavier, we can neglect their motion. They merely provide a background of positive charge to keep the plasma neutral. We further assume the plasma to be *cold* – which means that the electrons have no thermal motions and move only under the influence of the electric field of the wave. The readers may look at §12.3 and §12.4 of Choudhuri (1998) to learn about the effects of thermal motions, which make the analysis more complicated.

   Let **v**, **E** and **B** respectively denote the velocity of the electron fluid, the electric field and the magnetic field. The equation of motion of an electron is given by

$$m_e \frac{\partial \mathbf{v}}{\partial t} = -e\mathbf{E}. \qquad (8.72)$$

It is easy to show that the magnetic force $\mathbf{v} \times \mathbf{B}$ is much smaller if $|\mathbf{v}|$ is small compared to c. We need to combine (8.72) with the following two Maxwell's equations:

$$\nabla \times \mathbf{B} = -\mu_0 n_e e\mathbf{v} + \epsilon_0 \mu_0 \frac{\partial \mathbf{E}}{\partial t}, \qquad (8.73)$$

$$\nabla \times \mathbf{E} = -\frac{\partial \mathbf{B}}{\partial t}, \qquad (8.74)$$

where we have written $-n_e e\mathbf{v}$ for the current $\mathbf{j}$. Here n_e is the number density of electrons.

Since we want to consider an electromagnetic wave, let us assume that the time dependence of all the quantities is of the form $\exp(-i\omega t)$ so that we can everywhere replace $\partial/\partial t$ by $-i\omega$. We then get from (8.72)

$$\mathbf{v} = \frac{e}{i\omega m_{\mathrm{e}}}\mathbf{E}. \tag{8.75}$$

On substituting this for $\mathbf{v}$ in (8.73), we have

$$\nabla \times \mathbf{B} = -\frac{i\omega}{c^2}\left(1 - \frac{\omega_{\mathrm{p}}^2}{\omega^2}\right)\mathbf{E}, \tag{8.76}$$

where we have written $1/c^2$ for $\epsilon_0\mu_0$ and

$$\omega_{\mathrm{p}} = \sqrt{\frac{n_{\mathrm{e}}e^2}{\epsilon_0 m_{\mathrm{e}}}} \tag{8.77}$$

is known as the *plasma frequency*. On taking a time derivative of (8.76) and using (8.74), we end up with

$$\frac{\omega^2}{c^2}\left(1 - \frac{\omega_{\mathrm{p}}^2}{\omega^2}\right)\mathbf{E} = \nabla \times (\nabla \times \mathbf{E}). \tag{8.78}$$

Since the background plasma is homogeneous, we may look for solutions of the perturbed quantities which are sinusoidal in space. In other words, we assume all perturbations to be of the form $\exp(i\mathbf{k}.\mathbf{x} - i\omega t)$. On substituting in (8.78), we get

$$\mathbf{k} \times (\mathbf{k} \times \mathbf{E}) = -\frac{\omega^2}{c^2}\left(1 - \frac{\omega_{\mathrm{p}}^2}{\omega^2}\right)\mathbf{E}. \tag{8.79}$$

Without any loss of generality, we can choose our z axis in the direction of the propagation vector $\mathbf{k}$, i.e. we write $\mathbf{k} = k\mathbf{e}_z$. On substituting this in (8.79), we obtain the following matrix equation

$$\begin{pmatrix} \omega^2 - \omega_{\mathrm{p}}^2 - k^2c^2 & 0 & 0 \\ 0 & \omega^2 - \omega_{\mathrm{p}}^2 - k^2c^2 & 0 \\ 0 & 0 & \omega^2 - \omega_{\mathrm{p}}^2 \end{pmatrix}\begin{pmatrix} E_x \\ E_y \\ E_z \end{pmatrix} = \begin{pmatrix} 0 \\ 0 \\ 0 \end{pmatrix}. \tag{8.80}$$

It is clear from (8.80) that the x and y directions are symmetrical, as we expect. The z direction, being the direction along $\mathbf{k}$, is distinguishable. This indicates that we may have two physically distinct types of oscillatory modes. They are discussed below. The existence of these two modes in the plasma was first recognized by Tonks and Langmuir (1929).

8.13.1   Plasma oscillations

One solution of the matrix equation (8.80) is

$$E_x = E_y = 0, \qquad \omega^2 = \omega_p^2. \qquad (8.81)$$

Here the electric field is completely in the direction of the propagation vector **k**, and it follows from (8.72) that all the displacements are also in the same direction. We also note that the group velocity $(\partial\omega/\partial k)$ is zero. We therefore have a non-propagating longitudinal oscillation with its frequency equal to the plasma frequency ω_p. Such oscillations are known as *plasma oscillations*. They are often called *Langmuir oscillations*, after Langmuir who was the pioneer in the study of these oscillations (Langmuir, 1928; Tonks and Langmuir, 1929).

It is not difficult to understand the physical nature of these oscillations. Against a background of nearly immobile and hence uniformly distributed ions, there will be alternate layers of compression and rarefaction of the electron gas (unless $\mathbf{k} = 0$ so that the wavelength is infinite). The electrostatic forces arising out of such a charge imbalance drive these oscillations.

8.13.2   Electromagnetic waves

The only other possible solution of the matrix equation (8.80) is

$$E_z = 0, \qquad \omega^2 = \omega_p^2 + k^2 c^2. \qquad (8.82)$$

This clearly corresponds to a transverse wave. It is actually nothing but the ordinary electromagnetic wave modified by the presence of the plasma. If $\omega \gg \omega_p$, then we are led to limiting relation $\omega^2 = k^2 c^2$, which is the usual dispersion relation for electromagnetic waves in the vacuum. In other words, if the frequency of the wave is too high, even the electrons, which are much more mobile than the ions, are unable to respond sufficiently fast so that the plasma effects are negligible.

It is also to be noted from (8.82) that if $\omega < \omega_p$, then k becomes imaginary so that the wave is evanescent. If an electromagnetic wave of frequency ω is sent towards a volume of plasma with a plasma frequency ω_p greater than ω (if $\omega < \omega_p$), then the electromagnetic wave is not able to pass through this plasma and the only possibility is that it is reflected back.

The plasma frequency of the Earth's ionosphere is about 30 MHz. Radio waves from cosmic sources can penetrate through the ionosphere only if the frequency is higher than 30 MHz (or the wavelength is less than 10 m). Hence radio telescopes have to be operated at higher frequencies if we are to receive radio signals from cosmic sources, as pointed out in §1.7. On the other hand, if we want to communicate with faraway regions of the Earth's surface, then we may want to use radio waves of frequency less than 30 MHz which would be reflected back from the ionosphere.

Exercises

8.1 Consider a fluid flow pattern independent of time. Starting from the Euler equation, show that

$$\frac{1}{2}v^2 + \int \frac{dp}{\rho} + \Phi = \text{constant}$$

along a line of flow (Φ is the gravitational potential). This is known as *Bernoulli's principle* (Bernoulli, 1738).

8.2 Consider a constant initial magnetic field $\mathbf{B} = B_0 \mathbf{e}_y$ in a plasma of zero resistivity. Suppose a velocity field

$$\mathbf{v} = v_0 e^{-y^2} \mathbf{e}_x$$

is switched on at time $t = 0$. Find out how the magnetic field evolves in time. Make a sketch of the magnetic field lines at some time after switching on the velocity field.

8.3 Consider a cylindrical column of plasma with a current of uniform density $j \mathbf{e}_z$ flowing through it (the z direction being parallel to the axis of the cylinder). Find the magnetic field $B_\theta(r)$ resulting from this current (r, θ, z being the cylindrical coordinates). Show from (8.29) that the static equilibrium condition is given by

$$\frac{d}{dr}\left(P + \frac{B_\theta^2}{2\mu_0}\right) + \frac{B_\theta^2}{\mu_0 r} = 0$$

and determine how the gas pressure $P(r)$ varies inside the plasma column. It may be noted that this static equilibrium configuration is violently unstable, but we shall not get into a discussion of it here.

8.4 Suppose a uniform magnetic field $\mathbf{B}_0$ in a plasma with zero resistivity is perturbed. Assuming that the pressure and gravity forces are negligible compared to the magnetic force, i.e. writing the equation of motion as

$$\rho \frac{d\mathbf{v}}{dt} = \frac{(\nabla \times \mathbf{B}) \times \mathbf{B}}{\mu_0},$$

show that the perturbations give rise to wave motions moving along $\mathbf{B}_0$ with velocity given by

$$v_A = \frac{B_0}{\sqrt{\mu_0 \rho}}.$$

These waves are called *Alfvén waves* (Alfvén, 1942b) and v_A is called the *Alfvén speed*.

8.5 Consider a horizontal magnetic flux tube with magnetic field B and radius of cross-section a embedded in an isothermal atmosphere of perfect gas with constant gravity g. Let $\Lambda = P/\rho g$, which is a constant throughout an isothermal atmosphere. The flux tube rising due to magnetic buoyancy at speed v experiences a drag force per unit length given by

$$\frac{1}{2}C_D \rho v^2 a,$$

where C_D is a constant. Show that the flux tube eventually rises with an asymptotic speed

$$v_A \left(\frac{\pi a}{C_D \Lambda} \right)^{1/2},$$

where $v_A = B/\sqrt{\mu_0 \rho}$.

8.6 (For those who are good in special relativity.) Consider the one-dimensional problem of a relativistic particle being reflected from a set of reflectors moving with speeds either U or $-U$. Using special relativity, show that the average energy gained per collision is given by (8.44).

8.7 In the treatment of plasma oscillations given in §8.13, the motions of ions were neglected. Suppose ions with charge Ze and mass m_i also move in response to the electric field. Assuming the plasma to be a mixture of an electron fluid with velocity $\mathbf{v}_e$ and an ion fluid with velocity $\mathbf{v}_i$, show that the frequency of plasma oscillations will be

$$\omega = \omega_p \sqrt{1 + \frac{Z m_e}{m_i}},$$

where ω_p is given by (8.77).

8.8 From the dispersion relation (8.82) of electromagnetic waves propagating through a plasma, show that the group velocity is

$$v_{gr} = c \sqrt{1 - \frac{\omega_p^2}{\omega^2}}.$$

A radio signal starting from a pulsar and passing through the interstellar medium will reach an observer at a distance L in time

$$T_a = \int_0^L \frac{ds}{v_{gr}}.$$

Substitute the expression of v_{gr} in this and make a binomial expansion by assuming $\omega_p \ll \omega$. If signals with different frequencies started at the same time, show that they will be dispersed on reaching the observer, with the dispersion given by (6.68).

9

Extragalactic astronomy

9.1 Introduction

We have pointed out in §6.1 that astronomers in the early twentieth century thought that our Milky Way Galaxy is the entire Universe! Even a small telescope shows many nebulous objects in the sky. The great German philosopher Kant already conjectured in the eighteenth century that some of these nebulae could be island universes outside our Galaxy (Kant, 1755). However, astronomers at that time knew no way of either establishing or refuting this conjecture. In 1920 the National Academy of Sciences of USA arranged a debate on this subject – Shapley arguing that these nebulae are within our Galaxy and Curtis arguing that they are extragalactic objects (Shapley, 1921; Curtis, 1921). We discussed in §6.1.2 how the distances of Cepheid variable stars can be determined. Using the newly commissioned Mount Wilson telescope, which was much more powerful than any previous telescope, Hubble (1922) resolved some Cepheid variables in the Andromeda Galaxy M31 and estimated its distance, clearly showing that it must be lying far outside our Milky Way Galaxy. Our current best estimate of the distance of M31 is about 740 kpc. It soon became clear that many of the spiral nebulae are galaxies outside our Galaxy, heralding the subject of extragalactic astronomy and establishing that galaxies are the building blocks of the Universe.

9.2 Normal galaxies

Light coming from a typical simple galaxy seems like a composite of light emitted by a large number of stars. A galaxy of this kind is called a *normal galaxy*. We shall discuss the characteristics of such galaxies in this section. Galaxies with more complex properties will be taken up in §9.4.

Fig. 9.1 The spiral galaxy M51, photographed with the Hubble Space Telescope. Courtesy: NASA, ESA and Space Telescope Science Institute.

9.2.1 Morphological classification

Galaxies were first classified in the 1920s depending on their appearances through optical telescopes. Some galaxies appear to have beautiful spiral structures. They are called *spiral galaxies*. Figure 9.1 shows the Whirlpool Galaxy M51, which is a spiral galaxy. On the other hand, many galaxies seem to have featureless elliptical shapes. They are known as *elliptical galaxies*. Figure 9.2 is a photograph of such a galaxy. Apart from spiral and elliptical galaxies, there are some galaxies with irregular shapes which do not fit into either of these categories. They are simply called *irregular galaxies*.

Since all spiral galaxies are believed to be intrinsically shaped like circular disks, the apparent shape of a spiral galaxy in the sky gives an indication of its inclination with respect to the line of sight. On the other hand, different elliptical galaxies have different ellipticities, and their intrinsic ellipticities cannot be deduced easily from apparent shapes. For example, a highly flattened elliptical galaxy may appear fairly round in the sky if its short axis is turned towards us. Although the apparent shapes of elliptical galaxies may not be indicative of their real shapes, still elliptical galaxies are customarily classified according to their apparent shapes. The circular-looking elliptical galaxies are classified as E0. Then we go through a sequence of E1, E2, E3 . . . in order of increasing ellipticity, ultimately ending with E7 which are fairly flattened elliptical galaxies. Hubble (1936) developed a famous scheme of classifying galaxies, in which E7 elliptical galaxies are taken to be similar to spiral galaxies with very

Fig. 9.2 The elliptical galaxy NGC 1132, photographed with the Hubble Space Telescope. Courtesy: NASA and Space Telescope Science Institute.

closely wound spirals. There are some spiral galaxies with bars in the central regions, such as the galaxy shown in Figure 9.3. Hubble (1936) divided the spiral galaxies into ordinary spirals and barred spirals. Ordinary and barred spirals with very closely wound spirals are classified as S0 and SB0. Spiral galaxies with increasingly looser spiral structures are classified in the sequence Sa, Sb, Sc. Barred spirals are similarly classified in the sequence SBa, SBb, SBc. Figure 9.4 gives the famous *tuning fork diagram*, in which Hubble (1936) sequentially arranged all these galaxies.

Elliptical galaxies vary widely in luminosity – from very luminous giant galaxies to dwarf galaxies. The number density of elliptical galaxies with luminosity in the range L to $L + dL$ is approximately given by *Schechter's law* (Schechter, 1976)

$$\phi(L)\, dL \approx N_0 \left(\frac{L}{L_*} \right)^{\alpha} \exp(-L/L_*) \frac{dL}{L_*}, \qquad (9.1)$$

where $N_0 = 1.2 \times 10^{-2} h^3$ Mpc^{-3}, $\alpha = -1.25$ and $L_* = 1.0 \times 10^{10} h^{-2} L_\odot$. Here h appearing in the expressions of N_0 and L_* is obviously not Planck's

Fig. 9.3 The barred spiral galaxy NGC 1300, photographed with the Hubble Space Telescope. Courtesy: NASA and Space Telescope Science Institute.

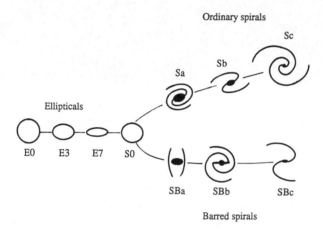

Fig. 9.4 Hubble's tuning fork diagram of galaxy classification.

constant but a very important dimensionless parameter in extragalactic astronomy, which will be introduced in §9.3. It is easy to see from (9.1) that dwarf elliptical galaxies greatly outnumber the giants. In contrast to elliptical galaxies, the spirals do not vary that much in size or luminosity. The typical fractions of spirals and ellipticals in a population of galaxies depend on the environment (Dressler, 1980). In the central regions of rich clusters of galaxies, only about 10% of the galaxies may be spirals. In contrast, the spirals may constitute nearly 80% of the bright galaxies in the low-density regions of the Universe.

The surface brightness of a galaxy is naturally the maximum at the centre and falls off as we go towards the outer edge. In the case of elliptical galaxies, the fall of surface brightness with distance from the centre can be fitted fairly well by the *de Vaucouleurs law* (de Vaucouleurs, 1948)

$$I(r) = I_e \exp\left\{-7.67\left[\left(\frac{r}{r_e}\right)^{0.25} - 1\right]\right\},\tag{9.2}$$

where r_e is called the *effective radius* within which half of the luminosity is contained (if the image of the galaxy happens to be circular), whereas $I_e = I(r_e)$. For the disk of a spiral galaxy, an exponential law gives a reasonably good fit for the fall in surface brightness:

$$I(r) = I_0 \exp\left(-\frac{r}{r_d}\right).\tag{9.3}$$

Here r_d is the distance at which the intensity falls to $0.37I_0$ and gives a measure of the size of the disk.

9.2.2 Physical characteristics and kinematics

Apart from the overall appearances, the physical characteristics of elliptical and spiral galaxies are also very different. We have introduced the concept of stellar populations in §6.4. A typical elliptical galaxy is very much like a Population II object. There is very little interstellar matter in an elliptical galaxy and star formation no longer takes place. So most of the stars are fairly old, giving a yellowish colour to the galaxy, in the absence of young bluish stars. Another property of Population II objects in our Galaxy, as we saw in §6.4, is that they have very little rotational velocity and are supported against gravity by random motions. Exactly similar considerations hold for elliptical galaxies as well. There is usually very little systematic rotation in an elliptical galaxy. The stars do not all collapse to the centre because of the random motions. The larger or the more luminous the elliptical galaxy is, the stronger is its gravity and the stars need to have more random motions in order to maintain a steady state. The velocity dispersion σ of an elliptical galaxy is related to its intrinsic luminosity by the *Faber–Jackson relation* (Faber and Jackson, 1976)

$$\sigma \approx 220 \left(\frac{L}{L_*}\right)^{0.25} \text{ km s}^{-1},\tag{9.4}$$

where L_* is the same luminosity as what appears in (9.1). It is clear from (9.4) that σ is larger for elliptical galaxies with higher luminosity L. Figure 9.5 shows velocity dispersions of several elliptical galaxies plotted against their luminosities. The observational data show a reasonably tight correlation corresponding to the Faber–Jackson relation, without too much scatter.

In addition to the differences in appearance and morphology, spiral galaxies differ from elliptical galaxies in the following basic characteristics: (i) spiral

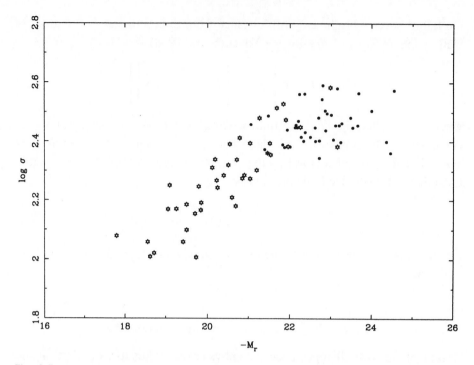

Fig. 9.5 The velocity dispersion σ of elliptical galaxies plotted against their absolute magnitudes. Remembering that the absolute magnitude is 2.5 times the logarithm of luminosity, one obtains the Faber–Jackson relation from this plot. From Oegerle and Hoessel (1991). (©American Astronomical Society. Reproduced with permission from *Astrophysical Journal*.)

galaxies contain considerable amounts of interstellar matter (ISM); and (ii) both stars and the ISM move in roughly circular orbits around the centre of a spiral galaxy such that the gravitational attraction towards the centre is balanced by the centrifugal force. Because of the presence of ISM, star formation goes on inside the disks of spiral galaxies, making them appear bluer than elliptical galaxies. We also receive synchrotron radiation (see §8.11) from the disks of spiral galaxies, which shows that spiral galaxies have magnetic fields as well as cosmic ray particles spiralling around them as in our Galaxy.

We discussed in §6.5 that the emission at the 21-cm line helped in mapping the distribution and kinematics of the ISM in our Galaxy. The ISMs of external spiral galaxies can also be studied by analysing the emission at the 21-cm line. If the galaxy is moving with respect to us, then we will of course find this line Doppler shifted. Additionally, in the case of a rotating disk, we expect the ISM to be moving towards us on one side of the galaxy and moving away from us on the other side (unless the line of sight is exactly perpendicular to the disk). The Doppler shifts of the 21-cm line should accordingly be different on the opposite sides of the spiral galaxy. This is indeed seen and one can use this variation of Doppler shift to determine how the circular speed v_c of the ISM varies with

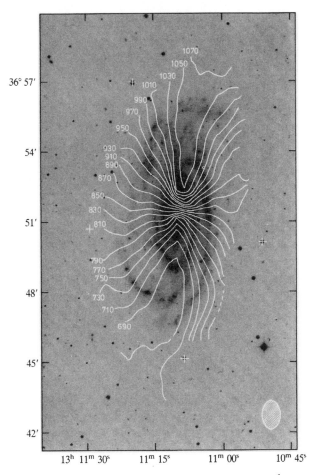

Fig. 9.6 The contours of constant Doppler shift (given in km s^{-1}) in the 21-cm line emission superposed on the negative optical image (i.e. the galaxy is shown in black against a light background) of the galaxy NGC 5033. From Bosma (1978). Courtesy: A. Bosma.

distance from the centre of the galaxy. Figure 9.6 shows the contours of constant Doppler shift in a spiral galaxy superposed on the optical image of the galaxy. The contour lines go well beyond the optical image, since the 21-cm emission of a typical spiral galaxy usually comes from a region much larger than the optical image. This implies that a spiral galaxy does not end at the edge of its optical image (primarily due to stars) and the disk of non-luminous ISM must be extending well beyond where stars are found.

From a figure like Figure 9.6, one can determine how the rotation speed v_c varies with the distance from the centre inside a galaxy. A plot of the circular speed v_c as a function of the radius of a galaxy is known as a *rotation curve*. Before presenting observationally determined rotation curves, let us first discuss what we expect on theoretical grounds. If v_c is the circular speed at a radial distance r from the centre, then equating the centrifugal force with the gravitational force gives

$$\frac{v_c^2}{r} = \frac{GM(r)}{r^2},$$ (9.5)

where $M(r)$ is the mass within the radius r. We should point out that (9.5) is strictly valid only for a spherically symmetric distribution of matter. In the case of a spiral galaxy, we expect (9.5) to give only an approximate qualitative idea of how v_c varies with r. If we take $M(r) \propto r^3$ in the central region of the galaxy, as we would expect in the case of a uniform spherical distribution, then it follows from (9.5) that

$$v_c \propto r$$ (9.6)

in the central region of the galaxy. If most of the mass is confined within a certain region, then the circular speed beyond that region, on the other hand, must be given by

$$v_c = \sqrt{\frac{GM_{\text{total}}}{r}},$$ (9.7)

where M_{total} is the total mass. In other words, we expect v_c to fall as $r^{-1/2}$ in the outer regions of the galaxy. Now we show in Figure 9.7 the rotation

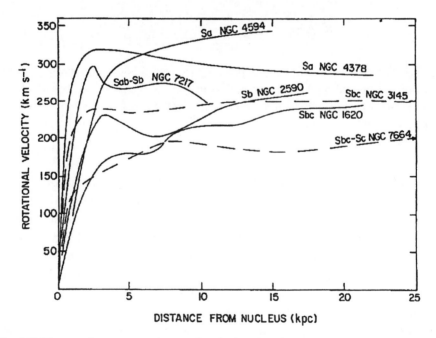

Fig. 9.7 The rotation curves of several galaxies showing how v_c varies with radius. From Rubin, Ford and Thonnard (1978). (©American Astronomical Society. Reproduced with permission from *Astrophysical Journal Letters*.)

curves of several spiral galaxies determined from the Doppler shift of the 21-cm line. It seems that v_c rises in the central regions of galaxies roughly as we expect from (9.6). However, the rotation curves become asymptotically flat and the values of v_c thereafter remain nearly constant with increasing radial distance. We do not see a fall in v_c as suggested by (9.7). This came as a very big surprise to astronomers when rotation curves of a few spiral galaxies were determined for the first time (Rubin and Ford, 1970; Huchtmeier, 1975; Roberts and Whitehurst, 1975; Rubin, Ford and Thonnard, 1978). Since the 21-cm emission is detected from regions of galaxies beyond the visible disk, it could be ascertained that the ISM keeps on going in circular orbits with constant v_c well beyond the regions emitting visible light in the galaxies.

What stops v_c from falling as $r^{-1/2}$ as suggested in (9.7)? The most plausible suggestion is that mass distribution continues beyond the visible stellar disk of the galaxy and even beyond the regions from where we receive 21-cm emission. That is why (9.7) based on the assumption that we are at the outer periphery of mass distribution is not applicable. It follows from (9.5) that $M(r) \propto r$ in a region where v_c is constant. So it is difficult to estimate the total mass of a galaxy if we are not able to detect the fall-off of v_c at larger distances. It appears that the total mass of a typical spiral galaxy is at least a few times the total mass of stars emitting light. In other words, most of the matter in a spiral galaxy does not emit light and is usually referred to as *dark matter*. Determining the nature of dark matter is one of the major challenges of modern astronomy. One important component of dark matter is obviously the ISM which exists in the form of a disk extending beyond the disk of stars. Since we do not see a fall-off of v_c till the edge of the region where atomic hydrogen (emitting the 21-cm line) is found, it is obvious that there must be matter even beyond this region and this matter is not atomic hydrogen. We have no information about the nature of this matter or its distribution. Does this dark matter lie in the disk beyond the disk of neutral hydrogen or does it form a halo around the galaxy? We do not know the full answer (see the discussion of gravitational lensing in §13.3.2).

The asymptotic circular speed v_c in the flat portion of the rotation curve would certainly depend on the mass of the spiral galaxy. We expect a higher v_c for a more massive galaxy. Since a more massive galaxy is expected to be more luminous as well, we anticipate a correlation between the asymptotic v_c and the intrinsic brightness of a spiral galaxy. Tully and Fisher (1977) discovered such a correlation. In the infrared 2.2 micron K band, the *Tully–Fisher relation* can be written as

$$v_c \approx 220 \left(\frac{L}{L_*} \right)^{0.22} \text{ km s}^{-1} \tag{9.8}$$

(Aaronson *et al.*, 1986), where L_* is the characteristic galaxy luminosity. This is reminiscent of the Faber–Jackson relation (9.4) of elliptical galaxies.

9.2.3 Open questions

We have seen in Chapters 3 and 4 that we have a reasonably complete theoretical understanding of various observed properties of stars. A similar theoretical understanding of galaxies has so far eluded astrophysicists. In Chapter 7 we have given an introduction to the theoretical methods for studying the dynamics of a collection of self-gravitating stars. Whereas an elliptical galaxy may seem like a collection of self-gravitating stars, a spiral galaxy is a more complex system due to the presence of the ISM. But why do these collections of stars take up only the morphologies which the galaxies seem to have? What rules out other possible morphologies which an imaginative person should be able to think of? We have no satisfactory theoretical answer to this question. For example, we still have no full theory to explain Hubble's tuning fork diagram. What determines whether a galaxy will turn out to be a spiral galaxy or an elliptical galaxy? Do the initial conditions during galaxy formation determine this, or does it depend on the environment? Our understanding of galaxy formation still being rather primitive, we do not know much about the initial conditions which may determine the nature of the galaxy. Since spiral galaxies have more angular momentum, the amount of angular momentum in the proto-galactic cloud may have some importance in determining the nature of the galaxy formed. The fact that the fraction of spiral galaxies is considerably less within dense clusters of galaxies gives an indication that the environment also must play a role, but we are not quite sure yet about the exact nature of this role. One possible explanation for the scarcity of spiral galaxies in rich clusters is that some of them had been converted into elliptical galaxies. We shall discuss in §9.5 that there are mechanisms by which a galaxy in a rich cluster may lose its ISM. However, if a spiral galaxy loses its ISM, is that sufficient to convert it into an elliptical galaxy? The current evidence points to a tentative answer 'yes', but we certainly do not understand the details of how this happens. We shall also point out in §9.5 that there are some indications that two colliding spiral galaxies may result in a large elliptical galaxy.

There are many other properties of galaxies which we understand only at a very rough qualitative level. For example, we expect the surface brightness of any galaxy to decrease as we move away from its centre. However, nobody has succeeded in giving detailed theoretical derivations from first principles to explain (9.2) or (9.3) quantitatively. Again, we certainly expect larger elliptical galaxies to have more velocity dispersion and larger spiral galaxies to have higher asymptotic v_c. So relations like the Faber–Jackson relation (9.4) and the Tully–Fisher relation (9.8) are certainly expected on the basis of broad qualitative arguments. But we do not have quantitative theoretical explanations why these relations have the precise mathematical forms which they seem to have. Since galaxies are often regarded as the building blocks of the Universe, it is certainly a very unsatisfactory situation that we understand galaxies so little.

9.3 Expansion of the Universe

Suppose a source emitting sound is moving away from you at speed v and the speed of sound through air is c_s. It is easy to show that the frequency ν_{obs} of the sound wave measured by you as an observer will be related to the frequency ν_{em} at which the sound is emitted by

$$\frac{\nu_{obs}}{\nu_{em}} = \frac{c_s}{c_s + v} \qquad (9.9)$$

(see, for example, Halliday, Resnick and Walker, 2001, §18-8). This is the well-known Doppler effect. For the case of light, it is a little bit more subtle, since the speed of light c is the same for all observers and the relative speed v between a light-emitting source and an observer cannot exceed c. However, if v is small compared to c, then (9.9) holds approximately for the Doppler effect of light as well, with c replacing c_s (Halliday, Resnick and Walker, 2001, §38-10). The wavelengths of light in the frames of the emitter and observer are then related by

$$\frac{\lambda_{obs}}{\lambda_{em}} = \frac{\nu_{em}}{\nu_{obs}} = 1 + \frac{v}{c}. \qquad (9.10)$$

Since $\lambda_{obs} > \lambda_{em}$ for a source moving away from us, a spectral line in the spectrum of the moving source will appear shifted towards the red side of the spectrum. Astronomers usually denote the redshift by the symbol z defined as

$$\frac{\lambda_{obs}}{\lambda_{em}} = 1 + z. \qquad (9.11)$$

It follows from (9.10) and (9.11) that

$$v = zc. \qquad (9.12)$$

Thus, measuring the redshifts of spectral lines in the spectrum of a receding source, one can find the speed of recession v.

From the spectrum of a galaxy, it is easy to measure the redshift (or the blueshift) of spectral lines which gives the speed at which the galaxy is moving away from us (or moving towards us). Slipher (1914) noted that the spectra of most galaxies showed redshift, indicating that they are moving away from us. After estimating distances of several galaxies by studying the Cepheid variables in them, Hubble (1929) discovered that more faraway galaxies are receding from us at higher speeds. Hubble (1929) proposed a linear relationship

$$v = H_0 l \qquad (9.13)$$

between the recession velocity v and the distance l. This is now known as *Hubble's law* and H_0 is known as the *Hubble constant*, perhaps the most important quantity in cosmology. As we shall see in the next chapter, it is possible for this constant of proportionality between v and l to evolve with time. In other words,

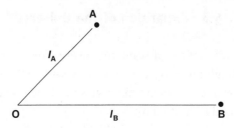

Fig. 9.8 An expanding plane with an observer O.

the Hubble constant is not a constant in time. It relates v and l at a particular epoch in the evolution of the Universe and is a constant for that epoch. In the next chapter, we shall use the symbol H to denote the Hubble constant at any arbitrary epoch, whereas H_0 denotes the Hubble constant at the present time.

On hearing about Hubble's law for the first time, the natural question which comes to one's mind is whether our Galaxy is situated in some special central location of the Universe from which the other galaxies are flying away. A little reflection shows that this linear law merely implies that the Universe is undergoing a uniform expansion and a law like Hubble's law will appear to be true at any point in the Universe. For simplicity, let us consider a plane surface which is expanding uniformly with some marks on it (Figure 9.8). Suppose O is an observer who is watching the two marks at A and B. The distances to A and B from O are respectively l_A and l_B such that

$$l_B = \alpha \, l_A. \tag{9.14}$$

After the plane has expanded uniformly for time dt, the changed distances of A and B will also satisfy a similar relation

$$l_B + dl_B = \alpha(l_A + dl_A)$$

by virtue of the uniformity of expansion. Hence

$$\frac{dl_B}{dt} = \alpha \frac{dl_A}{dt}.$$

Since the time derivative of distance is nothing but the recession velocity from O, we can write this as

$$v_B = \alpha \, v_A. \tag{9.15}$$

It follows from (9.14) and (9.15) that

$$\frac{v_B}{v_A} = \frac{l_B}{l_A}.$$

Thus various points in the plane will appear to move away from O obeying Hubble's law that the recession velocity is proportional to the distance. Since O is an arbitrary point in the plane, we expect a similar recession law to hold with respect to any point in the plane, provided the plane is expanding uniformly. Hubble's law, therefore, suggests a uniformly expanding Universe in which all points are equivalent.

In the example of the uniformly expanding plane, let us now assume that A and B are not static marks, but represent ants which are moving on the plane while the plane is expanding. Then the recession velocities of A and B will consist of two parts: (i) a part due to the expansion of the plane which will satisfy Hubble's law; and (ii) a part due to the motions of the ants with respect to the expanding plane. Exactly the same considerations apply to receding galaxies as well. If galaxies are moving away from us strictly as a part of the general flow of the expanding Universe, then Hubble's law would be exact. On the other hand, if galaxies also move around with a random velocity with respect to this general flow of expansion, then their relative velocities with respect to us can have another part which is random. So we expect the velocity of a galaxy to be given by

$$v = H_0 l + \delta v, \qquad (9.16)$$

where δv is the random part which can be of order $1000 \, \mathrm{km \, s^{-1}}$, since this is the typical velocity of a galaxy with respect to the background flow of expansion. The first term $H_0 l$ increases with distance, whereas the second term δv is of the same order for galaxies at different distances. Hence we expect $H_0 l$ to be the dominant term for galaxies at large distances for which (9.16) will approach Hubble's law (9.13). For nearby galaxies, however, one expects large departures from Hubble's law. In fact, the Andromeda Galaxy M31, which is one of our nearby large galaxies, shows blueshift implying that it is moving towards us. We shall see in §9.5 that both our Galaxy and the Andromeda Galaxy are part of a cluster of galaxies known as the *Local Group*. Typically a cluster of galaxies is a gravitationally bound system and does not expand with the expansion of the Universe. One typically has to look at galaxies at distances larger than the size of a galaxy cluster (a few Mpc) for Hubble's law to hold.

Determination of the Hubble constant

Determining the Hubble constant accurately is one of the major challenges in cosmology. The recession velocity v can be found easily from a measurement of the redshift z. So it is the measurement of the distance l which is the main source of uncertainty. The distance of a galaxy can be found if, within that galaxy, we can identify any object of known intrinsic luminosity. Such objects are often referred to as *standard candles* in astronomy. As we already pointed out, Hubble (1922) determined the distances of some nearby galaxies by studying Cepheid variables in them, which were taken as standard candles. For galaxies still further away for which Cepheid variables are not discernable, one can take the brightest stars as standard candles, assuming that the brightest stars in all galaxies are of approximately the same luminosity. We pointed out in §4.7 that Type Ia supernovae always have the same peak intensity. So, if a Type Ia supernova, which can be much brighter than the brightest stars and can be observed in very faraway galaxies in which even the brightest stars are not distinguishable, is observed in a galaxy, that can be used as a standard candle. Since one cannot always detect a supernova in a galaxy, one can use the Faber–Jackson relation (9.4) of elliptical galaxies or the Tully–Fisher relation (9.8) for spiral galaxies to determine the intrinsic luminosity of a galaxy, from which a measurement of the apparent luminosity gives the distance. All the distance measurement methods, however, become more and more uncertain as we go to larger and larger distances. Since an accurate determination of the Hubble constant requires measurements of both distance and cosmological recession velocity, we have a problematic situation. For nearby galaxies, the distances can be measured accurately, but the possible presence of a random component makes the measurement of cosmological recession velocities uncertain. On the other hand, recession velocities of distant galaxies may be primarily due to the cosmological expansion, but their distances are difficult to measure accurately. For several decades, the Hubble constant remained a quantity which had an unacceptably high error bar. One of the major aims of the Hubble Space Telescope was to narrow down this error bar by determining the distances of galaxies accurately. The project undertaken for this purpose was named the *Hubble Key Project*, which finally succeeded in pinning down the value of the Hubble constant to a great extent.

Before quoting the value of the Hubble constant, let us point out the unit used for it. Since velocities of galaxies are usually given in km s^{-1} and distances given in Mpc, it has become customary to give the value of the Hubble constant in the unit km s^{-1} Mpc^{-1}, although it has the dimension of inverse time. Hubble (1929) had found a rather large value of about 500 in this unit. During much of the second half of the twentieth century, different groups claimed values in the range 50 to 100. It became customary to write the Hubble constant in the following way:

$$H_0 = 100h \text{ km s}^{-1} \text{ Mpc}^{-1}, \tag{9.17}$$

where h was supposed to be a constant of which the value would be fixed by future measurements. Since many important quantities in cosmology depend on the Hubble constant, there were obvious advantages of substituting (9.17) for calculating the values of these quantities. Then one could clearly see how these quantities were affected by the uncertainties in the determination of H_0. For example, let us consider how we can find the distance l of a galaxy from its redshift z. From (9.12), (9.13) and (9.17), it follows that

$$l = 3000zh^{-1} \text{ Mpc}. \tag{9.18}$$

Once we know the exact value of h, (9.18) would give us the exact distance l. As another example, let us try to estimate the age of the Universe. If we assume that a galaxy moving away from us at speed v always moved at this speed v from the beginning of the Universe (which is not a correct assumption as we shall see in the next chapter), then the time taken by it to move through a distance l is $l/v = H_0^{-1}$. All the galaxies must have been on top of each other at the approximate time H_0^{-1} before the current epoch. This is called the *Hubble time* and gives an order of magnitude estimate of the age of the Universe. On using (9.17) and converting Mpc to km, we find the Hubble time to be

$$H_0^{-1} = 9.78 \times 10^9 h^{-1} \text{ yr.} \tag{9.19}$$

If h is assumed to lie in the range 0.5 to 1.0 (which was the case a few years ago), then one can find the uncertainty of the Hubble time from the above expression.

After analysing the various techniques of distance measurement carefully, the Hubble Key Project team finally announced a value of the Hubble constant (Freedman *et al.*, 2001):

$$H_0 = 72 \pm 8 \text{ km s}^{-1} \text{ Mpc}^{-1}. \tag{9.20}$$

This can be taken as the most reliable value of the Hubble constant that we have at the present time. Figure 9.9 plots recession velocities of some nearby galaxies against their Cepheid distances, to give an idea how good Hubble's law is within distances for which the distance measurement can be regarded as accurate.

Astronomers are aware of objects having redshift z of order or larger than 1, where z is calculated from (9.11) by using the observed wavelength

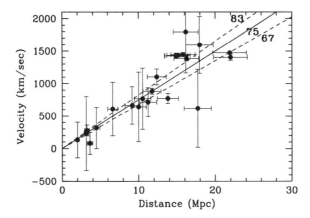

Fig. 9.9 Recession velocities of galaxies plotted against their distances measured from Cepheid variables. The Cepheid distance can be measured only if the galaxy is not too far away. The values of the Hubble constant corresponding to the three straight lines are indicated. From Freedman *et al.* (2001). (©American Astronomical Society. Reproduced with permission from *Astrophysical Journal*.)

of some known spectral line. Certainly (9.10) and (9.12), which hold only in the approximation $|v| \ll c$, cannot be true in such a situation, since (9.12) would imply a speed larger than c when $z > 1$. The full special relativistic expression for the Doppler shift allows for z to be larger than 1 when v approaches c. However, even this special relativistic interpretation is not very helpful in understanding the physics of what is happening when z is larger than 1. We need to apply general relativity for a proper study of the Universe at large redshift. We shall discuss the general relativistic interpretation of redshift in §10.3 and shall derive it in §14.3. This chapter is restricted to a discussion of only that part of the extragalactic Universe for which $z < 1$. The later chapters will discuss objects having $z > 1$.

9.4 Active galaxies

A normal galaxy is made up of stars and interstellar matter. Sometimes, however, it is found that a galaxy may additionally have a compact nucleus at its centre giving out copious amounts of radiation in several bands of electromagnetic spectrum from the radio to X-rays. Such a galaxy is called an *active galaxy* and its nucleus is called an *active galactic nucleus*, abbreviated as *AGN*.

9.4.1 The zoo of galactic activity

The study of active galaxies followed a rather chequered path. Objects which we now recognize to be very similar were often not realized to have anything in common when they were first discovered. As a result, the nomenclature in this field is heavily loaded with historical baggage. The names of different types of active galaxies give no clue as to how these different types of active galaxies may be related to each other.

The subject began when Seyfert (1943) noted that some spiral galaxies had unusually bright nuclei. The spectra of these nuclei were found to be totally different from the spectra of stars and had strong emission lines. So a typical galactic nucleus of this kind could not simply be a dense collection of stars. Depending on whether the emission lines were broad or narrow, these galaxies are now put in two classes. Galaxies with nuclei emitting very broad lines are called *Seyfert 1* galaxies. On the other hand, if the emission lines are narrow, then the galaxies are called *Seyfert 2* galaxies.

The next impetus to this field came from radio astronomy. It was found that some galaxies emitted radio waves. As resolutions of radio telescopes improved with the development of interferometric techniques, it became possible to study the detailed natures of these so-called *radio galaxies*. Jennison and Das Gupta (1953) discovered that the radio emission of the galaxy Cygnus A comes from two lobes located on two sides of the galaxy lying quite a bit outside the optical image of the galaxy. It was soon found that this was quite a common occurrence amongst radio galaxies rather than an oddity. Many radio galaxies consist of radio-emitting lobes lying outside the galaxy on two sides. The obvious question which bothered astronomers was: what could be the source of energy for powering radio emission from these lobes lying so far outside the galaxies?

At first sight, radio galaxies seemed to have nothing in common with Seyfert galaxies. Seyfert galaxies are spiral galaxies with bright nuclei. On the other hand, the radio galaxies, which were mostly found to be elliptical galaxies, have the radio emissions coming from lobes lying outside the galaxies. The fact that they could have something in common became apparent only when astronomers started probing the source of energy in radio galaxies. With improved radio telescopes, it was found that often oppositely directed radio-emitting jets were squirted out of the central regions of radio galaxies. These jets, which are presumably made of plasma flowing out at very high speed, made their ways by pushing away the intergalactic medium surrounding the galaxies. The lobes are located where the jets are finally stopped by the intergalactic medium. Figure 9.10 shows the radio image of Hercules A, which has almost symmetrical radio jets ending in radio-emitting lobes. It appears that the ultimate source of energy of a radio galaxy lies in its nucleus that produces the jets. The jets basically act as conduits for carrying the energy from the nucleus to the lobes (Blandford and Rees, 1974). Like Seyfert galaxies, radio galaxies are also galaxies which have active nuclei. The radio emissions from the jets and lobes of radio galaxies seem to be of the nature of synchrotron radiation, since they have the power-law spectra characteristic of synchrotron radiation. As discussed

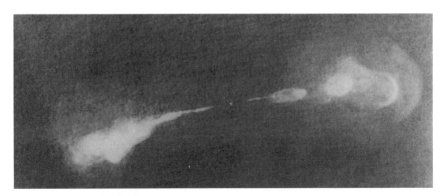

Fig. 9.10 The radio image of the radio galaxy Hercules A produced with the radio telescope VLA (Very Large Array). The radio-emitting jets and lobes are clearly seen. From Dreher and Feigelson (1984). (©Nature Publishing Group. Reproduced with permission from *Nature*.)

in §8.11, the emission of synchrotron radiation implies that the jets and lobes must have magnetic fields around which relativistic electrons are gyrating.

Quasars, which are the most extreme examples of active galaxies, came to the attention of astronomers in a dramatic fashion. These are radio sources of very compact size, some of which could be identified with optical sources looking very much like stars. However, these optical sources were found to have broad emission lines in the spectra, which seemed mysterious and non-identifiable at first. Finally Schmidt (1963) identified the spectral lines of the quasar 3C 273 to be nothing other than the ordinary spectral lines of hydrogen redshifted by an amount $z = 0.158$, which was considered an unbelievably large redshift at that time. Spectral lines of many other quasars were soon identified to be ordinary lines which had undergone even larger redshifts. If these redshifts are caused by recession velocities due to the expansion of the Universe, then (9.18) implied that these quasars must be lying at enormous distances – beyond the distances of most ordinary galaxies known at that time. If the quasars were really at such distances and still appeared so bright, then the typical luminosity of a quasar should be of order 10^{39} W, making it more than 100 times brighter than an ordinary galaxy. What baffled astronomers further is that the emissions from some quasars were found to be variable in time, the time scale of variation being sometimes of the order of days. If t is the typical time scale of variation, then the size of the emitting region cannot be larger than ct. It was inferred that some quasars were emitting their huge energies from very small nuclear regions, which could not be much larger than the solar system. This seemed so incredible at the first sight that many astronomers wondered if quasars were actually not located so far away and hence not intrinsically so luminous. If that were the case, then the redshifts of quasars should not be due to the cosmological expansion and one has to give an alternative explanation for the redshifts. One suggestion was that the quasars may be objects which were ejected at high velocities from the centre of our Galaxy due to some violent explosion there. Only gradually, over several years, astronomers came to accept the fact that quasars are really far-away objects and must be awesome energy-producing machines. In the case of at least a few nearby quasars, it became possible to show that they reside inside galaxies and must be nuclei of galaxies. Such observations are not easy to do, since the nuclei are many times brighter than the host galaxies and the remaining parts of the galaxies get obscured by the glare of the nuclei.

Figure 9.11 shows the typical spectrum of a quasar shifted to rest wavelengths (i.e. after dividing the observed wavelengths by $1 + z$). It may be noted that the spectra of central regions of Seyfert 1 galaxies look very similar, whereas the emission lines in the spectra of Seyfert 2 galaxies are narrower. The similarity in spectra suggests that Seyfert galaxies and quasars may be similar kinds of active galaxies, the Seyfert galaxies being the milder form of such active galaxies, whereas the quasars are the more extreme and rarer form

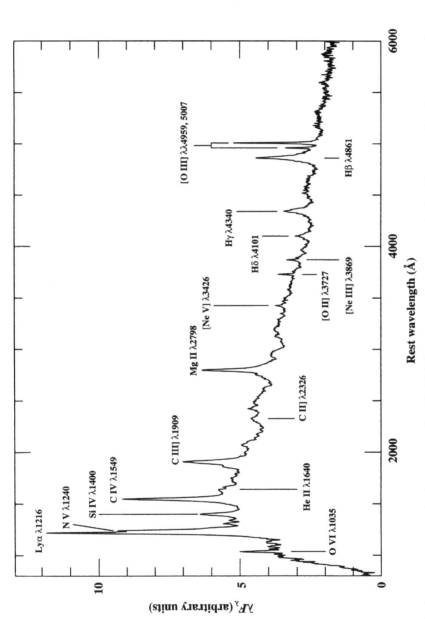

Fig. 9.11 The typical quasar spectrum in the rest wavelength. This is a composite spectrum obtained by averaging the spectra of many quasars. From Francis *et al.* (1991). (©American Astronomical Society. Reproduced with permission from *Astrophysical Journal*.)

which can be detected at very large distances where Seyfert galaxies would not be observable. It is seen from Figure 9.11 that quasars emit very strongly in UV. One efficient method of searching for quasars is to look for stellar-looking objects for which the UV brightness compared to the optical brightness is much higher than what we expect in the case of a star. As a result of such searches, a large number of sources could be found having all the other properties of quasars, except that they did not emit in the radio wavelengths. These radio-quiet quasars are often called *quasi-stellar objects*, abbreviated as QSOs. Since some authors use the terms 'quasar' and 'QSO' almost interchangeably, we shall use the terms radio-loud and radio-quiet quasars to denote quasars which do and do not emit in the radio. Radio-quiet quasars seem much more numerous than radio-loud quasars. Only a few percent of all quasars seem to be radio-emitters.

9.4.2 Superluminal motion in quasars

The only things in common between a radio galaxy and a radio-loud quasar may seem that they are both emitters of radio waves. Otherwise, they may at first sight appear to be very different kinds of objects, with quasars being very compact in appearance and radio galaxies being extended sources with huge jets and lobes. The fact that they may actually be the same objects became clear only when VLBI (see §1.7.2) was used to produce high-resolution maps of quasars showing moving parts inside them. Figure 9.12 shows images of 3C 273 taken in different years. It is clear that a radio-emitting blob is moving away from the central region. On multiplying the angular velocity of separation by the distance of the quasar, one finds a linear velocity larger than c. This phenomenon is referred to as *superluminal motion*. Does this mean that things can move faster than c in the world of very distant galaxies? Even before such superluminal motions were observationally discovered, Rees (1966) realized the possibility of apparent superluminal motions if something is moving towards the observer at a speed v comparable to c making a very small angle θ with respect to the line of sight. This is illustrated in Figure 9.13, where a source of radiation has moved from A to B in time δt such that $AB = v\,\delta t$. If the source emitted a signal when it was at A (setting $t = 0$ when the source was at A), then that signal would reach the observer O at a distance D away at time

$$t_{AO} = \frac{D}{c}.$$

If another signal is emitted after time δt when the source is at B, it is easy to see that this signal will reach the observer at time

$$t_{BO} = \delta t + \frac{D - v\,\delta t\cos\theta}{c}.$$

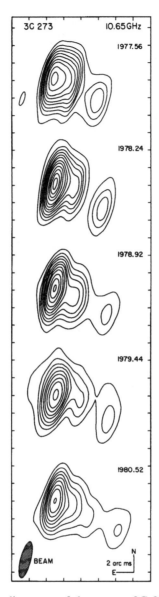

Fig. 9.12 High-resolution radio maps of the quasar 3C 273 obtained by VLBI (Very Large Baseline Interferometry) at different epochs. From Pearson *et al.* (1981). (©Nature Publishing Group. Reproduced with permission from *Nature*.)

Fig. 9.13 A sketch illustrating how superluminal motion arises.

So the observer will receive the two signals at times differing by

$$t_{BO} - t_{AO} = \delta t \left(1 - \frac{v}{c} \cos \theta \right).$$

Since the observer will find that the source has moved by an amount $v \sin \theta \, \delta t$ in the sky, it will appear to the observer that the transverse speed of the source perpendicular to the line of sight is

$$v_\perp = \frac{v \sin \theta \, \delta t}{t_{BO} - t_{AO}} = \frac{v \sin \theta}{1 - (v/c) \cos \theta}. \tag{9.21}$$

It is easy to see that $v_\perp$ can be larger than c if v is close to c and θ is small.

The existence of superluminal motion makes it clear that quasars often have parts moving towards the observer with relativistic speeds. The most plausible hypothesis is that the moving part is a jet ejected from the nucleus. If this is correct, one would regard a radio galaxy and a radio-loud quasar to be the same kind of object viewed from different angles. The radio galaxy has jets at large inclinations to the line of sight so that the jets are seen as extended objects in the sky. On the other hand, if the jet is directed towards the observer at a small angle to the line of sight, then the source is seen as a quasar. We have pointed out in §8.11 that radiation from a relativistically moving source appears to an observer to be beamed in the forward direction. The radio emission from the jet of a quasar (presumably by the synchrotron process) would be beamed in the direction of the jet and an observer lying in that direction would receive the beamed radiation, whereas an observer lying at a large angle with respect to the direction of the jet would get much less radiation. It is because of this relativistic beaming that the radio emission from a quasar appears amplified and quasars can be observed at very large distances where radio galaxies would be too faint to be detected. It may be pointed out that many radio galaxies have jets only on one direction. Since jets seem to be made up of relativistically moving plasma, relativistic beaming provides a natural explanation for this one-sidedness. If a jet has a component of velocity towards the observer, relativistic beaming makes it brighter than the jet in the opposite direction. We have seen in Figure 9.10 that radio galaxies also have radio-emitting lobes where the radio jets end. Presumably the lobes are not moving relativistically and radiation from the lobes is not relativistically beamed. That is why lobes of very distant quasars are not detectable and we receive relativistically beamed radiation coming primarily from the jet moving towards the observer.

9.4.3 Black hole as central engine

We now come to the central question: what powers an active galactic nucleus, making it able to produce huge amounts of energy in a very small volume? We saw in §5.6 that binary X-ray sources are powered by material falling in the deep potential wells of neutron stars or black holes through accretion disks and by the conversion of a part of the lost gravitational potential energy to other forms. Zeldovich and Novikov (1964) and Salpeter (1964) suggested that active galaxies must have black holes at their centres and matter losing gravitational potential energy in accretion disks around them should be providing the energy. It is easy to estimate the mass of the central black hole from (5.38) if we assume that the energy output from the active nucleus is close to the Eddington luminosity. In order to produce a luminosity of 10^{39} W, (5.38) tells us that we need a black hole of mass $\approx 10^8 M_\odot$. In §13.3 we shall introduce the concept of the Schwarzschild radius. Once matter falls within the Schwarzschild radius, it cannot send a signal to the outside world any more. So most of the radiation produced in an accretion disk around a black hole should be coming from regions a little bit beyond the Schwarzschild radius. We shall show in §13.3 that the Schwarzschild radius is given by

$$r_S = \frac{2GM}{c^2} = 3.0 \times 10^{11} M_8 \text{ m}, \tag{9.22}$$

where M_8 is the mass of the black hole in units of $10^8 M_\odot$. This is of the order of the Sun–Earth distance if $M_8 \approx 1$. So a black hole of mass $10^8 M_\odot$ can produce huge amounts of energy within a region of size comparable to the solar system.

We can estimate the temperature of the energy-producing region by equating the Eddington luminosity to $4\pi r_S^2 \sigma T^4$. Using (5.38) and (9.22), we get

$$\frac{c \, GM m_H}{\sigma_T} = \left(\frac{2GM}{c^2}\right)^2 \sigma T^4,$$

from which

$$T \approx 3.7 \times 10^5 M_8^{-1/4}. \tag{9.23}$$

It thus follows that more massive black holes are associated with smaller temperatures, which may seem counter-intuitive at first sight. In the case of larger black holes, however, the gravitational potential energy is lost over a larger radial distance, leading to a reduction in temperature. For a black hole of mass $10^8 M_\odot$, (9.23) suggests that the temperature would be appropriate for producing radiation in the extreme UV. We now estimate the mass infall rate required to produce the typical luminosity 10^{39} W of an AGN. We have seen in §5.6 that accretion is an efficient energy conversion mechanism in which a large fraction η of the rest mass energy can be converted into heat and radiation. Under favourable circumstances η can be as large as 0.1. If $\dot{M}$ is the mass accretion rate, then we must have

$$\eta \dot{M} c^2 \approx 10^{39} \text{ W},$$

from which

$$\dot{M} \approx 1.5 M_\odot \text{ yr}^{-1} \tag{9.24}$$

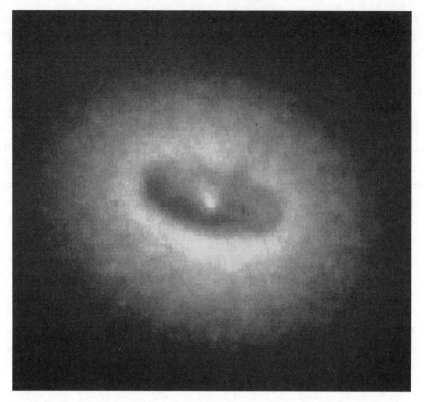

Fig. 9.14 A disk of gas and dust at the centre of NGC 4261 which presumably is feeding a black hole located inside the disk. Photographed with the Hubble Space Telescope. From Jaffe *et al.* (1993). (©Nature Publishing Group. Reproduced with permission from *Nature*.)

on taking $\eta = 0.1$. We see that many of the observed characteristics of active galaxies are explained by assuming that there is a black hole of mass $10^8 M_\odot$ in the galactic nucleus and that matter is falling into it through an accretion disk, the infall rate being given by (9.24). More direct evidence that black holes may power active galaxies came when Jaffe *et al.* (1993) succeeded in using the Hubble Space Telescope to photograph a disk of gas and dust at the centre of the active galaxy NGC 4261. This disk, shown in Figure 9.14, is presumably the colder outer region of the accretion disk which is feeding the black hole located inside.

One of the important characteristics of many active galaxies is the radio-emitting jets. A tiny fraction of the material falling into the black hole has to be ejected in the form of jets, which presumably come out perpendicular to the accretion disk. We still have a very incomplete theoretical understanding of how this happens. This is a very complex subject outside the scope of this elementary book. It may be mentioned that the active galaxy NGC 4261 has radio-emitting jets and the disk shown in Figure 9.14 is approximately perpendicular to the axis of the jets (Jaffe *et al.*, 1993). Many of the jets appear highly collimated, like the inner regions in the jets shown in Figure 9.10. It is believed that magnetic fields play an important role in the collimation of the jets. The emission of synchrotron radiation from the jets shows that these relativistic plasma jets must have magnetic fields and accelerated electrons. If the electrons were accelerated in the galactic nucleus and then carried with the magnetized plasma flowing into the jets, simple estimates show that the electrons may cool by the time they reach the extremities of the jets (see Exercise 9.4). Hence *in situ* particle acceleration has to take place inside jets.

9.4.4 Unification scheme

If all active galaxies derive their power from a central black hole, is it possible that all active galaxies are essentially the same kinds of objects? We now recognize many similarities between different kinds of active galaxies, which were thought to have nothing in common when they were first discovered. Seyfert galaxies and quasars now appear to be similar kinds of objects, the only difference being that quasars have much more powerful central engines. Presumably Seyfert galaxies have less massive black holes in their central nuclei. Radio galaxies and radio-loud quasars also seem to be the same objects viewed from different viewing angles. Is it possible that all active galaxies simply form a simple sequence according to the strength of their central engines, but appear as different kinds of objects to observers due to the different viewing angles?

Osterbrock (1978) proposed that a Seyfert galaxy appears as Seyfert 1 with broad spectral lines if we are located roughly perpendicular to the accretion torus so that we can get a view of the central region, whereas it appears as Seyfert 2 with narrow spectral lines if we are viewing it from an angle so that the accretion torus obscures the central region. The typical widths of Seyfert 1 spectral lines correspond to velocities larger than even 10^3 km s^{-1} (i.e. $\delta\lambda > 10$ Å). If this is due to thermal broadening, then the emitting gas has to be at the unrealistically high temperature of more than 10^8 K. A more reasonable assumption is that there are fast-moving gas blobs called *broad-line regions* (abbreviated as *BLR*s) near the central engine. Only in the case of Seyfert 1, radiation emitted by these reaches the observer and the emission lines appear broadened due to the Doppler-broadening of these fast-moving gas blobs. In the case of Seyfert 2, we receive radiation only from gas blobs much further away from the central engine which are moving more slowly and are known as *narrow-line regions* (abbreviated as *NLR*s), giving rise to narrower emission lines.

One of the important differences amongst active galaxies to keep in mind is that Seyfert galaxies are spiral and radio galaxies are elliptical. Because of the great distances of quasars, it is usually not easy to ascertain the nature of the host galaxies of quasars. It appears, however, that most radio-loud quasars are associated with elliptical galaxies, whereas most radio-quiet quasars are associated with spiral galaxies. This suggests that an active nucleus lodged in a spiral galaxy is not likely to become a radio source (as in the case of a Seyfert galaxy or a radio-quiet quasar), whereas an active nucleus lodged in an elliptical galaxy is likely to produce jets and lobes which emit radio waves (as in the case of a radio galaxy or a radio-loud quasar). We at present have no good theoretical explanation of why active nuclei in elliptical galaxies are more likely to produce radio jets.

To sum up, the current ideas about unification of various kinds of active galaxies are the following. A central engine lodged in a spiral galaxy produces a Seyfert or a radio-quiet quasar depending on whether the central engine is weak or strong. The two types of Seyferts are merely the same objects seen from different angles. A central engine lodged in an elliptical galaxy is seen as a radio-loud quasar if the viewing angle is close to the radio jet and is seen as a radio galaxy if the viewing angle is larger. Although this appears like an attractive scheme, we must stress that at present this scheme cannot be taken as completely established. The best way of establishing a unification scheme like this is to carry out various statistical tests. We expect the active galaxies to be oriented randomly at all possible angles with respect to the line of sight. It is found that about a quarter of all Seyfert galaxies are Seyfert 1. If the opening solid angle of the accretion torus on each side is about $\pi/2$ (i.e. one-eighth of the total solid angle 4π around a point in space), then the unification idea based on the viewing angle provides a natural explanation of why Seyfert 1 galaxies make up about a quarter in the total population of Seyferts. If the radio jet of an active galaxy has to be within a certain angle with respect to the line of sight in order for the galaxy to appear as a radio-loud quasar and if it appears as a radio galaxy otherwise, then one can think of carrying out similar statistical tests on the populations of radio-loud quasars and radio galaxies. Unfortunately the relativistic beaming effect makes this statistical test very complicated and we do not have definitive results yet. Similarly, we expect a distribution law for the strengths of the central engines and, if radio-quiet quasars are merely extreme forms of Seyfert galaxies, then one would think that the populations of Seyfert galaxies and radio-quiet quasars may form different regimes of the same distribution law. Even this is not easy to show, since most of the quasars are found at large distances where Seyfert galaxies would not be visible. We expect that more research in the coming years will put the unification scheme of active galaxies on a firmer footing.

9.5 Clusters of galaxies

Like gregarious animals, galaxies also seem to like living in herds. Our Galaxy is a member of a cluster of about 35 galaxies called the *Local Group*. The

Andromeda Galaxy M31 is the most prominent member of the Local Group and our Galaxy is the second most prominent member. The members of the Local Group are spatially distributed in an irregular fashion without any symmetry. Rich clusters with many galaxies, however, tend to have more symmetric and regular appearance. A typical rich cluster can have more than 100 galaxies distributed within a range of order 1 Mpc. Abell (1958) catalogued a few thousand clusters of galaxies, all lying within several hundred Mpc from us. A cluster of galaxies is usually a gravitationally bound system and does not expand with the expansion of the Universe. So it is the clusters of galaxies rather than galaxies themselves which are moving away from each other with the expansion of the Universe. In fact, there are indications that the Local Group may have an infall velocity of a few hundred km s^{-1} towards the Virgo cluster, the nearest large cluster, superposed on the Hubble expansion (Aaronson *et al.*, 1982). Different standard candles suggest slightly different distances for the Virgo cluster, the average being slightly lower than 20 Mpc. It appears that there can be departures from the Hubble flow over such distance scales. Although the Virgo cluster is a rich cluster, its appearance is not very regular. The nearest fairly regular cluster is the Coma cluster, of which the redshift $z = 0.023$ suggests a distance of $69h^{-1}$ Mpc.

Figure 9.15 shows the Virgo cluster. It is seen that a large elliptical galaxy M87 is at the centre of the Virgo cluster. It is quite common to find a large elliptical galaxy at the centre of a rich cluster of galaxies. Such an elliptical galaxy belongs to a special class of galaxies known as cD galaxies. It is believed that such galaxies form due to the merger of several galaxies (Ostriker and Tremaine, 1975). Since stars occupy only a tiny fraction of volume inside a galaxy, physical collisions between stars inside a galaxy is an extremely unlikely event. On the other hand, individual galaxies occupy a much larger fraction of volume inside galaxy clusters and astronomers have succeeding in catching several galaxies in the act of collision. Figure 9.16 shows a pair of colliding galaxies, which interestingly have produced long tail-like structures. Toomre and Toomre (1972) were able to model such features theoretically through simulations of galaxy-galaxy collision. Such colliding galaxies often merge together. Presumably a cD galaxy at the centre of a cluster is the result of a merger of a few large galaxies. We have pointed out in §7.4 that a star moving inside a star cluster can be slowed down by dynamical friction. A galaxy moving in a galaxy cluster also can be slowed down by dynamical friction in exactly the same way. Once a cD galaxy forms, other galaxies moving through nearby regions may lose their energy due to dynamical friction and then fall into the cD galaxy. This process is called *galactic cannibalism*.

Galaxies inside a rich cluster typically have velocity dispersions of order 1000 km s^{-1}. We have seen in §7.2 that the virial theorem can be applied to estimate the mass of a star cluster. If a galaxy cluster is relaxed, then one can apply (7.9) to estimate the mass of the galaxy cluster as well. On substituting $v \approx 1000$ km s^{-1} and $R \approx 1$ Mpc, the mass of a galaxy cluster typically turns

Fig. 9.15 The Virgo cluster of galaxies. The large elliptical galaxy at its centre is M87. Photographed with the Hubble Space Telescope. Courtesy: NASA and Space Telescope Science Institute.

Fig. 9.16 The colliding galaxies NGC 4038 and NGC 4039 photographed from Antilhue Observatory, Chile. Courtesy: Daniel Verschatse.

out to be of order $M_{gc} \approx 10^{15} M_{\odot}$. On the other hand, the total luminosity of a galaxy cluster typically is of order $L_{gc} \approx 10^{13} L_{\odot}$. For a typical galaxy cluster, we then have

$$\frac{M_{gc}}{L_{gc}} \approx 100 \frac{M_{\odot}}{L_{\odot}}. \tag{9.25}$$

If the galaxies in a cluster were made up of stars like the Sun and there was no other matter inside the cluster, then one would have M_{gc}/L_{gc} equal to $M_{\odot}/L_{\odot}$. If a galaxy has lots of stars less massive than the Sun and much less efficient in producing energy (due to the mass–luminosity relation (3.37)), then it is possible that M/L of the stellar component can be of order $10 M_{\odot}/L_{\odot}$. A factor of 100 in (9.25) implies that it is not possible for the stars inside the galaxies to account for more than 10% of the mass of a typical galaxy cluster. Most of the matter in a galaxy cluster has to be dark matter, i.e. matter which does not emit radiation, but makes its presence felt only by the gravitational field it produces (leading to the high random velocities of the galaxies in the cluster). The existence of huge amounts of dark matter in galaxy clusters was first realized by Zwicky (1933). We have pointed out in §9.2.2 that flat rotation curves suggest the presence of dark matter in spiral galaxies. So a part of the dark matter implied by (9.25) may be attached to the galaxies. Careful analyses indicate that at most 30% of the dark matter in a galaxy cluster may be attached to galaxies. The rest of the dark matter is somehow distributed within the cluster. Nobody at present has a good clue about how this dark matter is distributed in a galaxy cluster or what it is made of. A further discussion of this topic will be presented in §11.5.

Hot X-ray emitting gas in galaxy clusters

Although we do not know much about most of the material in the galaxy cluster lying outside galaxies, we know that a small fraction of it exists in the form of hot thin gas, because this gas emits X-rays. The first extragalactic X-ray source to be detected was M87 in the Virgo cluster (Byram *et al.*, 1966). The Uhuru X-ray satellite established many galaxy clusters as X-ray sources (Giacconi *et al.*, 1972). Figure 9.17 shows contours of X-ray brightness superposed on the optical image of the Coma cluster. The approximately spherical X-ray image suggests that the hot gas emitting X-rays is distributed fairly uniformly and symmetrically throughout the galaxy cluster. The primary emission mechanism is believed to be bremsstrahlung discussed in §8.12. Various aspects of the X-ray emission of galaxy clusters can be explained by assuming that it is bremsstrahlung emitted by a hot plasma of temperature 10^8 K with particle number density $n_e \approx 10^3$ m^{-3} (Felten *et al.*, 1966). On substituting these values in (8.71) and assuming that the total volume of the gas is of order 1 Mpc^3, it is easy to show that the total X-ray luminosity from the cluster would be 10^{37} W. This is indeed the typical X-ray luminosity of galaxy clusters. The total mass

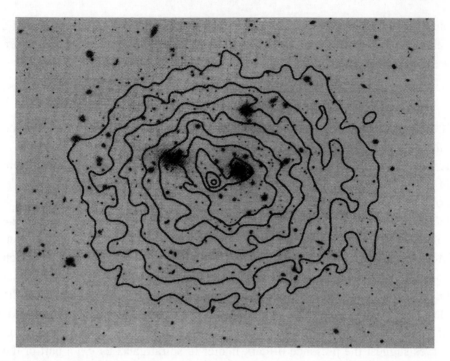

Fig. 9.17 An X-ray image (shown as contours of equal intensity) of the Coma cluster, obtained with the *Einstein Observatory* (1978–81, a NASA mission of X-ray observations), superposed on the negative optical image of the cluster in which galaxies are seen as dark blobs. As given by Sarazin (1986), who obtained the figure from C. Jones and W. Forman. Courtesy: W. Forman.

of the hot gas is of order $10^{13} M_\odot$, indicating that the hot gas contributes only a small fraction of the mass of the galaxy cluster. Before addressing the question of where this hot gas came from, let us show that the cooling time of the hot gas is very large. A number density $n_e \approx 10^3 \, \mathrm{m}^{-3}$ over a volume of 1 Mpc^3 gives a total of about $N \approx 5 \times 10^{70}$ particles, after multiplying the number of electrons by a factor of 2 to account for the protons (assuming that the gas is primarily hydrogen). The total thermal energy of the gas is then

$$N \kappa_B T \approx 7 \times 10^{55} \, \mathrm{J}$$

on taking $T \approx 10^8$. Dividing this total thermal energy by the typical X-ray luminosity 10^{37} W, we get a cooling time of order 2×10^{11} yr which is larger than the age of the Universe as estimated in (9.19), though not excessively larger. Hence, once the hot gas is put in the galaxy cluster, it will remain hot over time scales in which we are interested, even in the absence of any additional heating mechanism to keep the gas hot. A more careful calculation, however, suggests that the effects of cooling may not be completely negligible in the core of the galaxy cluster where the gas density is highest and hence cooling is expected to be fastest, since the bremsstrahlung emission is proportional to the square of density as seen in (8.71). If the gas in the central core becomes colder and causes a decrease in pressure there, it is expected that a radially inward flow of the hot gas known as the *cooling flow* will be induced (Cowie and Binney, 1977).

Let us now come to the questions (i) from where the gas in the galaxy cluster came and (ii) why it is so hot. As we shall point out in §11.9, galaxy formation is still a rather ill-understood subject. Presumably, when the galaxy clusters formed, some of the primordial gas got trapped within the clusters. The trapped gas can become heated simply by falling in the gravitational potential well of the cluster. Suppose a proton of mass m_p falls in the gravitational potential of the cluster and a fraction η of the potential energy lost gets converted to thermal energy. Then

$$\eta \frac{GM_{gc}}{R_{gc}} m_p \approx \kappa_B T.$$

Using $M_{gc} \approx 10^{15} M_\odot$ and $R_{gc} \approx 1$ Mpc, we get

$$T \approx 5 \times 10^8 \eta \text{ K}.$$

It thus seems that even a relatively low value $\eta \approx 0.2$ can explain the high temperature of the gas in galaxy clusters.

Although the X-ray spectra from galaxy clusters can be explained quite well by assuming the emission to be due to bremsstrahlung, the spectra of some clusters show an emission line at 7 keV, which is identified as a line due to highly ionized iron (Mitchell *et al.*, 1976). Figure 9.18 shows the X-ray spectrum from the Coma cluster. From the intensity of this line, one concludes that the ratio of the number of iron atoms to the number of hydrogen atoms in a typical galaxy cluster is of order

$$\frac{\text{Fe}}{\text{H}} \approx 2 \times 10^{-5},$$

which is about half the solar value. As far as we know, heavy elements like iron can be synthesized only in the interiors of massive stars. The ISM of a galaxy gets contaminated with such heavy elements when supernovae spew out materials from the interiors of massive stars into the ISM. The presence of iron in the gas in galaxy clusters indicates that this gas could not be pure primordial gas, but material from the ISMs of the galaxies in the cluster also must have been mixed with the gas. This issue is connected with the issue pointed out in §9.2.1 that the fraction of spiral galaxies in the central regions of rich clusters is much less than that fraction in the low-density regions of the Universe. The most plausible explanation would be that some of the spiral galaxies in the cluster have been converted into elliptical galaxies by losing their ISMs and this gas lost from the spiral galaxies has been mixed with the gas in the cluster, thereby causing the observed iron abundance in that gas.

How do spiral galaxies in a cluster lose their ISM? When two spiral galaxies collide, Spitzer and Baade (1951) argued that the stellar components could pass

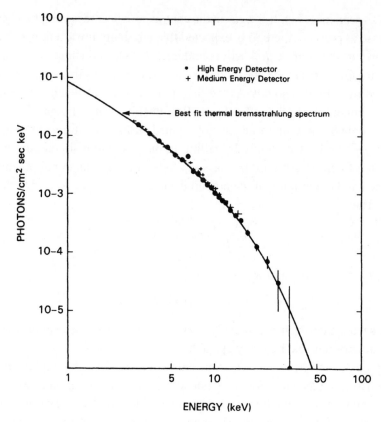

Fig. 9.18 The X-ray spectrum of the Coma cluster showing an emission line at energy 7 keV. From Henriksen and Mushotzky (1986). (©American Astronomical Society. Reproduced with permission from *Astrophysical Journal*.)

through each other, leaving behind the ISM. It is, however, possible for spiral galaxies to lose ISM in more non-violent fashion. When a spiral galaxy moves fast through the cluster, the gas in the cluster would seem to blow against it like a strong wind. A simple application of Bernoulli's principle (see, for example, Choudhuri, 1998, §4.5) shows that this wind blowing against the galaxy would exert a pressure $\frac{1}{2}\rho v^2$ on the surface of the galaxy where the wind is braked. This is called the *ram pressure*. Gunn and Gott (1972) suggested that this ram pressure can force the ISM out of the galaxy, provided it is strong enough to overcome the gravitational force with which the ISM is bound in the galaxy.

   If spiral galaxies in rich clusters are continuously being converted into elliptical galaxies, one would expect that the clusters had more spiral galaxies at earlier times compared to what they have now. When we are looking at galaxy clusters at high redshift, we are basically looking at clusters at earlier times because light would have taken time to reach us from those clusters. Butcher and Oemler (1978) claimed to have found that clusters at redshifts of order $z \approx 0.4$ have more spiral galaxies than nearby clusters. This suggests that many spiral galaxies would have been converted into elliptical galaxies during the time light took to travel from clusters at $z \approx 0.4$ to us, which is of order $4 \times 10^9 h^{-1}$ yr.

9.6 Large-scale distribution of galaxies

Are the galaxy clusters the largest structures in the Universe or are there even bigger structures? To answer this question, it is necessary to map the three-dimensional distribution of many galaxies. We see galaxies distributed over the two-dimensional celestial sphere. Once we find the redshift of a galaxy, we can use (9.18) to find its distance and thereby obtain its position in three dimensions. We therefore need to measure redshifts of many galaxies to con-struct a three-dimensional map of galaxy distribution. This is a rather telescope-intensive project, since it takes a few minutes to obtain the spectrum of a faint galaxy to measure its redshift and redshifts of only a limited number of galaxies can be determined in one night. The first pioneering project of this type was the CfA Survey undertaken by de Lapparent, Geller and Huchra (1986), which was later followed by much bigger projects like the Las Campanas Survey (Shectman *et al.*, 1996), the Sloan Digital Sky Survey (York *et al.*, 2000) and the 2dF Survey (Colless *et al.*, 2001). Figure 9.19 shows a famous slice of the sky from the CfA Survey (de Lapparent, Geller and Huchra, 1986) along with a similar slice from the Las Campanas Survey (Shectman *et al.*, 1996) exactly a decade later. The radial coordinate is the redshift distance cz. Remember that $cz = 10,000$ km s^{-1} would correspond to a distance of $100h^{-1}$ Mpc and a redshift of $z = 0.033$. Note that the Las Campanas Survey covers galaxies at much larger distances than the CfA Survey. Even in the CfA Survey, the distances over the slice are much larger than the sizes of galaxy clusters. Still we do not find a uniform distribution. It seems that the galaxy clusters make up still larger wall-like structures which are called *superclusters of galaxies*. Lurking between the superclusters, we see *voids* typically of size $30h^{-1}$ Mpc which are regions free from galaxies. When we have matter distributed intermittently in space, we can think of two extreme kinds of distribution. One possibility is that clumps of matter appear distributed in space. The other possibility is that bubbles of empty space remain surrounded by matter. The large-scale distribution of galaxies seems to follow the second possibility. Any theory of galaxy formation has to take this fact into account.

It may be noted that there are several radially elongated structures in Figure 9.19, which are particularly prominent in the upper CfA Survey panel. These are artifacts in the redshift distance space which may not correspond to real structures in actual space. To understand how they arise, let us con-sider a cluster of galaxies which have large random velocities caused by the gravitational field of the cluster. Some galaxies would have random velocities away from us (superposed on the mean recession velocity of the cluster) and

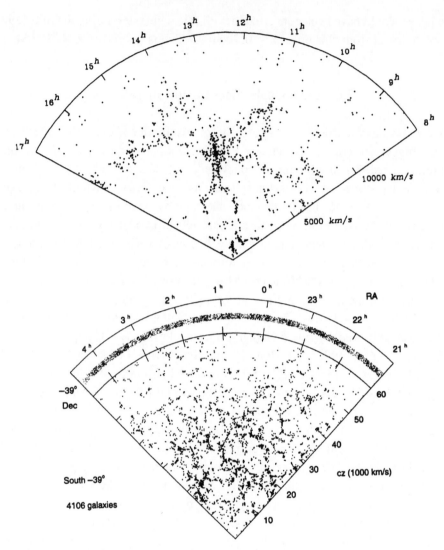

Fig. 9.19 The distribution of galaxies in slices of the sky, as obtained by (a) the CfA
Survey (de Lapparent, Geller and Huchra, 1986): upper panel; and (b) the Las Campanas
Survey (Shectman *et al.*, 1996): lower panel. The radial coordinate gives the redshift
distances of galaxies. (Remember that $cz = 10,000$ km s^{-1} corresponds to $100h^{-1}$
Mpc.) We are at the centre of the slices.

would appear in the redshift distance space to be radially further away than
the mean position of the cluster. On the other hand, galaxies with random
velocities towards us would appear nearer. Thus a cluster of galaxies having
large random velocities appears stretched in the radial direction in the redshift
distance space. Often astronomers refer to such radial elongations as 'fingers of
God' suggesting that one has to be cautious in interpreting redshift distances.

In the next chapter, we shall see that one of the guiding principles of cosmology is the cosmological principle, which assumes that matter is distributed
uniformly at sufficiently large scales, i.e. at scales larger than the sizes of local

non-uniformities. While it was not clear from the CfA Survey (the upper panel of Figure 9.19) whether this is indeed the case, the much deeper Las Campanas survey (the lower panel of Figure 9.19) involving many more galaxies and going to much higher redshifts shows that the superclusters may be the largest structures in the Universe and different patches of the Universe look very similar when we go to scales larger than $100h^{-1}$ Mpc. Other big observational support for the cosmological principle comes from the amazing uniformity of the cosmic background radiation to be introduced in §10.5. Since cosmological models based on the cosmological principle seem to give results in broad agreement with various kinds of observational data, no serious cosmologist doubts the validity of the cosmological principle at a sufficiently large scale, which now seems reasonably well established from galaxy distribution studies.

9.7   Gamma ray bursts

We end our discussion of extragalactic astronomy with a few words about one of the most intriguing and ill-understood objects in astronomy – *gamma ray bursts*, abbreviated as GRBs. These are bursts of γ-rays typically lasting for a few seconds. As of now, there is not yet a consensus amongst astrophysicists how these bursts are produced.

The discovery of GRBs occurred in a dramatic way. They were first detected by American *Vela* satellites designed to detect γ-rays from any secret nuclear tests carried out by the Soviet Union or other countries. Whether these satellites actually detected any secret nuclear tests or not is classified information, but it quickly became clear that some of the signals detected were of extraterrestrial origin (Klebesadel, Strong and Olson, 1973). Initially it was thought that GRBs are produced within our Galaxy. After the launch of the Compton Gamma Ray Observatory in 1991 (see §1.7.4), many GRBs were detected, and it became apparent that their distribution was isotropic and not confined to the galactic plane, suggesting an extragalactic origin. It was eventually possible to detect a fading optical source immediately after the occurrence of a GRB at exactly the same point in the sky (van Paradijs *et al.*, 1997). Afterwards such optical 'afterglows' have been found to follow many GRBs. The optical counterparts appear to be faint galaxies for many of which redshifts have been measured, leading to estimates of the distance. One can try to calculate the energy emitted in a GRB from a knowledge of its distance, although such calculations become uncertain if the emitting material is moving relativistically towards the observer and the relativistic beaming effect is involved (see §8.11). Still it appears that GRBs may be the most energetic explosions in the Universe after the Big Bang!

Apart from the fact that GRBs most probably involve relativistically moving materials, astronomers do not agree about the details of the physical mechanism by which they are produced. It is not even clear if all the GRBs are produced in the same way or if there are two or three different mechanisms, suggesting that all the GRBs may not arise in the same way. One popular theoretical idea used to be that GRBs are caused by collisions between two neutron stars. We have discussed the binary pulsar in §5.5.1. Two neutron stars orbiting each other would continuously lose energy due to gravitational radiation and would come closer to each other, eventually leading to a collision. One can make some statistical estimates of how often such collisions are likely to occur in a galaxy and how often we may expect to see a GRB. GRBs are detected too frequently for this theory to be true as the cause of all GRBs. Are the GRBs then some kind of extreme supernovae? Or are they connected with AGNs? Probably we shall know the answers to these questions in a few years.

Exercises

9.1 Suppose an elliptical galaxy appears circular in the sky with the fall in surface brightness given by the de Vaucouleurs law (9.2). Show that the total light coming from this galaxy given by $\int_0^\infty I(r)\, 2\pi r\, dr$ is equal to $7.22\pi r_e^2 I_e$. Show also that the light coming from within r_e is exactly half this amount.

9.2 On the basis of Figure 9.5, argue that $\sigma \propto L^\alpha$ and estimate the power-law index α.

9.3 Suppose a plasma jet is coming from a quasar at a relativistic speed $0.98c$. At what angle with respect to the line of sight must it lie to cause the maximum superluminal motion? What is the value of the maximum apparent transverse velocity?

9.4 Assuming that magnetic fields inside extragalactic jets are of the order of magnetic fields in the ISM (i.e. about 10^{-10} T), estimate the value of γ for relativistic electrons which produce synchrotron radiation in the radio frequencies. Use (8.59) to estimate the cooling time of these electrons due to synchrotron emission. Compare this with the time taken by some material to start from the central galactic nucleus and reach the outer lobes in a jet of size of 1 Mpc moving with speed $0.1c$. Do you think that electrons have to be accelerated inside the jets, or electrons accelerated in the nucleus of the galaxy would still be able to produce synchrotron radio emission after reaching the lobes?

9.5 Estimate the typical time a galaxy would take to cross a cluster of galaxies and the relaxation time. A cluster of galaxies is clearly not a relaxed system, but it has to be virialized for the virial theorem to be applicable. Do you think that typical galaxy clusters are virialized?

10

The spacetime dynamics of the Universe

10.1 Introduction

How the Universe began is a fundamental question which has occupied the minds of men from prehistoric times. However, only after Einstein's formulation of general relativity (Einstein, 1916), did it become possible to build theoretical models for the evolution of the Universe. Very simple considerations show that the Newtonian theory of gravity is inadequate for handling the Universe. Consider an infinite Universe uniformly filled with matter of density ρ. If we try to find the gravitational field $\mathbf{g}$ at a point on the basis of the Newtonian theory, we immediately run into inconsistencies. Since all directions are symmetric in an infinite uniform Universe, we expect the gravitation field to be zero at any point because there is no preferred direction in which the gravitational field vector can point. However, Newtonian theory of gravity leads to the Poisson equation

$$\nabla.\mathbf{g} = -4\pi G\rho,$$

which also must be satisfied. If $\mathbf{g}$ is zero at all points, then the left-hand side of this equation has to be zero, so that a non-zero ρ clearly leads to contradictions.

In all theories of physics before the formulation of general relativity, spacetime was supposed to provide an inert background against which one could study the dynamics of various systems. In other words, spacetime itself was not supposed to have any dynamics. However, general relativity allowed spacetime also to have its own dynamics. We have discussed the recessive motions of galaxies in §9.3. The common sense interpretation of Hubble's law (9.13) is that galaxies are rushing away from each other through the empty space of the Universe. However, general relativity provides a radically different viewpoint. We assume that the galaxies stay put in space, but space itself is expanding – thereby moving apart the galaxies which are embedded in space.

General relativity requires the machinery of tensor analysis for its mathematical formulation. Is it not possible to learn cosmology, the science dealing with the origin and evolution of the Universe, without acquiring a mastery over the mathematical machinery of tensors? Yes and no! A deep understanding of cosmology certainly requires a technical knowledge of general relativity based on tensor analysis. However, some aspects of cosmology can be studied without learning general relativity first. The dynamics of spacetime, of course, can be handled properly only through general relativity. However, we also need to analyse various physical phenomena taking place in the expanding space of the Universe, in order to understand why the Universe is what it is today. Much of this can be done without a detailed technical knowledge of general relativity. Even the dynamics of expanding space can be studied to some extent without introducing general relativity. If we assume that we are at the centre of the Universe and the Universe is expanding around us in a spherically symmetric fashion, then the equation of motion derived from Newtonian mechanics turns out to be essentially the same as the equation that follows from the detailed general relativistic analysis. This is an amazing coincidence, which allows us to make considerable progress in cosmology without general relativity. This approach is often called *Newtonian cosmology* (Milne and McCrea, 1934). Many issues certainly remain unclear at a conceptual level in this approach and some assumptions have to be taken merely as ad hoc hypotheses without trying to justify them.

The aim of this chapter is to study the dynamics of expanding space through the approach of Newtonian cosmology. Then the next chapter will discuss various physical phenomena taking place in the expanding space. We present a qualitative introduction to general relativity in §10.2, to give readers an idea of what is left out by not doing a fully relativistic treatment. Readers not wishing to learn general relativity at a technical level will have some idea of cosmology from this chapter and the next. Then Chapter 12 provides elementary introductions to tensor analysis and general relativity for those readers who would like to learn this subject at a technical level. Finally, Chapter 14 is devoted to relativistic cosmology, where we cover some important topics which are beyond the scope of Newtonian cosmology and are, therefore, not discussed in this chapter.

10.2   What is general relativity?

General relativity provides a field theory of gravity. To explain what is meant by a field theory, let us consider the example of the other great classical field theory with which the reader must be familiar – the theory of electromagnetic fields. We consider two charges q_1 and q_2 with $\mathbf{r}_{12}$ as the vectorial distance of

q_2 from q_1. According to Coulomb's law, the electric force on q_2 due to q_1 is given by

$$F_{12} = \frac{1}{4\pi\epsilon_0}\frac{q_1 q_2}{r_{12}^3}r_{12}. \tag{10.1}$$

In the action-at-a-distance approach, we only concern ourselves with the charges and the forces acting on them, without bothering about the surrounding space. On the other hand, the field approach suggests that the charge q_1 creates an electric field around it which is given at the distance r_{12} by the expression

$$E = \frac{1}{4\pi\epsilon_0}\frac{q_1}{r_{12}^3}r_{12} \tag{10.2}$$

and that the charge q_2 experiences the force

$$F_{12} = q_2 E \tag{10.3}$$

by virtue of being present in the electric field. It may appear that we are merely splitting (10.1) into two equations (10.2) and (10.3), and nothing new is gained in the process. As long as the charges are at rest, it is true that the field approach does not give us anything new compared to the action-at-a-distance approach. However, the situation changes if the charges are in motion.

Suppose the charge q_1 suddenly starts moving at an instant of time. What should we now substitute for r_{12} in the expression (10.1)? You may think that we have to consider the simultaneous locations of the two charges at an instant of time and obtain r_{12} from that. However, we know from special relativity that simultaneity in different regions of space is a subtle concept and depends on the frame of reference. If we can somehow suitably define simultaneity in some frame of reference, then there is a more serious problem with (10.1). Since the separation between q_1 and q_2 changes as soon as q_1 starts moving, the force F_{12} on q_2 also should change immediately according to (10.1). This means that the information that the charge q_1 has started moving has to be communicated to the other charge q_2 at infinite speed, which contradicts special relativity. These difficulties disappear if the electromagnetic fields are treated with the help of Maxwell's equations. It can be shown from Maxwell's equations that the information about the motion of q_1 propagates at speed c and the charge q_2 starts getting affected only when the information reaches its location. We thus see that the action-at-a-distance approach fails to provide a consistent theory of charges in motion and we need a field approach.

Exactly the same considerations exist in the case of the gravitational field. When two masses are at rest, Newton's law of gravitation gives the force between them. But, when the masses start moving, we have the same difficulties which we have with moving charges. After developing special relativity, Einstein realized that Newton's theory of gravity is not consistent with special relativity. We need to develop a field approach, which will ensure that gravitational information does not propagate at a speed faster than c. In contrast to two types of electric charges, we have only one type of mass. It might, therefore, seem at first sight that a field theory of gravity would be simpler than the field theory of electromagnetism. In reality, however, the field theory of gravity is much more complicated. Let us point out some reasons for this. A particle of charge q and mass m placed in an electric field $\mathbf{E}$ will have acceleration $(q/m)\mathbf{E}$. This acceleration will in general be different for different particles. In contrast, all particles placed in a small region of gravitational field (over which the variations of the field can be neglected) have the same acceleration. As a result, a gravitational field is equivalent to an accelerated frame. Einstein realized that this *equivalence principle* would have to be an important part of a field theory of gravity. A second complication for a field theory of gravity is that it has to be a nonlinear theory, in contrast to the theory of the electromagnetic field which is linear. One can give a simple argument for why the field theory of gravity has to be nonlinear. Any field such as a gravitational field has some energy associated with it and energy is equivalent to mass according to special relativity. Hence a gravitational field, having some equivalent mass associated with it, can itself act as a source of gravitational field. This is not the case in electromagnetic theory, where charges and currents are the sources of the electromagnetic field and the field itself cannot be its own source. Since a field theory of gravity somehow has to incorporate the equivalence principle as well as some nonlinear aspects, it necessarily has to be more complex than the theory of electromagnetic fields.

Einstein (1916) had the profound insight of viewing gravity as a curvature of spacetime. In a region without gravitational fields, the spacetime is flat and a body moves in a straight line. However, in a region where the spacetime is curved (which implies the presence of gravity), a body is deflected from a rectilinear path. Thus, instead of saying that the gravitational force deflects the body, we say that the curved spacetime makes it move in a curved path. This geometrical interpretation of gravity automatically provides an explanation for the equivalence principle. The curved path which a body would take in a region is determined merely by the curvature of spacetime and should be independent of the mass of the body. Hence bodies of different masses should follow the same curved path.

To develop a field theory of gravity, we therefore need a mathematical machinery to handle the curvature of spacetime. Let us first consider the curvature of a two-dimensional surface. There are two ways of looking at the curvature: extrinsic and intrinsic. When we perceive a surface as curved in a common sense way, we essentially perceive the surface embedded in three-dimensional space to be curved. This extrinsic way of looking at curvature is not so useful when we consider the curvature of four-dimensional spacetime. Do we have to think of it as being embedded in a five-dimensional something within which it is curved? To explain the alternative intrinsic way of viewing

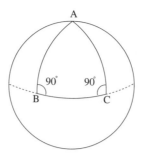

Fig. 10.1 Sketch of a triangle *ABC* on a sphere, made with two circles of longitude and the equator.

curvature, let us again consider the example of the two-dimensional surface. Let there be some intelligent ants or other creatures living on this two-dimensional surface who have no conception of three-dimensional space. Can they make some measurements within this two-dimensional surface to find out if it is curved or not? If a surface is plane, we know that any arbitrary triangle will have the sum of three angles equal to 180°. But this is not necessarily true for a curved surface. In the case of a curved surface, a straight line has to be replaced by a geodesic, which is the shortest path between two points of the surface lying wholly on the surface. Great circles on a sphere are geodesics. We can think of a triangle on the Earth's surface made with two circles of longitude and the equator, as shown in Figure 10.1. Clearly this triangle made up of three geodesics has the sum of its angles larger than 180°. So the intelligent ants living on the two-dimensional surface can determine whether the surface is curved by considering some arbitrary triangles and measuring the sums of their angles. This intrinsic view of curvature does not require any consideration whether the curved surface is embedded in a higher-dimensional space. When we consider the curvature of four-dimensional spacetime, we naturally have to look at the curvature from an intrinsic point of view.

The intrinsic view of curvature usually conforms to our everyday notion of curvature – but not always! For example, we normally think of the surface of a cylinder as a curved surface. But this surface can be unrolled to a plane and has the same geometric properties as a plane surface. All triangles drawn on a cylindrical surface have the sum of their angles equal to 180°. Therefore, when we are taking an intrinsic point of view, there is nothing to distinguish a cylindrical surface from a plane and we have to consider the surface of a cylinder as a plane or flat surface – a point of view which appears puzzling at first. Now we come to the crucial question: what determines if a two-dimensional surface is flat? If the surface can be spread on a plane *without stretching or shrinking* anywhere, then it should be regarded as flat. This cannot be done with a spherical surface and that is why a spherical surface is not flat. When geographers have to represent the map of the whole Earth on a flat sheet of paper, they have to apply some projections which do drastic things like making Greenland appear as large as Africa or making some continents highly distorted.

When we stretch or shrink some portions of a surface, we change distances between various pairs of points on the surface. On the other hand, when we try to spread the surface on a plane without any stretching or shrinking, we essentially apply transformations which keep distances between all possible pairs of points invariant. Whether such transformations would enable us to spread the surface on a plane or not depends on how the distances between various pairs of points are related to each other. If we introduce the standard (θ, ϕ) coordinates on a spherical surface of radius a, the distance ds between two nearby points (θ, ϕ) and $(\theta + d\theta, \phi + d\phi)$ is given by

$$ds^2 = a^2(d\theta^2 + \sin^2\theta \, d\phi^2). \tag{10.4}$$

In general, the distance between two neighbouring points on a surface is given by an expression of the form

$$ds^2 = \sum_{\alpha,\beta} g_{\alpha\beta} \, dx_\alpha \, dx_\beta, \tag{10.5}$$

where $g_{\alpha\beta}$ is called the metric tensor. It is this tensor which determines how distances between various possible pairs of points are related. Hence, it is this tensor $g_{\alpha\beta}$ which decides whether it would be possible to spread the surface on the plane, i.e. whether the surface is flat or not. These considerations carry over to spaces of higher dimensions as well. The distances between nearby points in a higher dimensional space also can be written in the form (10.5). It is again the metric tensor $g_{\alpha\beta}$ for the space which determines whether the space is flat or curved. In §12.2 we shall develop the appropriate mathematical tools of tensor analysis which can be applied to a metric tensor $g_{\alpha\beta}$ to calculate the curvature of the space.

On using polar coordinates, the metric tensor of a plane surface is given by

$$ds^2 = dr^2 + r^2 \, d\theta^2. \tag{10.6}$$

Using the same coordinates (x_1, x_2), the two metrics (10.6) and (10.4) can be written as

$$ds^2 = a^2(dx_1^2 + x_1^2 \, dx_2^2), \tag{10.7}$$

$$ds^2 = a^2(dx_1^2 + \sin^2 x_1 \, dx_2^2). \tag{10.8}$$

Another possible metric for a two-dimensional surface is

$$ds^2 = a^2(dx_1^2 + \sinh^2 x_1 \, dx_2^2). \tag{10.9}$$

When we apply the technique of curvature calculation to be developed in §12.2.4, we find that the metric (10.7) has zero curvature (i.e. it corresponds to a flat surface), whereas the metrics (10.8) and (10.9) correspond to uniform

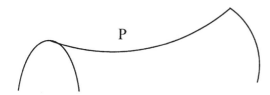

Fig. 10.2 Sketch of a saddle-like surface.

surfaces of constant curvature $2/a^2$ and $-2/a^2$ respectively. Certainly a sphere, of which (10.8) is the metric, has a surface of uniform curvature. But what kind of surface does (10.9) correspond to? Figure 10.2 shows a saddle-like surface. It can be shown that the saddle point P has negative curvature. However, the saddle surface is not a uniform surface, but has its geometric properties changing from point to point. The metric (10.9) corresponds to a uniform surface of which every point is a saddle point. Certainly there is no real two-dimensional surface embedded in three-dimensional space which has this property. However, one can mathematically postulate such a surface and study its properties by analysing (10.9).

In order to learn general relativity at a technical level, one first has to learn the mathematical machinery of calculating curvature. That is presented in §12.2. Here we merely point out that the theoretical structure of general relativity has certain analogies with the theory of electromagnetic fields. The basic idea of electromagnetic theory is that charges and currents produce electromagnetic fields. We have Maxwell's equations which tell us how the electromagnetic field can be obtained if we know the distribution of charges and currents. In an analogous way, general relativity suggests that mass and energy can give rise to the curvature of spacetime, and the central equation, known as Einstein's equation (to be introduced in §12.4.2), describes how the curvature of spacetime is related to mass-energy. Hence, given the distribution of mass-energy, one can in principle find out the metric of spacetime from Einstein's equation and thereby determine the structure of spacetime. To complete electromagnetic theory, we need another equation describing how a charge moves in an electromagnetic field, which is the Lorentz equation

$$m\frac{d\mathbf{v}}{dt} = q(\mathbf{E} + \mathbf{v} \times \mathbf{B}). \tag{10.10}$$

We require an analogous equation in general relativity, which will tell us how a mass will move in the curved spacetime. The basic idea of general relativity is that a mass moves along a geodesic of spacetime if non-gravitational forces are absent. Hence, in the place of (10.10), we have the equation of the geodesic (to be discussed in §12.2.5) in general relativity. Table 10.1 presents a comparison between electrodynamics and general relativity. In general relativity, we basically have to replace the Newtonian theory of gravity by the idea that mass and energy create curvatures in spacetime and a particle moves along geodesics in this curved spacetime.

Table 10.1 Analogy between electrodynamics and general relativity.

	ELECTRODYNAMICS	*GENERAL RELATIVITY*
Basic field equations	**Maxwell's equations:** {Charge, current} ⇒ {Electromagnetic field}	**Einstein's equation:** {Mass, energy} ⇒ {Curvature of spacetime}
Equation of motion in the field	The Lorentz equation (10.10)	Motion along geodesics

10.3 The metric of the Universe

After the brief qualitative introduction to general relativity in the previous section, we now discuss the possible structure of spacetime in the Universe. In §9.6, we have introduced the *cosmological principle*, which states that space is homogeneous and isotropic. Now, it is possible for space to be homogeneous and isotropic only if it has uniform curvature everywhere. We have seen that (10.7), (10.8) and (10.9) are the only possible metrics for a two-dimensional surface which is uniform (i.e. which has a constant curvature everywhere). We now have to write down similar metrics for a uniform three-dimensional space. Using spherical coordinates (r, θ, ϕ), the distance between two nearby points in a flat space is given by

$$ds^2 = dr^2 + r^2(d\theta^2 + \sin^2\theta \, d\phi^2). \tag{10.11}$$

One can explicitly calculate the curvature of this metric (by methods to be introduced in §12.2.4) and show that it is zero. Writing $r = a\chi$, this metric takes the form

$$ds^2 = a^2(d\chi^2 + \chi^2 \, d\Omega^2), \tag{10.12}$$

where

$$d\Omega^2 = d\theta^2 + \sin^2\theta \, d\phi^2. \tag{10.13}$$

It may be noted that the three-dimensional metric (10.12) looks very similar to the two-dimensional metric (10.7). In analogy with (10.8) and (10.9), we can consider the following three-dimensional metrics

$$ds^2 = a^2(d\chi^2 + \sin^2\chi \, d\Omega^2), \tag{10.14}$$

$$ds^2 = a^2(d\chi^2 + \sinh^2\chi \, d\Omega^2), \tag{10.15}$$

where $d\Omega^2$ is always given by (10.13). If we apply the techniques of curvature calculation to be discussed in §12.2.4, we indeed find that the two metrics (10.14) and (10.15) have uniform curvatures $6/a^2$ and $-6/a^2$ respectively. Just as all points of a sphere are equivalent, all points in the space described by metrics (10.14) and (10.15) must be equivalent, since the curvature has the same value at all points in either of these metrics. In fact, (10.12), (10.14) and (10.15) are the only possible forms of three-dimensional metrics for which all points are equivalent. Hence, if the cosmological principle has to be satisfied, then the spatial part of the metric of the Universe has to have one of these three forms.

It is not difficult to show that the space described by the metric (10.14) must have finite volume. Let us consider an element of volume with its sides along the three coordinate directions. Keeping in mind that $d\Omega^2$ is given by (10.13), it follows from (10.14) that the sides of the volume element have lengths $a\,d\chi$, $a\sin\chi\,d\theta$ and $a\sin\chi\sin\theta\,d\phi$. Hence the volume of this volume element is

$$dV = a^3 \sin^2\chi\,d\chi\,\sin\theta\,d\theta\,d\phi.$$

To get the total volume of the space, we have to integrate this over all possible values of χ, θ (0 to π) and ϕ (0 to 2π). What is the range of values of χ? The factor $\sin^2\chi$ appearing in the metric takes the same values for $\chi = 0$ and $\chi = \pi$, beyond which there is a repetition of the same range. Hence the total volume of the space is given by

$$V = a^3 \int_{\chi=0}^{\chi=\pi} d\chi\,\sin^2\chi \int_{\theta=0}^{\theta=\pi} d\theta\,\sin\theta \int_{\phi=0}^{\phi=2\pi} d\phi = 2\pi^2 a^3. \qquad (10.16)$$

Just as a sphere has a surface of finite area without any edges, this space also similarly has a finite volume without any bounding surface. It is easy to show that the spaces given by the metrics (10.12) or (10.15) have infinite volumes.

Let us now introduce a slightly different notation. We substitute $\chi = r$ in (10.12), $\sin\chi = r$ in (10.14) and $\sinh\chi = r$ in (10.15). It is straightforward to show that (10.12), (10.14) and (10.15) now become

$$ds^2 = a^2(dr^2 + r^2\,d\Omega^2),$$

$$ds^2 = a^2\left(\frac{dr^2}{1-r^2} + r^2\,d\Omega^2\right),$$

$$ds^2 = a^2\left(\frac{dr^2}{1+r^2} + r^2\,d\Omega^2\right).$$

These three equations can be written together in the combined compact form

$$ds^2 = a^2\left(\frac{dr^2}{1-kr^2} + r^2\,d\Omega^2\right), \qquad (10.17)$$

where k can have values 0, +1 and −1, respectively corresponding to uniform space with zero, positive and negative curvature.

So far we have considered the metric of three-dimensional space. To describe the spacetime of the Universe, we need the metric of four-dimensional spacetime. We expect (10.17) to provide the spatial part of this metric. We now have to add the time part. Special relativity can guide us how to do this. We know that the metric of special relativity is given by

$$ds^2 = -c^2dt^2 + dx^2 + dy^2 + dz^2.$$

Since $dx^2 + dy^2 + dz^2$ is the spatial part of the metric, it is clear that we get the full spacetime metric by adding $-c^2dt^2$ to it. In exactly the same fashion, we expect to get the four-dimensional spacetime metric of the Universe by adding $-c^2dt^2$ to (10.17). This gives

$$ds^2 = -c^2dt^2 + a(t)^2 \left(\frac{dr^2}{1 - kr^2} + r^2 d\Omega^2 \right). \qquad (10.18)$$

Substituting for $d\Omega^2$ from (10.13), the full four-dimensional form of the metric is given by

$$ds^2 = -c^2dt^2 + a(t)^2 \left[\frac{dr^2}{1 - kr^2} + r^2(d\theta^2 + \sin^2\theta \, d\phi^2) \right]. \qquad (10.19)$$

This is known as the *Robertson–Walker metric* (Robertson, 1935; Walker, 1936) and is the only possible form of the metric for a Universe satisfying the cosmological principle (i.e. having the spatial part uniform). The three values 0, $+1$ and -1 of k give the three possible kinds of Universe with flat, positively curved and negatively curved space. Sometimes it is useful to write the Robertson–Walker metric in terms of the variable χ used in (10.12), (10.14) and (10.15) rather than r. This is easily seen to be

$$ds^2 = -c^2dt^2 + a(t)^2 \left[d\chi^2 + S^2(\chi)(d\theta^2 + \sin^2\theta \, d\phi^2) \right], \qquad (10.20)$$

where the function $S(\chi)$ has to be χ, $\sin\chi$ or $\sinh\chi$ corresponding to the values 0, $+1$ or -1 of k.

One very important point to note is that we have written a in (10.19) and (10.20) in the form $a(t)$ to make it explicit that $a(t)$ can in general be a function of time. To understand the physical significance of this, let us look at the metric (10.4) for the surface of a sphere. There the parameter a was the radius of the sphere and would increase with time if the sphere expanded. In exactly the same spirit, we can regard $a(t)$ appearing in (10.19) as a measure of the size of the Universe. It is called the *scale factor* of the Universe. A time evolution equation of $a(t)$ will tell us how the Universe evolves with time. Such an equation can be obtained by substituting (10.19) into Einstein's equation, the basic equation of general relativity. We shall carry out this exercise in §14.1. However, we have already pointed out in §10.1 that, by a miraculous coincidence, exactly the same equation for the evolution of the Universe can be obtained from Newtonian mechanics with some suitable assumptions. We shall discuss this in the next section.

Let us now point out another useful analogy with the metric (10.4) for the spherical surface. Suppose we consider some marks on the spherical surface.

The coordinates (θ, ϕ) for any particular mark will not change with time if the sphere expands. But the marks will move away from each other because the radius of the sphere is increasing and the distance between any two marks (along the great circle connecting them) is proportional to the radius. In relativistic cosmology, we take a similar point of view that galaxies are moving from each other because the scale factor of the Universe is increasing, but the spatial coordinates (r, θ, ϕ) of a galaxy would not change (provided we neglect any motion of the galaxy with respect to the Hubble expansion). In other words, a galaxy stays put at a point in space while space is expanding. A coordinate system in which galaxies do not change their coordinates with the expansion of the Universe is called a *co-moving coordinate system*. The Robertson–Walker metric is usually assumed to be a metric corresponding to a co-moving coordinate system. Suppose we take our Galaxy to be the origin of our coordinate system and we want to find the distance of a galaxy located at (r, θ, ϕ). One way of obtaining a measure of this distance is to integrate the spatial part of ds between us and that galaxy. If we are at the origin, this integration will clearly be in the radial direction and we easily conclude from (10.19) that this measure of distance to the galaxy is given by

$$l = a(t) \int_0^r \frac{dr'}{\sqrt{1 - kr'^2}}.$$ (10.21)

We should point out that the concept of distance in general relativity involves some subtleties. We shall present an analysis of length measurement in §13.1. How different kinds of galaxy distances can be inferred from observational data will be discussed in §14.4. However, these different distances as well as the distance measure given by (10.21) converge if the redshift z of the galaxy is small compared to 1 such that the curvature of the Universe is not important within the distance to the galaxy. Taking l given by (10.21) as the distance to the galaxy, the recession velocity of the galaxy with the expansion of the Universe is obviously

$$v = \dot{a}(t) \int_0^r \frac{dr'}{\sqrt{1 - kr'^2}},$$ (10.22)

where a dot represents differentiation with t throughout this chapter. It follows from (9.13) that the Hubble constant is given by

$$H = \frac{v}{l} = \frac{\dot{a}}{a}.$$ (10.23)

We have introduced some elementary concepts of relativistic cosmology. Now we shall use only Newtonian mechanics in the rest of this chapter and see how far we can go with it. We shall find that it is possible to study the mathematical equations for the evolution of the Universe without getting into the details of general relativity. However, our treatment will lack self-consistency and will not

Table 10.2 Conceptual differences between relativistic cosmology and Newtonian cosmology.

	RELATIVISTIC COSMOLOGY	NEWTONIAN COSMOLOGY
1	All points in space are equivalent	We are at the centre of the Universe
2	Space is expanding with galaxies	Galaxies are moving away in space
3	Redshift caused by stretching of light wavelength due to expansion	Redshift caused by Doppler effect due to recession of galaxies

be satisfactory at a deep conceptual level if we shy away from general relativity. Table 10.2 lists the main differences between relativistic cosmology and the so-called Newtonian cosmology. Just as all points on a spherical surface are equivalent, all points in a uniform three-dimensional space described by the metric (10.19) are equivalent. In Newtonian cosmology, however, we shall take the point of view that we are at the centre and the Universe is expanding radially outwards with us at the centre. In the co-moving coordinate system introduced in relativistic cosmology, space is expanding and carrying the galaxies with it. On the other hand, we shall have to assume in Newtonian cosmology that galaxies are moving away in space, which is regarded as the inert background without any dynamics. Another problematic aspect of Newtonian cosmology is the interpretation of the redshift of spectral lines, which is regarded as a simple Doppler shift due to the recession of the galaxies. When the redshift z defined in (9.11) is of the order of 1 or larger (which is the case for many objects found through the most powerful telescopes), this interpretation does not make very good sense, as pointed out at the end of §9.3. When we discuss the relativistic theory of light propagation in an expanding Universe in §14.3, we shall see that the wavelength of light gets stretched with the expansion of the Universe. Thus, if a were the scale factor of the Universe when light started from a distant galaxy and if a_0 is the present scale factor (which must be larger than a for an expanding Universe), it follows from relativistic considerations that

$$1 + z = \frac{\lambda_{\text{obs}}}{\lambda_{\text{em}}} = \frac{a_0}{a}. \tag{10.24}$$

We shall see that the expressions of most of the observable quantities in cosmology will involve the ratio of scale factors rather than the scale factor itself, as in (10.24).

10.4 Friedmann equation for the scale factor

We now want to derive an equation describing how the scale factor $a(t)$ appearing in (10.19) evolves with time. We shall use some simple considerations

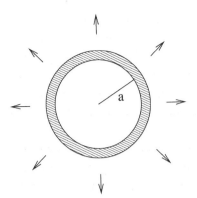

Fig. 10.3 Sketch of a spherical shell in a region of spherical expansion.

of Newtonian mechanics. As we already pointed out, we have to assume the Universe to be spherically symmetric around us located at the centre of a uniform expansion, as sketched in Figure 10.3, and the equation we shall get miraculously turns out to be the same as what one gets by putting the Robertson–Walker metric in Einstein's equation.

Let us consider a spherical shell of radius a indicated in Figure 10.3. The kinetic energy per unit mass of the shell is $\frac{1}{2}\dot{a}^2$ and the potential energy per unit mass is $-GM/a$, where M is the mass enclosed within this shell. Assuming a uniform density ρ, the potential energy turns out to be

$$-\frac{\frac{4}{3}\pi G\rho a^3}{a} = -\frac{4}{3}\pi G\rho a^2$$

so that the total mechanical energy E, which is a constant of motion, is given by

$$E = \frac{1}{2}\dot{a}^2 - \frac{4}{3}\pi G\rho a^2. \tag{10.25}$$

If we know how ρ depends on a, then this equation can be solved to find how a will evolve with time. As we shall see in §14.1, on substituting the Robertson–Walker metric into Einstein's equation, we get essentially the same equation as (10.25) with E given by

$$E = -\frac{kc^2}{2}, \tag{10.26}$$

where k is the same k that appears in the Robertson–Walker metric (10.19) and can have values $+1$, -1 or 0. On substituting (10.26) in (10.25), we get

$$\frac{\dot{a}^2}{a^2} + \frac{kc^2}{a^2} = \frac{8\pi G}{3}\rho, \tag{10.27}$$

which is known as the *Friedmann equation* (Friedmann, 1924).

It may be pointed out that Einstein's equation allows the possibility of an extra term (called the *cosmological constant*), which causes an extra acceleration of the Universe (Einstein, 1917). As we shall discuss in §14.5, the recent redshift data of distant supernovae indicate that the Universe may be accelerating, which suggests that the Friedmann equation (10.27) may not be the complete equation and an additional cosmological constant term may be present. Even if that is true, for purely pedagogical purposes, it is useful to study the consequences of (10.27) before getting into a discussion of the effects of the cosmological constant. All the discussions in this chapter will assume a zero cosmological constant and will be based on (10.27). In Chapter 14 we shall discuss the modifications of the theory necessitated by the cosmological constant. Since the cosmological constant was believed to be zero until a few years ago, most of the standard textbooks of cosmology written till about 2000 assumed (10.27) to be the complete equation and discussed its solutions. We shall see in §14.2 that the cosmological constant becomes more dominant as the Universe becomes older. It appears that the Universe is right now passing through the phase when the cosmological constant term has become as large as the other terms in the equation. For a study of the Universe when it was young, (10.27) is more than adequate.

For a projectile moving against gravity, we know that a positive total energy E would imply that it will move forever and escape to infinity, whereas a negative energy implies that it will eventually fall back due to the attraction of gravity. We expect similar considerations to hold here also. Noting from (10.26) that k has a sign opposite of E, we can at once draw a very important conclusion. If $k = -1$, then the Universe will expand forever. On the other hand, if $k = +1$, then the expansion of the Universe will eventually be halted, making the Universe fall back and collapse (provided the cosmological constant is zero). Such a Universe will last for a finite time before it ends in a big crunch. We have already seen in §10.3 that a Universe with positive curvature (i.e. $k = +1$) has a finite volume, whereas a Universe with negative curvature (i.e. $k = -1$) is infinite. This leads to an interesting statement. A Universe finite in space (with $k = +1$) should last for a finite time. On the other hand, a Universe infinite in space (with $k = -1$) will last for infinite time.

Whether the Universe will expand forever or not must depend on the density of the Universe, which determines the strength of the gravitational attraction. The value of density for which the Universe will lie exactly on the borderline between these two possibilities (with $k = 0$) is called the *critical density* and is denoted by ρ_c. On putting $k = 0$ in (10.27) and using (10.23), the critical density is given by

$$\rho_c = \frac{3H^2}{8\pi G}. \tag{10.28}$$

On substituting the present-day value of the Hubble constant as given by (9.17), the present-day value of the critical density turns out to be

$$\rho_{c,0} = 1.88 \times 10^{-26} h^2 \text{ kg m}^{-3}. \tag{10.29}$$

If the average density of our present Universe is less than this, then it should expand forever. On the other hand, a higher average density is expected to make the Universe eventually fall back. We shall come to the question of the value of average density of the Universe in the next section where we discuss the contents of the Universe. The ratio of the density to the critical density is called the *density parameter* and is denoted by Ω, i.e.

$$\Omega = \frac{\rho}{\rho_c}. \tag{10.30}$$

On making use of (10.23), (10.28) and (10.30), we can write (10.27) in the form

$$\frac{kc^2}{a^2 H^2} = \Omega - 1. \tag{10.31}$$

It should be noted that a, H and Ω all evolve with time. Their values at any particular epoch t have to be related by (10.31). Their values at the present epoch are denoted by a_0, H_0 and Ω_0. It should be clear from (10.30) and (10.31) that $k = 0$, $k = +1$ and $k = -1$ respectively correspond to the cases $\rho = \rho_c$, $\rho > \rho_c$ and $\rho < \rho_c$. This is consistent with the conclusion we have already drawn that a Universe with $k = +1$ should eventually fall back, whereas a Universe with $k = -1$ should expand forever. Only when the density is more than the critical density ρ_c, will the gravitational force be strong enough to pull back the Universe eventually ($k = +1$ case). On the other hand, a density less than the critical density ρ_c corresponds to a weak gravitational pull that cannot halt the expansion of the Universe ($k = -1$ case).

In the Newtonian expression (10.25) of the spherical shell, it is possible for E to have any real value. However, it follows from (10.26) that general relativistic considerations constrain E to have only three values corresponding to the three values of k. This may seem surprising at first sight. It is to be noted that the variables like a, $\dot{a}$ and ρ appearing in (10.27) can have continuous possible values. The three possible values of k basically force a, $\dot{a}$ and ρ to satisfy three possible relationships amongst themselves which follow from (10.27) on substituting the values of k. Since it is the mass-energy which creates the curvature of spacetime in general relativity, we do expect such relationships. On the other hand, when we apply purely Newtonian considerations to the spherical shell of Figure 10.3, its radius a does not have to be related to the density ρ. We could consider spherical shells of different radii and write down equations of the form (10.25) for them. The value of E can be different for them. In relativistic cosmology, on the other hand, a is like the radius of curvature of the Universe and general relativity certainly imposes some extra constraints which would not be present in the Newtonian formulation of the expanding shell. While Newtonian considerations lead us to the equation (10.25) having the same form as what we get from general relativity, these subtle differences should be kept in mind.

We shall discuss in the next section how ρ varies with a. Here we merely point out how we proceed to solve (10.27) when we have ρ as a function of a. It is particularly easy to handle (10.27) when $k = 0$. Since $\dot{a}$ has to be positive for an expanding Universe, we get

$$\dot{a} = \sqrt{\frac{8\pi G \rho}{3}} a. \tag{10.32}$$

This is very easy to integrate when we have ρ as a function of a. When $k = \pm 1$, it is useful to change from t to another time-like variable η defined through

$$c\,dt = a\,d\eta. \tag{10.33}$$

On using this variable η, the Robertson–Walker metric (10.19) would have the form

$$ds^2 = a(\eta)^2 \left[-d\eta^2 + \frac{dr^2}{1 - kr^2} + r^2(d\theta^2 + \sin^2\theta\,d\phi^2) \right]. \tag{10.34}$$

Keeping in mind that a dot denotes a differentiation with respect to t, a differentiation with respect to η has to be indicated explicitly. We find from (10.33) that

$$\dot{a} = \frac{c}{a}\frac{da}{d\eta}.$$

Substituting in (10.27), we get

$$\frac{c^2}{a^4}\left(\frac{da}{d\eta}\right)^2 + \frac{kc^2}{a^2} = \frac{8\pi G}{3}\rho,$$

from which

$$\frac{da}{d\eta} = \pm\sqrt{\frac{8\pi G}{3c^2}\rho a^4 - ka^2}.$$

This can be put in the form of a quadrature

$$\eta = \pm \int \frac{da}{\sqrt{\frac{8\pi G}{3c^2}\rho a^4 - ka^2}}. \tag{10.35}$$

Once ρ is given as a function of a, one can work out this quadrature to find how a varies with the time-like variable η. If one is interested in determining the variation of a with t, then it is further necessary to relate η to t by solving (10.33) after obtaining a as a function of η. We shall carry out some calculations of this type explicitly in §10.6.

10.5   Contents of the Universe. The cosmic blackbody radiation

As pointed out in §10.4, we need to specify how the density ρ of the Universe varies with the scale factor a in order to solve for a as a function of time. We have to consider the contents of the Universe for this purpose. Before we get into the discussion of the specific contents, let us consider a fluid filling the Universe with density ρ and equivalent energy density ρc^2. Since ρc^2 has the dimension of pressure, we can write the pressure due to this fluid as

$$P = w\rho c^2. \tag{10.36}$$

We consider a volume a^3 of the Universe which is increasing with time. The total internal energy inside this volume is $\rho c^2 a^3$. Assuming the expansion to be adiabatic, the first law of thermodynamics $dQ = dU + P\,dV$ leads to

$$d(\rho c^2 a^3) + w\rho c^2\, d(a^3) = 0,$$

from which it follows that

$$\rho \propto \frac{1}{a^{3(1+w)}}. \tag{10.37}$$

If we know the appropriate value of w appearing in (10.36) for a particular component of the Universe, (10.37) tells us how the density of that component will vary with a. One standard result of the kinetic theory of gases is that the pressure of a gas is given by

$$P = \frac{1}{3}\rho \overline{v^2},$$

where v is the molecular velocity (see, for example, Saha and Srivastava, 1965, §3.12). Comparing with (10.36), we find that

$$w = \frac{1}{3}\frac{\overline{v^2}}{c^2}. \tag{10.38}$$

For a non-relativistic gas, we have

$$w \approx 0,\ \rho \propto \frac{1}{a^3}, \tag{10.39}$$

whereas for a gas of relativistic particles all moving around with speed c,

$$w \approx \frac{1}{3},\ \rho \propto \frac{1}{a^4}. \tag{10.40}$$

Although matter in the Universe is distributed in a hierarchy of structures, we pointed out in §9.6 that the matter distribution starts looking homogeneous when we go to scales larger than about $100h^{-1}$ Mpc, in accordance with the cosmological principle. If the luminous stars constituted all the matter in the

Universe, then a careful analysis of observational data indicates that the density parameter Ω defined in (10.30) would be of order

$$\Omega_{\text{Lum}} \approx 0.01. \qquad (10.41)$$

However, we pointed out in §9.2.2 that rotation curves of galaxies suggest a significant amount of dark matter beyond the stellar disks of galaxies. Then we discussed in §9.5 that the application of the virial theorem to galaxy clusters suggests even larger amounts of dark matter. The density parameter estimated from the virial masses of galaxy clusters turns out to be independent of the uncertainties in the Hubble constant (see Exercise 10.1) and is of order

$$\Omega_{\text{M},0} \approx 0.3. \qquad (10.42)$$

The subscript 0 implies that this is the present value of the density parameter due to matter, which can have different values at other epochs. We shall discuss other independent arguments in §14.5 that $\Omega_{\text{M},0}$ indeed has this value. From (10.29) and (10.30), the present matter density should be

$$\rho_{\text{M},0} = 1.88 \times 10^{-26} \Omega_{\text{M},0} h^2 \text{ kg m}^{-3}. \qquad (10.43)$$

Using (10.39), we can now write down the matter density at an arbitrary epoch in the form

$$\rho_{\text{M}} = \rho_{\text{M},0} \left(\frac{a_0}{a} \right)^3, \qquad (10.44)$$

where a_0 is the value of the scale factor at the present epoch and a its value at that arbitrary epoch.

Hubble's law (9.13) implies that all the galaxies were on top of each other at a certain epoch, often called the *Big Bang*. In other words, the physical parameters of the Universe like its density and its temperature were infinite at that epoch, according to straightforward theoretical considerations, so that it is not possible to extrapolate our currently understood physical laws to times earlier than the epoch of the Big Bang. We know that hot matter emits radiation. Since the early Universe must have been dense and hot, it would have been filled with radiation existing in thermodynamic equilibrium with matter. As we discussed in §2.2.4 and §2.3, radiation in equilibrium with matter has to be blackbody radiation. As the Universe expanded and its density fell, at some stage the Universe became transparent to radiation and the radiation ceased to be in equilibrium with matter. We shall discuss this decoupling of radiation and matter in greater depth in §11.7. As the Universe kept expanding after this decoupling, the radiation would undergo adiabatic expansion because it ceases to interact with matter any more. One of the important results from the thermodynamics of blackbody radiation (see, for example, Saha and Srivastava, 1965, §15.25) is that blackbody radiation continues to remain blackbody radiation under adiabatic expansion, although its temperature keeps decreasing with expansion. As we shall show in §14.3, the general relativistic analysis of light propagation in the expanding Universe also leads to the same conclusion that the radiation continues to remain blackbody radiation even though it is not interacting with matter any more. We thus expect on theoretical grounds that the Universe should still be filled with a blackbody radiation background, which will be cooling with the expansion of the Universe. Alpher and Herman (1948) were the first to point this out and predicted that the present temperature of this blackbody radiation should be of order 10 K. Without being aware of this theoretical prediction, Penzias and Wilson (1965) accidentally discovered this radiation which seemed to have a temperature of 3 K. Since much of this blackbody radiation at 3 K lies in the microwave range, it is called the *cosmic microwave background radiation*, abbreviated as *CMBR*.

The discovery of CMBR was perhaps the most important milestone in the development of observational cosmology after the discovery of Hubble's law. Lemaitre (1927) was the first person to argue that the Universe must have begun from a hot Big Bang. The CMBR is a remnant of this Big Bang and its existence provided reasonably compelling proof (at least compelling enough to most astrophysicists) that there was really a hot Big Bang from which the Universe was born. Penzias and Wilson (1965) were able to measure only a small part of the CMBR spectrum. The satellite COBE (Cosmic Background Explorer) was launched in 1989 to study the CMBR in detail. Figure 10.4 shows the spectrum of CMBR measured by COBE, as reported by Mather *et al.*

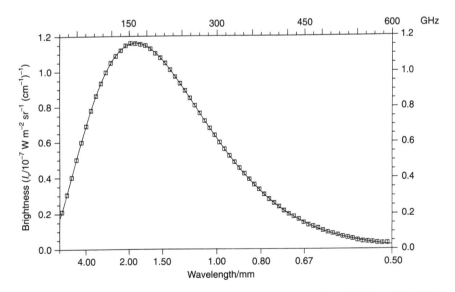

Fig. 10.4 The spectrum of cosmic microwave background radiation (CMBR) as obtained by COBE. From Mather *et al.* (1990). (©American Astronomical Society. Reproduced with permission from *Astrophysical Journal*.)

(1990). It is a fantastic fit to the Planck spectrum (2.1) for blackbody radiation at temperature

$$T_0 = 2.735 \pm 0.06 \text{ K}. \tag{10.45}$$

We know that the energy density of blackbody radiation at temperature T is given by $a_B T_0^4$, where a_B is Stefan's constant (see Exercise 2.1). Hence the contribution of CMBR to the density of the Universe is

$$\rho_\gamma = \frac{a_B}{c^2} T^4. \tag{10.46}$$

We shall discuss in §11.4 that the Universe is expected to have a background of neutrinos in addition to the background of photons in the CMBR. The detailed properties of the neutrino background will be worked out in §11.4. For the time being, let us only mention that the total density of relativistic background particles (photons and neutrinos together) at the present epoch will be shown to be

$$\rho_{R,0} = 1.68\rho_{\gamma,0}, \tag{10.47}$$

where $\rho_{\gamma,0}$ is the present-day contribution of CMBR to the density of the Universe. Since the CMBR photons and the neutrinos are both relativistic gases, we expect (10.40) to hold for both so that we can write

$$\rho_R = \rho_{R,0} \left(\frac{a_0}{a}\right)^4. \tag{10.48}$$

Since the CMBR density ρ_γ separately would also fall as a^{-4}, it easily follows from (10.46) that

$$T \propto \frac{1}{a}, \tag{10.49}$$

which tells us how the temperature of the CMBR must be falling with the expansion of the Universe. We shall show in §11.4 that the neutrino temperature also should fall in the same way, although its value will be different from the value of CMBR temperature at the same instant.

By summing up (10.44) and (10.48), the total density of the Universe can be written as

$$\rho = \rho_{M,0} \left(\frac{a_0}{a}\right)^3 + \rho_{R,0} \left(\frac{a_0}{a}\right)^4. \tag{10.50}$$

In the following discussion, we shall refer to both photons and neutrinos as 'radiation'. Some obvious conclusions can be drawn from the expression (10.50) for the density. The radiation density (falling as a^{-4}) falls more rapidly than the matter density (falling as a^{-3}) with the expansion of the Universe. Going backwards in time, as we approach the Big Bang closer and closer, the radiation density must be rising faster than the matter density with increasingly smaller a. At the present time, we have

$$\rho_{M,0} \gg \rho_{R,0}.$$

However, there must be a past epoch when the matter and radiation densities were equal, before which the radiation was dominant. If a_{eq} was the value of the scale factor at that epoch of *matter-radiation equality*, on equating (10.44) and (10.48) we get

$$\frac{a_0}{a_{eq}} = \frac{\rho_{M,0}}{\rho_{R,0}} = \frac{\rho_{M,0}c^2}{1.68 a_B T_0^4} \tag{10.51}$$

where we have made use of (10.46) and (10.47). Substituting from (10.43) and (10.45), we get

$$\frac{a_0}{a_{eq}} = 2.3 \times 10^4 \Omega_{M,0} h^2. \tag{10.52}$$

We can divide the history of the Universe into two distinct periods. At times earlier than the matter-radiation equality when the scale factor a was smaller than a_{eq} given by (10.52), the Universe is said to be *radiation-dominated*. On the other hand, after the matter-radiation equality the Universe has become *matter-dominated*.

To study the evolution of the Universe, we now need to solve the Friedmann equation (10.27) after substituting for ρ from (10.50). The calculations, however, become much simpler without introducing any significant error if we assume $\rho = \rho_M$ given by (10.44) when we study the matter-dominated period of the Universe and assume $\rho = \rho_R$ given by (10.48) when we study the radiation-dominated period of the Universe. It is possible to get analytical solutions in these simpler cases. The next two sections will be devoted to solving the Friedmann equation for the matter-dominated and the radiation-dominated Universe respectively.

10.6   The evolution of the matter-dominated Universe

We shall now solve the Friedmann equation (10.27) by taking $\rho = \rho_M$ given by (10.44) appropriate for the matter-dominated Universe. We need to consider the three cases $k = -1, 0, +1$.

Let us first consider the case $k = 0$, for which the Friedmann equation leads to (10.32). On substituting from (10.44) in (10.32), we have

$$\dot{a} = \sqrt{\frac{8\pi G \rho_{M,0} a_0^3}{3}} a^{-1/2},$$

of which the solution is

$$\frac{2}{3}a^{\frac{3}{2}} = \sqrt{\frac{8\pi G\rho_{M,0}a_0^3}{3}}\,t \tag{10.53}$$

on setting $t = 0$ at the epoch of the Big Bang when $a = 0$. For the $k = 0$ case, the density is equal to the critical density given by (10.28) so that

$$\rho_{M,0} = \frac{3H_0^2}{8\pi G}.$$

On substituting this in (10.53), we get

$$\frac{a}{a_0} = \left(\frac{3}{2}H_0t\right)^{2/3}. \tag{10.54}$$

We thus reach the very important conclusion that the size of the Universe increases with time as $t^{2/3}$. This solution for the $k = 0$ case is often called the *Einstein–de Sitter model*, since it was studied by Einstein and de Sitter (1932).

For the cases $k = \pm 1$, we use the quadrature formula (10.35) obtained from the Friedmann equation (10.27). On substituting (10.44) into (10.35), we get

$$\eta = \pm \int \frac{da}{\sqrt{\frac{8\pi G\rho_{M,0}a_0^3}{3c^2}a - ka^2}}. \tag{10.55}$$

10.6.1   The closed solution ($k=+1$)

As already discussed in §10.4 and as should be clear from (10.31), this is the case where the density is larger than the critical density, i.e. $\Omega_{M,0} > 1$. When $k = +1$, the quadrature (10.55) can be worked out to give

$$a = \frac{4\pi G}{3c^2}\rho_{M,0}a_0^3(1 - \cos\eta).$$

This can be written as

$$\frac{a}{a_0} = \frac{1}{2}\left(\frac{8\pi G\rho_{M,0}}{3H_0^2}\right)\frac{a_0^2 H_0^2}{c^2}(1 - \cos\eta).$$

On making use of (10.28), (10.30) and (10.31), it becomes

$$\frac{a}{a_0} = \frac{\Omega_{M,0}}{2(\Omega_{M,0} - 1)}(1 - \cos\eta), \tag{10.56}$$

where $\Omega_{M,0} = \rho_{M,0}/\rho_{c,0}$ is the density parameter at the present epoch due to matter. The evolution of a as a function of the time-like variable η is given by (10.56). In order to bring in t, we use (10.33) which gives

$$t = \frac{1}{c} \int a \, d\eta = \frac{\Omega_{M,0} a_0}{2c(\Omega_{M,0} - 1)}(\eta - \sin \eta)$$

on substituting from (10.56) for a. Multiplying by H_0 and making use of (10.31), we get

$$H_0 t = \frac{\Omega_{M,0}}{2(\Omega_{M,0} - 1)^{3/2}}(\eta - \sin \eta). \qquad (10.57)$$

The two equations (10.56) and (10.57) together give an implicit solution of a as a function of t. It follows from (10.56) that a goes to zero when η increases to 2π. In other words, this is a solution which corresponds to a Universe eventually ending up in a big crunch. We have already used arguments based on simple Newtonian mechanics in §10.4 to conclude that a Universe with $k = +1$, which has finite volume and has density more than the critical density, should eventually collapse. This is now explicitly seen in the solution. Figure 10.5 shows a/a_0 as a function of t.

10.6.2   The open solution ($k = -1$)

This is the case corresponding to $\Omega_{M,0} < 1$. With $k = -1$, instead of (10.56) and (10.57), we have

$$\frac{a}{a_0} = \frac{\Omega_{M,0}}{2(1 - \Omega_{M,0})}(\cosh \eta - 1), \qquad (10.58)$$

$$H_0 t = \frac{\Omega_{M,0}}{2(1 - \Omega_{M,0})^{3/2}}(\sinh \eta - \eta). \qquad (10.59)$$

Again these two equations together give an implicit solution of a as a function of t. This solution of a/a_0 as a function of t is also plotted in Figure 10.5,

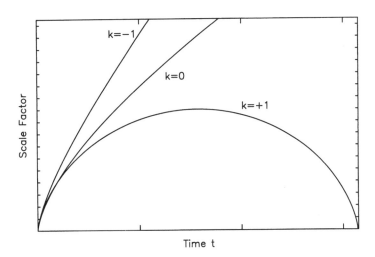

Fig. 10.5 The evolution of the scale factor a/a_0 as a function of the time t for the cases $k = -1, 0, +1$.

which shows the solution for the $k = 0$ case as well. It is seen that the solution for $k = -1$ increases forever at a rate faster than the rate of increase of the critical solution $k = 0$. We had already anticipated in §10.4 that the $k = -1$ solution, which corresponds to an infinite Universe with density less than the critical density, will expand forever. The explicit solution now confirms this.

As we pointed out in §10.5, it appears that $\Omega_{M,0}$ is about 0.3. It would then seem that the open solution is the appropriate solution for our Universe. However, so far in our discussion, we have not included the cosmological constant which would contribute another term to the Friedmann equation (10.27). In §14.5 we shall discuss the observational evidence that this cosmological constant may actually be non-zero, leading to an acceleration of the Universe. The cosmological constant term, however, becomes more important with time. At the present epoch, this term is comparable to the matter density term. At earlier epochs, the cosmological term was less important. So the open solution is presumably the appropriate solution for the evolution of the Universe at early epochs and can be used up to the present epoch in many calculations without introducing too much error. We shall show explicitly in §14.2 that the solution with non-zero cosmological constant (given by (14.20)) reduces at early times to the approximate solution we are going to discuss now.

10.6.3 Approximate solution for early epochs

We see in Figure 10.5 that the solutions for all the three values of k behave very similarly at sufficiently early times. We can simplify the solutions for $k = \pm 1$ when $\eta \ll 1$. Both (10.56) and (10.58) reduce to

$$\frac{a}{a_0} \approx \frac{\Omega_{M,0}}{2|1 - \Omega_{M,0}|} \frac{\eta^2}{2}$$

when η is small. Similarly both (10.57) and (10.59) reduce to

$$H_0 t \approx \frac{\Omega_{M,0}}{2|1 - \Omega_{M,0}|^{3/2}} \frac{\eta^3}{6}.$$

Eliminating η between these two equations, we get

$$\frac{a}{a_0} \approx \left(\frac{3}{2}\Omega_{M,0}^{1/2} H_0 t\right)^{2/3}. \tag{10.60}$$

This reduces to the critical solution (10.54) when the density parameter $\Omega_{M,0}$ is 1. Although (10.60) may not be strictly true at the present epoch, it is still very useful in many quick calculations. Since $a = a_0$ at the present epoch $t = t_0$, it follows from (10.60) that

$$t_0 \approx \frac{2}{3} H_0^{-1} \Omega_{M,0}^{-1/2}. \tag{10.61}$$

Then (10.60) can be written as

$$\frac{a}{a_0} \approx \left(\frac{t}{t_0}\right)^{2/3}. \tag{10.62}$$

It may be pointed out that the ratio a/a_0 is a more observationally relevant quantity than the scale factor a itself, since this ratio is related to the redshift z through (10.24). If a source is at redshift z, then light from it started at time t to reach us at the present epoch t_0. The relation between z and t can be obtained by combining (10.24) with (10.62):

$$\frac{t}{t_0} \approx (1+z)^{-3/2}. \tag{10.63}$$

When we look at a galaxy at redshift $z = 1$, we essentially see the galaxy as it existed when the age of the Universe was $2^{-3/2}$ times its present age.

As we already discussed, the very early Universe was radiation-dominated. So (10.60), based on the assumption that the Universe is matter-dominated, should not hold at those early times. However, the Universe became matter-dominated fairly early. From (10.52) and (10.62), the epoch t_{eq} of matter-radiation equality is given by

$$t_{eq} = 2.9 \times 10^{-7} \Omega_{M,0}^{-3/2} h^{-3} t_0. \tag{10.64}$$

Since t_0 is believed to be of the order of 10^{10} years, the Universe would have become matter-dominated a few thousand years after the Big Bang. From then onwards, equations like (10.60) and (10.62) should hold till the present time.

10.6.4   The age of the Universe

An approximate expression for the age of the Universe is given in (10.61), where we see that the age is shorter if $\Omega_{M,0}$ is larger. One can understand this result quite easily. A larger $\Omega_{M,0}$ means a stronger deceleration, which implies that the expansion rate of the early Universe would have been faster if $\Omega_{M,0}$ was larger. This faster expansion leads to a shorter age. Instead of using the approximate expression (10.61), one can easily find out the exact value of the age t_0 for a given $\Omega_{M,0}$. Since $a = a_0$ at the present epoch, the age t_0 of the Universe is given by the value of t which makes $a = a_0$. If $\Omega_{M,0} > 1$, then the numerical value of $H_0 t_0$ can be found from (10.56) and (10.57). On the other hand, we have to use (10.58) and (10.59) if $\Omega_{M,0} < 1$. Figure 10.6 plots the numerical values of $H_0 t_0$ as a function of $\Omega_{M,0}$. The expansion rate of the Universe would have

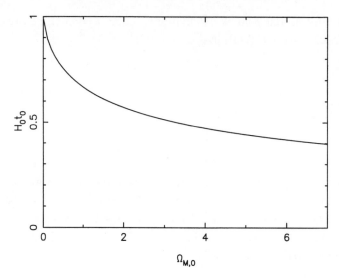

Fig. 10.6 The plot of $H_0 t_0$ against $\Omega_{M,0}$, showing how the age t_0 of the Universe depends on $\Omega_{M,0}$.

been unchanged if $\Omega_{M,0}$ were equal to zero, making t_0 equal to the Hubble time H_0^{-1}. We see in Figure 10.6 that t_0 becomes a smaller and smaller fraction of the Hubble time as $\Omega_{M,0}$ is increased, giving the value $(2/3)H_0^{-1}$ when $\Omega_{M,0} = 1$, in accordance with (10.61).

10.7 The evolution of the radiation-dominated Universe

We shall now discuss how the Universe evolved at the early times before t_{eq} given by (10.64) when the Universe was dominated by radiation. In the case of the matter-dominated Universe, it should be clear from Figure 10.5 that the solutions for $k = -1, 0, +1$ converge at sufficiently early times. The reason for this is not difficult to understand. It is the curvature term kc^2/a^2 in the Friedmann equation (10.27) which is responsible for making the three solutions different. At sufficiently early times, this term becomes negligible compared to the mass density term which goes as a^{-3} and becomes more dominant when a is sufficiently small. During the radiation-dominated epoch, the radiation density term going as a^{-4} becomes even more dominant and we can ignore the curvature term kc^2/a^2, so that (10.32) obtained by putting $k = 0$ in the Friedmann equation (10.27) is applicable. On substituting $\rho = \rho_R$ as given by (10.48), we obtain from (10.32) that

$$a\dot{a} = \sqrt{\frac{8\pi G \rho_{R,0}}{3} a_0^2},$$

of which the solution is

$$\frac{a}{a_0} = \left(\frac{32\pi G\rho_{R,0}}{3}\right)^{1/4} t^{1/2}. \tag{10.65}$$

The radiation-dominated Universe expanded with time as $t^{1/2}$ in contrast to the early matter-dominated Universe which expanded as $t^{2/3}$ according to (10.62).

We have pointed out in (10.49) that the temperature of the CMBR falls as the inverse of a. So we expect

$$\frac{a}{a_0} = \frac{T_0}{T}.$$

Using this and substituting for $\rho_{R,0}$ from (10.46), we get from (10.65) that

$$T = \left(\frac{3c^2}{32\pi Ga_B}\right)^{1/4} t^{-1/2}. \tag{10.66}$$

On substituting the values of c, G and a_B, we obtain

$$T \text{ (in K)} = \frac{1.52 \times 10^{10}}{\sqrt{t}}. \tag{10.67}$$

Here T is in kelvin as indicated and t has to be in seconds. It may be noted that (10.66) and (10.67) are derived by assuming that photons were the only relativistic particles in the early Universe. As we shall see in Chapter 11, there were other relativistic particles and these equations have to be suitably modified in more accurate calculations.

A typical photon in blackbody radiation at temperature T has energy $E = \kappa_B T$. It can be easily shown that an energy E in eV corresponds to temperature

$$T = 1.16 \times 10^4 E \text{ (in eV)}. \tag{10.68}$$

It is sometimes useful to express temperature in units of energy like eV. If we do this, then (10.67) becomes

$$T \text{ (in MeV)} = \frac{1.31}{\sqrt{t}}. \tag{10.69}$$

This is a very important equation which gives an indication of the typical energy a photon (or any other kind of particle) would have at time t after the Big Bang. We shall make extensive use of (10.69) in the next chapter.

Since we have neglected the curvature term kc^2/a^2, we now argue that this term really becomes negligible compared to the other terms in the Friedmann equation (10.27) as we go close to the Big Bang. It follows from (10.31) that

$$|\Omega - 1| = \frac{c^2}{a^2 H^2} = \frac{c^2}{\dot{a}^2}$$

on using (10.23). Since a goes as $t^{1/2}$, we find from this that

$$|\Omega - 1| \propto t. \tag{10.70}$$

In other words, as $t \to 0$, the density parameter Ω approaches 1 arbitrarily closely and the curvature term becomes totally insignificant compared to other terms, in spite of the fact that the curvature term kc^2/a^2 by itself becomes infinite as a goes to zero.

Exercises

10.1 Since the recession velocity of a galaxy can be determined reasonably accurately from the redshift in its spectrum, an application of Hubble's law to estimate distance makes the estimated distance uncertain as h^{-1} if the value of the Hubble constant is uncertain. If the average density of the Universe is determined by estimating the masses of galaxy clusters by the application of the virial theorem, show that the density parameter Ω determined therefrom will be independent of the uncertainties in the Hubble constant.

10.2 Consider quasars with redshifts $z = 1$ and $z = 4$. We want to find out the age of the Universe (as a fraction of its present age) when light started from these quasars. First do the straightforward estimates for the case $\Omega_{M,0} = 1$. Then estimate the percentile errors you would make if it turned out that $\Omega_{M,0} = 0.5$ and $\Omega_{M,0} = 1.5$.

10.3 Assuming the Universe to be matter-dominated, numerically calculate the age of the Universe for different values of $\Omega_{M,0}$. Make the plot to produce Figure 10.6.

10.4 Suppose the Universe had only radiation and no matter. Solve the Friedmann equation for the values 0, $+1$ and -1 of k. Make plots of the scale factor a as functions of time for all the three cases.

10.5 According to the steady state theory of cosmology (Bondi and Gold, 1948; Hoyle, 1948), the average density of the Universe remains constant due to continuous creation of matter as the Universe expands. Using (9.17) and (10.43), estimate the number of hydrogen atoms which, on an average, have to be created per year in a volume of 1 km^3 to keep the average density of the Universe constant.

11

The thermal history of
the Universe

11.1 Setting the time table

The present uniform expansion of the Universe suggests that there was an epoch in the past when the Universe was in a singular state with infinite density. Since most of the known laws of physics become inapplicable to such a singular state, we cannot extrapolate to earlier times before this epoch of singularity. We therefore concern ourselves only with what happened after this epoch of singularity, which is called the *Big Bang*. In the solutions discussed in §10.6 and §10.7, the time t was measured from the Big Bang.

The spacetime dynamics discussed in Chapter 10 sets the stage of the Universe. Now we shall look at the *dramatis personae* who were involved in the grand drama which unfolded and is still unfolding against this background stage of spacetime. How the temperature of the early Universe varied with time is given by (10.67) and (10.69). At times earlier than 1 s after the Big Bang, typical photons had energies somewhat larger than 1 MeV. Since such photons are known to produce electron-positron pairs, the Universe at these early times must have been full of electrons and positrons which would have been approximately as abundant as photons. When photons had energies larger than 2 GeV at still earlier times, they would have given rise to proton-antiproton pairs and neutron-antineutron pairs along with pairs of many other particles listed in elementary particle physics textbooks and their antiparticles. At even higher energies existing at still earlier times, the quarks making up the elementary particles would have been free. The Universe at this very early time consisted of quarks and leptons along with their antiparticles and the bosons which mediate the various fundamental interactions. One often says that the early Universe was the best laboratory of particle physics that ever existed. However, if all the records of a laboratory have completely disappeared, then that laboratory would not be of great interest to most of us today even if it was the best laboratory once

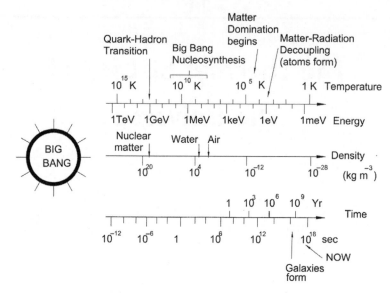

Fig. 11.1 Landmarks in the thermal history of the Universe. Adapted from Kolb and Turner (1990, p. 73).

upon a time! We shall see in this chapter that many things which we observe now have connections with what happened in the early Universe.

Figure 11.1 shows a timetable with time plotted in a logarithmic scale along with temperature in K and in eV. When the temperatures were higher than 1 GeV, we enter the particle physics era. Since theoretical considerations of these very early times are often of rather speculative nature and do not have much direct connections with observational data, we shall not discuss much about times earlier than the particle physics era in this elementary textbook. Typical nuclear reactions involve energies of order MeV. During the epoch from about $t = 1$ s to $t = 10^2$ s, particles in the Universe had energies appropriate for nuclear reactions and various kinds of nuclear reactions must have taken place. We shall discuss the issue of primordial nucleosynthesis at some detail later in this chapter. It follows from (10.64) that the Universe was a few thousand years old when it changed from being radiation-dominated to being matter-dominated. This important epoch is indicated in Figure 11.1. Another landmark epoch came a little bit later when T fell to about 1 eV and formation of atoms took place. Since typical ionization energies are of order eV, we would not expect atoms when T was larger than this and matter must have existed in the form of free electrons and bare nuclei. Although (10.69) is no longer valid at the epoch of atom formation when the Universe is already matter-dominated, we can get an approximate value of time t for this epoch by setting $T = 1$ eV in (10.69), which gives $t \approx 5 \times 10^4$ yr as the epoch of atom formation. There was a very important consequence of the formation of atoms. Radiation interacted with matter before this epoch primarily through Thomson scattering discussed in §2.6.1. When the atoms formed, all the electrons got locked inside them and there ceased to be any free electrons available to produce Thomson scattering. So radiation got decoupled from matter and the Universe suddenly became transparent to radiation. We shall discuss these things in §11.7. The aim of this chapter is to elaborate on the history of the Universe summarized in Figure 11.1 and to connect it to the present-day observations.

11.2 Thermodynamic equilibrium

Let us consider a reaction in which A, B, C, ...combine to produce L, M, N, ... :

$$A + B + C + \cdots \longleftrightarrow L + M + N + \cdots, \qquad (11.1)$$

which can proceed in both directions. Under normal circumstances, we expect this reaction to reach a *chemical equilibrium* when the concentrations of A, B, C, ..., L, M, N, ...are such that the backward rate balances the forward rate and the concentrations do not change any more. It may be noted that we shall be using the term chemical equilibrium even if (11.1) is strictly not a 'chemical' reaction, i.e. it can be a nuclear reaction involving nuclei. The condition for the chemical equilibrium of reaction (11.1) is

$$\mu_A + \mu_B + \mu_C + \cdots = \mu_L + \mu_M + \mu_N + \cdots, \qquad (11.2)$$

where the μ-s are the chemical potentials. This is a very well-known condition discussed in any standard textbook covering this topic (see, for example, Reif, 1965, §8.9) and we shall not get into a detailed discussion of this condition here.

Various kinds of nuclear and particle reactions were possible in the early Universe. The fundamental question we want to ask now is whether these reactions could reach chemical equilibrium. As the Universe expanded at the rate $H = \dot{a}/a$, the condition for chemical equilibrium kept changing. Only if a reaction proceeded at a rate faster than this expansion rate, can we expect the reaction to reach equilibrium. Suppose Γ is the interaction rate per particle, i.e. it is the number of interactions a particle is expected to have in unit time. Within time $\sim 1/\Gamma$, most of the particles will have one interaction and we may expect the system to reach equilibrium if it starts from a state not too far from equilibrium. If $\Gamma \gg H$, then the reaction would be able to reach chemical equilibrium and, as the Universe expanded, it would evolve through successive states in chemical equilibrium appropriate to the physical conditions at the successive instants of time. On the other hand, if $\Gamma \ll H$, then the reaction would not proceed fast enough to change the concentrations of the particles involved in the reaction. Let us write down

$$\frac{\Gamma}{H} \gg 1 \tag{11.3}$$

as the condition for a reaction to reach chemical equilibrium. Since the interaction rate Γ depends on the density of the Universe, we expect Γ to decrease
with the expansion of the Universe. Although H also decreases with time (for
example, H decreases as t^{-1} if a goes as some power t^n of t), the decrease
in Γ is typically faster. We then expect the condition $\Gamma \gg H$ to change over
to the condition $\Gamma \ll H$ as the Universe evolves. In such a situation, a reaction
which was initially able to remain in chemical equilibrium eventually falls out of
equilibrium. The concentrations of the particles involved in the reaction are not
expected to change after the reaction goes out of equilibrium. So we can assume
the concentrations to be frozen at the values which they had when $\Gamma \approx H$ and
the reaction was just going out of equilibrium. It should be emphasized that the
ideas presented in this paragraph should be taken as rough rules of thumb. To
obtain more accurate results, one has to carry out a detailed calculation of the
reaction in the expanding Universe (usually done numerically with computers).

When a reaction is able to reach chemical equilibrium, the concentrations
of the various species of particles involved in the reaction will be given by
the standard results of thermodynamic equilibrium. The number of particles
occupying a quantum state with momentum $\mathbf{p}$ is given by

$$f(\mathbf{p}) = \left[\exp\left(\frac{E(\mathbf{p}) - \mu}{\kappa_B T} \right) \pm 1 \right]^{-1}, \tag{11.4}$$

where we have to use the plus sign for fermions obeying Fermi–Dirac statistics
(Fermi, 1926; Dirac, 1926) and the minus sign for bosons obeying Bose–
Einstein statistics (Bose, 1924; Einstein, 1924), $E(\mathbf{p}) = \sqrt{p^2 c^2 + m^2 c^4}$ being
the energy associated with the momentum $\mathbf{p}$. To obtain the actual number
density from (11.4), we have to keep in mind that the six-dimensional phase
space volume element $dV\, d^3p$ (where dV is the ordinary volume element) has
$g\, dV\, d^3p/h^3$ quantum states in it. Here g is the degeneracy, which is 2 for both
electrons and photons, corresponding to the two spin states and two degrees of
polarization respectively. Hence the number density per unit volume must be
given by multiplying (11.4) by $g\, d^3p/h^3$ and then integrating over all possible
momenta, i.e.

$$n = \frac{g}{(2\pi)^3} \int f(\mathbf{p}) \frac{d^3p}{\hbar^3}. \tag{11.5}$$

The contribution to density made by this species of particles is

$$\rho = \frac{g}{(2\pi)^3} \int \frac{E(\mathbf{p})}{c^2} f(\mathbf{p}) \frac{d^3p}{\hbar^3}. \tag{11.6}$$

Let us make some comments on the chemical potential μ appearing in
(11.2) and (11.4). It is well known that $\mu = 0$ for photons which can be created

or destroyed easily. If photons produce particle-antiparticle pairs, then (11.2) is satisfied only if the sum of the chemical potentials of the particle and the antiparticle is zero. In other words, if the chemical potential of the particle is $+\mu$, the chemical potential of the antiparticle has to be $-\mu$. It follows from (11.4) that these differences in chemical potential would make the number densities of particles and antiparticles different.

We now consider the case $\kappa_B T \gg mc^2$ such that most particles would have energies much larger than the rest mass energy and would be relativistic, i.e. we can write $E \approx pc$. At such a high temperature, the Universe is expected to be full of these particles and their antiparticles, which should be present in comparable numbers. This is possible only if μ is much smaller compared to $\kappa_B T$. Writing $d^3p = 4\pi p^2 \, dp$, $E = pc$ and $\mu = 0$, we find from (11.4) and (11.5) that

$$n = \frac{g}{2\pi^2 \hbar^3} \int_0^\infty \frac{p^2 dp}{e^{pc/\kappa_B T} \pm 1}. \tag{11.7}$$

It similarly follows from (11.4) and (11.6) that

$$\rho = \frac{g}{2\pi^2 c \hbar^3} \int_0^\infty \frac{p^3 dp}{e^{pc/\kappa_B T} \pm 1}. \tag{11.8}$$

Both (11.7) and (11.8) can be evaluated analytically (see Exercise 11.1). For the cases of bosons and fermions, (11.7) can be integrated to give

$$n = \begin{cases} \frac{\zeta(3)}{\pi^2} g \left(\frac{\kappa_B T}{\hbar c} \right)^3 & \text{(boson)}, \\ \frac{3}{4} \frac{\zeta(3)}{\pi^2} g \left(\frac{\kappa_B T}{\hbar c} \right)^3 & \text{(fermion)}, \end{cases} \tag{11.9}$$

where $\zeta(3) = 1.202$ is the Riemann zeta function (see, for example, Abramowitz and Stegun, 1964, Chapter 23). Similarly from (11.8) we get

$$\rho = \begin{cases} \frac{g}{2c^2} a_B T^4 & \text{(boson)}, \\ \frac{7}{8} \frac{g}{2c^2} a_B T^4 & \text{(fermion)}. \end{cases} \tag{11.10}$$

Here

$$a_B = \frac{\pi^2 \kappa_B^4}{15 \hbar^3 c^3} \tag{11.11}$$

is the Stefan constant. If we take $g = 2$ for photons, then the energy density ρc^2 according to (11.10) is $a_B T^4$, which is the standard expression for energy density of blackbody radiation at temperature T (see Exercise 2.1). It is now instructive to determine the entropy density of this gas of relativistic bosons or fermions. For this purpose, we begin with the standard thermodynamic relation

$$T \, dS = dU + P \, dV, \tag{11.12}$$

where the internal energy U inside a volume V is given by $\rho c^2 V$. If s is the entropy density, then we can write $S = sV$. On making these substitutions, (11.12) gives

$$TVds = Vc^2 d\rho + [(\rho c^2 + P) - Ts]dV. \tag{11.13}$$

We see in (11.10) that ρ is a function of T alone. We expect the entropy density s also to be a function of T alone. If both s and ρ are independent of V, then consistency requires that the coefficient of dV in (11.13) must be zero. This means

$$s = \frac{\rho c^2 + P}{T}. \tag{11.14}$$

On making use of the fact that $P = (1/3)\rho c^2$ for a relativistic gas (see the discussion in §10.5) and substituting for ρ from (11.10), we get

$$s = \begin{cases} \frac{2g}{3} a_B T^3 & \text{(boson)}, \\ \frac{7}{8}\frac{2g}{3} a_B T^3 & \text{(fermion)}. \end{cases} \tag{11.15}$$

After devoting the last paragraph to the case $\kappa_B T \gg mc^2$ when most particles are relativistic, we now take a brief look at the opposite case $\kappa_B T \ll mc^2$. We can take

$$E(\mathbf{p}) \approx mc^2 + \frac{p^2}{2m}$$

in (11.4) in this case. Substituting (11.4) in (11.5), the particle density in the non-relativistic limit is found to be

$$n = \frac{g}{\hbar^3} \left(\frac{m\kappa_B T}{2\pi} \right)^{3/2} \exp\left(-\frac{mc^2 - \mu}{\kappa_B T} \right). \tag{11.16}$$

Note that the exponential factor is supposed to be quite small in the non-relativistic limit. So the number density of non-relativistic particles under thermodynamic equilibrium would be negligible compared to the number density of relativistic particles.

We already pointed out in §10.5 that the CMBR still has a spectral distribution appropriate for thermodynamic equilibrium even though the photons in the CMBR are no longer in equilibrium with matter. So we can still use (11.9) to calculate the number density of photons in the CMBR at present. Using T as given by (10.45) and taking $g = 2$, we find from (11.9) that the present photon number density is

$$n_{\gamma,0} = 4.14 \times 10^8 \, \text{m}^{-3}. \tag{11.17}$$

It is instructive to compare this with the present number density of baryons (i.e. the total number of protons and neutrons in unit volume). If $\Omega_{B,0}$ is the contribution of baryons to the density parameter, then the number density $n_{B,0}$

of baryons should clearly be $\Omega_{B,0}\rho_{c,0}/m_p$, where m_p is the mass of a proton. On using (10.29), we get

$$n_{B,0} = 11.3\,\Omega_{B,0}h^2\ \text{m}^{-3}. \tag{11.18}$$

The ratio of baryon to photon number densities is an interesting dimensionless quantity which has the value

$$\eta = \frac{n_{B,0}}{n_{\gamma,0}} = 2.73 \times 10^{-8}\Omega_{B,0}h^2. \tag{11.19}$$

As the Universe expands, the photon number density n_γ falls as T^3, which, in combination with (10.49), suggests that n_γ falls as a^{-3}. The baryon number density also is obviously falling as a^{-3}. Hence the ratio of these number densities does not change with time. In other words, the value of this ratio given by (11.19) is not only its present value, but also its value at earlier or later times (as long as there are no reactions to change n_γ or n_B suddenly). Thus η is an important quantity in cosmology and we shall comment on its significance later.

11.3 Primordial nucleosynthesis

As indicated in Figure 11.1, the epoch from $1\,\text{s}$ to $10^2\,\text{s}$ after the Big Bang was suitable for nuclear reactions. In early work, Gamow (1946) suggested that most of the heavy nuclei were synthesized during this short epoch just after the Big Bang. As we pointed out in §4.3, there is no stable nucleus of mass 5 or 8 and hence it is not easy for nuclear reactions to synthesize nuclei heavier than helium. In the interiors of heavy stars, this mass gap is superseded by the triple alpha reaction (4.28). This reaction, however, involves a three-body process and cannot take place unless the number density of helium nuclei is sufficiently high. Detailed calculations show that this reaction was very unlikely under the conditions of the early Universe and heavier nuclei could not be synthesized there. We now believe that all heavy nuclei starting from ^{12}C are produced inside stars.

Realistic calculations of nucleosynthesis in the early Universe have to be done numerically. Peebles (1966) and Wagoner, Fowler and Hoyle (1967) developed the first codes soon after the discovery of the CMBR, which established that the Universe really began with a hot Big Bang. The code perfected by Wagoner (1973) came to be regarded as the 'standard code' on which most of the later computations are based. The results depend on η introduced in (11.19). A higher η implies that there were more baryons per unit volume and nuclear reactions were more likely to take place. Figure 11.2 shows the mass fractions of various nuclei produced in the early Universe as a function of η. Although this figure is based on results obtained by numerical simulation, we shall now show that at least some of the results can be understood from general arguments.

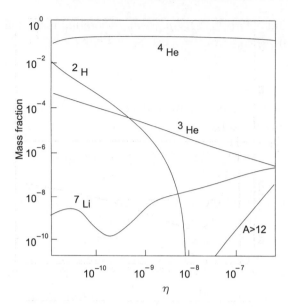

Fig. 11.2 Theoretically calculated primordial abundances of various light nuclei as a function of η. Adapted from Wagoner (1973).

Let us first consider the production of helium. Two protons and two neutrons have to combine to produce a helium nucleus. During the epoch of nuclear reactions, the protons and neutrons should take part in reactions like (5.29) and (5.30), which were listed as possible reactions inside neutron stars. These reactions are mediated by the weak interaction. The crucial question is whether these reactions would have been able to reach chemical equilibrium, for which the condition is given by (11.3). To answer this question, we need to find out the reaction rate Γ, which depends on the coupling constant of the weak interaction and can be calculated from a knowledge of the half-life of a free neutron. We know that free neutrons decay by the reaction (5.29) with a half-life of 10.5 minutes. Since not all readers may be familiar with the theory of weak interactions, we skip the derivation and merely quote the result, which is

$$\frac{\Gamma}{H} \approx \left[\frac{T \text{ (in MeV)}}{0.8 \text{ MeV}}\right]^{3}. \tag{11.20}$$

It follows from this that the condition of chemical equilibrium (11.3) would be satisfied when T (in MeV) $\gg 0.8$ MeV. Under this situation, we can apply (11.16) to calculate the number densities of protons or neutrons. Since chemical potentials are expected to be small compared to the rest masses, it follows from (11.16) that the ratio of these number densities will be

$$\frac{n_\text{n}}{n_\text{p}} = \exp\left[-\frac{(m_\text{n} - m_\text{p})c^2}{\kappa_\text{B} T}\right], \tag{11.21}$$

where the various symbols have obvious meanings. There is a very simple way of understanding (11.21). We can regard the proton as the ground state for a baryon and the neutron as the first excited state. Then (11.21) can be regarded merely as the Boltzmann distribution law (2.28) applied to this situation. As the temperature keeps decreasing, we expect (11.21) to be valid till $T \approx 0.8\,\text{MeV}$, after which the reactions will no longer be able to maintain thermodynamic equilibrium and the ratio n_n/n_p will be approximately frozen. Since $(m_n - m_p)c^2 = 1.29\,\text{MeV}$, this frozen ratio is given by

$$\frac{n_n}{n_p} \approx e^{-\frac{1.29}{0.8}} \approx 0.20. \tag{11.22}$$

After the neutron number is frozen, these neutrons may eventually be used up to synthesize helium. Let us assume that n_n neutrons in the unit volume combine with n_n protons to synthesize helium nuclei and the other $n_p - n_n$ protons remain as protons. Then the helium mass fraction should be given by

$$\frac{2n_n}{n_n + n_p} = \frac{2(n_n/n_p)}{1 + (n_n/n_p)} = 0.33 \tag{11.23}$$

on substituting from (11.22). It may be noted that the ratio n_n/n_p does not remain completely frozen during the time of order $100\,\text{s}$ when nucleosynthesis takes place, but decreases due to the decay of neutrons by the reaction (5.29) which is still possible even when thermodynamic equilibrium no longer prevails. So this ratio becomes somewhat less than 0.20 given in (11.22), leading to a smaller helium fraction compared to what is given in (11.23). Careful numerical simulations suggest a value of about 0.25 for the helium mass fraction. We see in Figure 11.2 that the helium mass fraction is nearly independent of η, although it becomes slightly smaller for low η. The reason behind this is not difficult to understand. A low η implies a low baryon number density, which means that the nuclear reactions which build up helium nuclei would proceed at a slower rate and more neutrons would decay before being bound up inside helium nuclei, thereby leading to a lower helium mass fraction.

Let us now point out what the observational data tell us. Although the mass fractions of higher elements are found to vary considerably in different astrophysical sources, the helium fraction is found to be very similar in widely different astrophysical systems, lying in the range 0.23–0.27. It is one of the triumphs of Big Bang cosmology that this observation can be explained very naturally if we assume that most of the helium was produced by primordial nucleosynthesis, with nuclear reactions in the interiors of stars contributing a little bit afterwards. The theoretically calculated value of the helium fraction is completely in agreement with observational data. The fact that the abundances of higher elements vary considerably in different astrophysical sources suggests that they were not produced in the early Universe, because such a production would imply a more uniform distribution today.

We now consider the abundance of deuterium ^{2}H. It is an intermediate product in the synthesis of helium, as can be seen from the pp chain reactions listed in §4.3. If η is high and the nuclear reaction rate is fast, then most of the deuterium would get converted into helium during the primordial nucleosynthesis era. On the other hand, a lower η and a slower reaction rate would imply that the synthesis of helium would not be so efficient in the nucleosynthesis era and some deuterium would be left over. We therefore see in Figure 11.2 that the deuterium fraction falls sharply with the increase in η. From observations, it is found that the cosmic deuterium abundance is not less than 10^{-5}. This puts an upper bound on η, which is $\eta < 10^{-9}$. A larger value of η would not allow the observed deuterium to be left over after the primordial nucleosynthesis.

This upper bound on η has a tremendously important significance. It follows from (11.19) that this upper bound on η translates into an upper bound on $\Omega_{B,0}$, which is $\Omega_{B,0} < 0.037h^{-2}$. Even if we take 0.64 as the lowest possible value of h in accordance with (9.20), we are still bound by the limit $\Omega_{B,0} < 0.09$. On the other hand, the value of $\Omega_{M,0}$ based on virial masses of clusters of galaxies was quoted to be about 0.3 in (10.42). We shall discuss in §14.5 that there is other evidence suggesting that $\Omega_{M,0}$ indeed has a value close to 0.3. How do we reconcile this result with the limit $\Omega_{B,0} < 0.09$? The only way of reconciling these results is to conclude that a large part of the matter in the Universe is non-baryonic in nature and hence is not included when we estimate $\Omega_{B,0}$ or η. This is a truly extraordinary conclusion. The different objects around us and we ourselves are made up of atoms with nuclei which consist of baryonic matter. If much of the matter in the Universe is non-baryonic, it means that this matter is not made up of ordinary atoms we are familiar with. As discussed in §10.5, we believe that most of the matter in the Universe does not emit light and hence is dark. We do not know much about the distribution of this dark matter in galaxies or galaxy clusters. Now we conclude that a major component of this dark matter is not even made up of ordinary atoms. What is it made up of then? We shall come to this question in §11.5.

11.4 The cosmic neutrino background

Various thermal processes in the early Universe created photons and that is why we still have a cosmic background of blackbody radiation. Exactly similarly we would expect a background of neutrinos because neutrinos were created in the early Universe by reactions like (5.29) and (5.30) which involved weak interactions and these neutrinos must be present today. When the weak interaction rate was faster than the rate of expansion of the Universe, these neutrinos must have been in thermodynamic equilibrium with matter and would have satisfied relations like (11.9), (11.10) and (11.15) for fermions. After the weak interaction rate became slower, the neutrinos would get decoupled from matter and thereafter must have evolved adiabatically. We know that photons in thermodynamic equilibrium (i.e. obeying the blackbody spectrum) continue to remain in thermodynamic equilibrium under adiabatic expansion, as we shall prove explicitly in §14.3. If neutrinos are either massless or continue to remain relativistic (i.e. the thermal energy continues to remain higher than rest mass energy $m_\nu c^2$), then exactly the same considerations should hold for the background neutrinos as well. In other words, even after the neutrinos have decoupled from matter, they should continue to have a distribution appropriate for thermodynamic equilibrium, with the temperature T falling as a^{-1}. In the early Universe, photons, neutrinos and matter particles must all have been in thermodynamic equilibrium and must have had the same temperature. Since the temperatures of both photons and neutrinos fall as a^{-1} after decoupling, should we still expect them to have the same temperature?

The background neutrinos would have the same temperature as the CMBR photons only if nothing had happened to change the temperature of photons after the neutrinos decoupled from the other particles. We believe that one important phenomenon had changed the temperature of the photons. When the neutrinos decoupled, the electrons were still relativistic. So the Universe must have been filled with electrons and positrons with number densities given by (11.9). When the temperature fell, the electrons and positrons would have annihilated each other creating photons, thereby putting more energy in the photon background and increasing its temperature. Let T_i be the temperature before the electron-positron annihilation (i.e. it would have been the temperature of the electrons and positrons as well as photons) and let T_f be the enhanced temperature of the photons after this annihilation. Since this is an adiabatic process, we expect the entropy to remain conserved. Equating the final entropy with the initial entropy, we shall now obtain a relation between T_i and T_f. To find the initial entropy density s_i, we have to add up the contributions of photons, electrons and positrons as given by (11.15). For all these particles, we have $g = 2$ – because of the two polarization states of photons and two spin states of electrons or positrons. Remembering that photons are bosons whereas electrons and positrons are fermions, we get

$$s_i = \left(\frac{4}{3} + \frac{7}{8} \times \frac{4}{3} + \frac{7}{8} \times \frac{4}{3}\right) a_{\mathrm{B}} T_i^3 = \frac{11}{4} \times \frac{4}{3} a_{\mathrm{B}} T_i^3. \qquad (11.24)$$

The final entropy density s_f is due only to the photons and must be

$$s_f = \frac{4}{3} a_{\mathrm{B}} T_f^3. \qquad (11.25)$$

Equating s_i and s_f as given by the above two equations, we get

$$\frac{T_f}{T_i} = \left(\frac{11}{4}\right)^{1/3}.$$

(11.26)

In other words, the photon temperature jumped by this factor after the electron-positron annihilation, but the neutrino temperature did not change because the neutrinos constituted a distinct system decoupled from other matter. Since both the photon temperature and the neutrino temperature afterwards fell as a^{-1}, the photon temperature would still be larger than the neutrino temperature by this factor $(11/4)^{1/3}$. The present neutrino temperature should be

$$T_{\nu,0} = \left(\frac{4}{11}\right)^{1/3} T_0,$$

(11.27)

where T_0 is the present temperature of the CMBR given by (10.45). On taking $T_0 = 2.735$ K, we get

$$T_{\nu,0} = 1.95 \text{ K}.$$

(11.28)

We can calculate the energy density of the neutrino background by using (11.10). Since there are three types of neutrinos (associated with electrons, muons and τ-particles) and each neutrino has an antineutrino, we take $g = 6$ so that the present energy density of the neutrino background, according to (11.10), is

$$\rho_{\nu,0} = \frac{7}{8}\frac{3}{c^2}a_B\left[\left(\frac{4}{11}\right)^{1/3}T_0\right]^4 = 0.68\frac{a_B}{c^2}T_0^4 = 0.68\rho_{\gamma,0},$$

(11.29)

where $\rho_{\gamma,0}$ is the present energy density of the CMBR. We have already made use of this in writing down (10.47), which is obtained by adding the energy density of neutrinos to the energy density of photons.

At the end of this discussion, we again stress one point which we mentioned at the beginning of this section. The contribution to density by neutrinos will be given by (11.29) only if the neutrinos are still relativistic. If the neutrinos have mass and if the thermal energies become smaller than $m_\nu c^2$ with the fall in temperature, then it is possible for neutrinos to contribute much more to the density because there would be a lower bound in the contribution to density made by a neutrino. A neutrino with mass m_ν has to contribute at least m_ν to density. We discuss this further in the next section.

11.5 The nature of dark matter

We have already pointed out that much of the matter in our Universe does not emit light. Evidence for dark matter comes from the rotation curves of spiral galaxies (§9.2.2) and from the application of the virial theorem to galaxy clusters (§9.5). In fact, luminous matter contributes only a small fraction to the total matter density of the Universe, as discussed in §10.5. Primordial nucleosynthesis calculations described in §11.3 lead us to the extraordinary conclusion that much of the dark matter is non-baryonic, i.e. not made up of ordinary atoms. We now briefly discuss the question of what could be the constituents of dark matter.

While neutrinos are very light particles, physicists wondered for a very long time whether they are massless or have very small mass. If neutrinos have mass and the masses of the three types of neutrinos are different, then it is possible for neutrinos to have oscillations in which one type of neutrino gets converted into another. Since these oscillations depend on the mass difference between different kinds of neutrinos, the discovery of neutrino oscillations led to the conclusion that $|\Delta m|^2$ should be of order $5 \times 10^{-5} \, \text{eV}^2$ (Ahmad *et al.*, 2002). While we do not know the individual masses of different types of neutrinos, we now know that neutrinos have mass. Is it then possible that the background neutrinos are responsible for the non-baryonic mass part of dark matter?

To answer this question, let us first find out the number density of neutrinos. As long as neutrinos are relativistic, the number density is given by (11.9). It turns out that (11.9) can still be used to calculate the number density of neutrinos even if they had become non-relativistic, provided we take T to be a quantity which falls off as a^{-1} in the expanding Universe. After learning general relativity, the reader is asked in Exercise 14.4 to analyse the dynamics of particles in the expanding Universe and to show this. On taking $g = 6$ and substituting the temperature given by (11.28), we obtain from (11.9) that

$$n_{v,0} = 3.36 \times 10^8 \, \text{m}^{-3}. \tag{11.30}$$

If m_v is the average mass of neutrinos, then we have to divide $m_v n_{v,0}$ by $\rho_{c,0}$ given by (10.29) to get the density parameter $\Omega_{v,0}$ due to neutrinos, which turns out to be

$$\Omega_{v,0} = 3.18 \times 10^{-2} h^{-2} \left(\frac{m_v}{\text{eV}}\right), \tag{11.31}$$

where m_v has to be in eV. We have seen in (10.42) that $\Omega_{M,0}$ should have a value of about 0.3. Since the neutrino contribution to the density parameter has to be less than this, we get the following limit by demanding that $\Omega_{v,0}$ given by (11.31) should be less than 0.3:

$$m_v < 9.4h^2 \, \text{eV}. \tag{11.32}$$

This is known as the *Cowsik–McClelland limit* and was a more stringent limit on the neutrino mass than the limit from laboratory experiments when it was first pointed out (Cowsik and McClelland, 1972). This is an example of how astrophysical considerations can be relevant in particle physics.

If we hypothesize that neutrinos have an average mass close to the Cowsik–McClelland limit (11.32), then we can solve the mystery of non-baryonic dark matter because the neutrinos will provide the estimated mass of this dark matter. There are, however, some serious difficulties with this hypothesis that dark matter is made up of neutrinos with mass just a little less than the limit given by (11.32). Although we do not have a good idea about the distribution of dark matter, observations indicate that it is not uniformly spread throughout the Universe. Estimates of the masses of spiral galaxies and galaxy clusters suggest that much of the dark matter should be gravitationally bound with these systems. For this to be possible, the typical kinetic energy of dark matter particles should not exceed the gravitational binding energy $m_\nu |\Phi|$, where Φ is the gravitational potential associated with structures like galaxies and galaxy clusters. This condition is hard to satisfy if the limit of m_ν is given by (11.32). If dark matter particles satisfy (11.32), then that type of dark matter is called *hot dark matter*. Such dark matter would tend to be distributed throughout the Universe without clumping in the gravitational structures like galaxies or galaxy structures. If we want dark matter to be bound in these gravitational structures, then we need to have *cold dark matter*, in which particles are more massive, move more slowly at a given temperature (since the thermal velocity is given by $\sqrt{2\kappa_B T/m}$) and can get gravitationally bound in galaxies or galaxy clusters. When we discuss structure formation in §11.9, we shall point out that cold dark matter helps to fulfil some requirements of structure formation as well.

In view of (11.32), is it possible to have cold dark matter with particles more massive than this limit? Note that we obtained (11.32) by using (11.9), which would be applicable only if the particles were relativistic at the time of decoupling. If the particles are so heavy that they were already non-relativistic at the time of decoupling, then the number density would be given by (11.16) rather than (11.9). Some supersymmetric theories of particle physics suggest the possibility of some particle with mass of about a few GeV, which acts with other particles only through the weak interaction. Such a particle is expected to get decoupled from the other constituents of the Universe when the temperature was of order MeV. This particle would remain in thermodynamic equilibrium before the decoupling and its number density at the time of decoupling would be given by (11.16). The exponential factor in (11.16) would ensure a low number density of this particle. If the particle is more massive than about 3 GeV, detailed calculations show that its number density would be sufficiently suppressed by this exponential factor and it would not make a contribution to the density parameter more than what is estimated from observations (i.e. $\Omega_{M,0} \approx 0.3$). The limit that the particles of cold dark matter have to be more massive than 3 GeV is known as the *Lee–Weinberg limit* (Lee and Weinberg, 1977).

We basically conclude that dark matter particles could not have mass in the range 10 eV to 3 GeV. They either have to be lighter than 10 eV (hot dark matter) or heavier than 3 GeV (cold dark matter). Current evidence suggests the second possibility that the dark matter in the Universe should be cold dark matter made up of particles heavier than 3 GeV. We shall discuss this point further in §11.9.

11.6   Some considerations of the very early Universe

In §11.3 we considered the nuclear reaction era when the temperature of the Universe was of order MeV ($\approx 10^{10}$ K). The next two sections followed the fate of particles which got decoupled during this era. This era roughly lasted from about 10^{-1} s to 10^2 s after the Big Bang, according to (10.67) or (10.69). With new developments in particle physics in the last few decades, there has been considerable interest in investigating what might have happened in the Universe at still earlier times when the temperature was of the order of GeV or higher. The Universe at this extremely early epoch is usually referred to as the *very early Universe*. The study of the very early Universe is still a rather speculative subject, without too many connections with present-day astrophysical data. It is beyond the scope of this book to get into that subject. We merely touch upon two topics which have some astrophysical relevance.

11.6.1   The horizon problem and inflation

Since the Universe began with a violent Big Bang, it is very unlikely that the Universe was created as a very homogeneous system. Why then is the Universe so homogeneous now? Suppose inhomogeneities are suddenly created inside a gas kept within a container. We expect that the gas from regions of higher density will move to regions of lower density to establish homogeneity again. If c_s is the sound speed, we expect that regions of size $c_s t$ will become homogeneous in time t. At the present time when the age of the Universe is t, we could not have received information from regions further away than ct because any information starting from those regions beyond ct would not be able to reach us by today. Hence a sphere of radius ct around us is our *horizon*. We can have causal contacts only with regions inside this horizon. We may expect the Universe to have become homogeneous over regions of horizon size, but not over larger regions. It may be noted that an accurate calculation of the horizon in an expanding Universe requires a more careful analysis of light propagation (to be discussed in §14.3). But we need not get into those details.

There is enough evidence to show that the Universe is actually homogeneous over regions much larger than the horizon. We have discussed the CMBR in §10.5. We shall point out in §11.7 that photons in the CMBR reach us after travelling through space for time comparable to the age of the Universe. Consider CMBR photons coming from two diametrically opposite regions in the sky. These two regions are causally connected to us, but are not causally connected to each other because the information from one of the regions had just time enough to reach us and did not yet have time to reach the other region. The isotropy of CMBR, however, suggests that these two regions which apparently had never been in causal contact have the same physical characteristics, since they produce CMBR of exactly the same nature. How is it possible that regions which are out of each other's horizons and which have never been in causal contact are so homogeneous? This is known as the *horizon problem* in cosmology.

Guth (1981) proposed a solution to the horizon problem. On the basis of some field theoretic arguments which are beyond the scope of this book, Guth (1981) suggested that there was a brief phase in the very early Universe when the Universe expanded very rapidly and became larger by several orders of magnitude. This is called *inflation*. If this is true, then the Universe before inflation must have been much, much smaller than what we would expect it to be if inflation had not taken place. Different parts of the Universe could have been causally connected if the Universe was very small before inflation and thereby the homogeneity of the Universe could have been established.

11.6.2 Baryogenesis

According to (11.19), the photon number density $n_{\gamma,0}$ is nearly eight orders of magnitude larger than the baryon number density $n_{B,0}$. As long as new photons or baryons are not created, both these numbers fall as a^{-3} and their ratio does not change. Even at the time when photons decoupled from matter, this ratio must have had this value. Why are there many more photons than baryons?

In the early Universe when the temperature was higher than a few GeV and baryon-antibaryon pairs could be formed, the number of either baryons or antibaryons would have been comparable to the number of photons, since all these numbers would have been given by (11.9). But the number of baryons must have been slightly larger than the number of antibaryons to ensure that some baryons were left over after the baryon-antibaryon annihilation which must have taken place when the temperature fell below GeV. If Δn_B was the excess in the number density of baryons compared to the number density of antibaryons before the annihilation, then we must have

$$\frac{\Delta n_B}{n_B} \approx 10^{-8} \tag{11.33}$$

if we want the baryon-to-photon ratio to have a value like this after annihilation.

Many physicists feel that it is esthetically more satisfying to assume that the Universe was created with equal numbers of baryons and antibaryons rather than to assume that the Universe was created with such a tiny imbalance. If the Universe really had equal numbers of baryons and antibaryons in the beginning, then the net baryon number Δn_B was initially zero and it had to change to a non-zero value later. We find the baryon number to be a conserved quantity in all particle interactions we study at the present time. If the Universe was created with equal numbers of baryons and antibaryons, then the small excess of baryons over antibaryons could arise only if baryon number conservation was violated in the very early Universe. This is known as the problem of *baryogenesis* and has been of some interest to theoretical particle physicists.

11.7 The formation of atoms and the last scattering surface

After discussing the physics of the nucleosynthesis era and its consequences in §11.3–§11.5, we took a brief digression in §11.6 to raise some theoretical issues pertaining to still earlier times. Now we again follow the evolutionary history of the Universe and look at some of the later landmarks in that evolutionary history. After the electron-positron annihilation discussed in §11.4, the Universe consisted of the basic constituents of ordinary matter – protons, helium nuclei and electrons – in addition to the non-baryonic matter and the relativistic particles (photons and neutrinos). We have seen in (10.52) that the Universe became matter-dominated when its size was $1 + z = 2.3 \times 10^4 \, \Omega_{\mathrm{M},0} h^2$ times smaller than its present size. This happened about 10^4 yr after the Big Bang, as indicated in (10.64). Since the temperature at that time was 2.735 K multiplied by this redshift factor, the Universe was still too hot for the formation of atoms.

We know that matter and radiation were in equilibrium in the early Universe. We now ask the question how long this coupling between matter and radiation continued. We have seen in §2.6.1 that photons interact with electrons through Thomson scattering. As long as electrons are free, we expect Thomson scattering to keep matter and radiation in equilibrium. Radiation gets decoupled from matter when atoms form and all electrons get locked up inside atoms. We can apply the Saha equation (2.29) to estimate the ionization fraction which gives a measure of the number of free electrons. The ionization potential for hydrogen is $\chi = 13.6$ eV, corresponding to a temperature of 1.5×10^5 K by (10.68). However, a simple application of (2.29) shows that the number of free electrons becomes insignificant only when the temperature falls to a much lower value of about 3000 K (see Exercise 11.4). We thus conclude that the Universe becomes transparent to photons when the temperature falls below 3000 K, causing radiation to get decoupled from matter. Interestingly, the plot of stellar opacity in Figure 2.8 shows that stellar material also becomes transparent when the temperature falls to about this value. Since the present CMBR temperature is 2.735 K, a simple application of (10.49) suggests that the Universe must have been about 1000 times smaller in size when the temperature was 3000 K. A more careful calculation gives the redshift $z_{\mathrm{dec}} \approx 1100$ as the era when matter-radiation decoupling took place. The Universe became transparent after this decoupling and photons no longer interacted with matter.

All the CMBR photons which reach us today last interacted with matter at the era of redshift $z_{dec} \approx 1100$. These photons started as blackbody radiation of temperature 3000 K. The redshift of 1100 has made them the present blackbody radiation of temperature 2.735 K. When we look at the Sun, we basically see photons which last interacted with matter at the solar surface, since the space between the solar surface and us is transparent to visible light. So the photons coming from the Sun show us the solar surface. In exactly the same way, the CMBR photons coming from all directions show us a surface of primordial matter surrounding us as it existed at redshift $z_{dec} \approx 1100$. This is called the *last scattering surface*. If the primordial matter at redshift $z_{dec} \approx 1100$ was completely homogeneous, then this last scattering surface would appear smooth and CMBR coming from it would be totally isotropic. On the other hand, if there were inhomogeneities in the last scattering surface, they would manifest themselves as angular anisotropies in the CMBR.

11.7.1 Primary anisotropies in CMBR

We believe that the matter distribution in the primordial Universe was reasonably homogeneous. A standard paradigm in cosmology is that there were some small initial perturbations in matter density which kept on being enhanced with time as the Universe expanded and eventually led to the formation of structures that we see today – stars, galaxies and galaxy clusters. If this paradigm is correct, then there must have been some density perturbations in the last scattering surface, causing anisotropies in the CMBR. We pointed out in §10.5 that the mission COBE showed the spectrum of CMBR to be a perfect blackbody spectrum (Mather *et al.*, 1990). COBE also kept looking for anisotropies in CMBR and finally discovered them (Smoot *et al.*, 1992). It was found that CMBR looks exactly like blackbody radiation in all directions, but the temperature of the blackbody radiation was found to vary slightly from direction to direction. The temperature variation was discovered to be of order

$$\frac{\Delta T}{T} \approx 10^{-5}. \tag{11.34}$$

The upper panel of Figure 11.3 shows a map of this temperature anisotropy as discovered by COBE (Smoot *et al.*, 1992). Since COBE had an angular resolution of about 7°, the temperature variation at smaller angular scales could not be determined by COBE. This was finally achieved by the later mission WMAP. The lower panel of Figure 11.3 is the temperature anisotropy obtained by WMAP (Bennett *et al.*, 2003). We shall discuss in §11.9 how these inhomogeneities in the last scattering surface grew to produce structures. The typical angular size of anisotropies can be used to put important constraints on various cosmological parameters. This topic will be taken up in §14.5. The anisotropies in CMBR resulting from irregularities in the last scattering surface are often

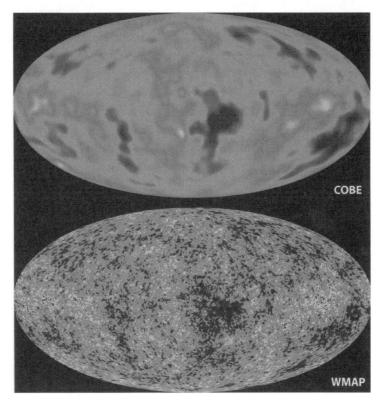

Fig. 11.3 Map of the CMBR temperature distribution in different directions of the sky, as obtained by (a) COBE (Smoot *et al.*, 1992): upper panel; and (b) WMAP (Bennett *et al.*, 2003): lower panel.

called *primary anisotropies*, to distinguish them from anisotropies which may arise during the passage of the CMBR photons from the last scattering surface to us.

11.7.2   The Sunyaev–Zeldovich effect

We have said that the photons from the last scattering surface reach us without interacting with matter any more. This is an almost correct statement, since most of the space between the last scattering surface and us is devoid of free electrons. There is, however, one important exception. We have discussed in §9.5 that galaxy clusters contain hot gas, which is ionized and has free electrons. So CMBR photons passing through galaxy clusters can interact with the free electrons in the hot gas. In normal Thomson scattering, photons scatter off electrons without any significant change in energy. Thomson scattering involving an energy exchange between photons and electrons is called the *Compton effect* (Compton, 1923). While the mathematical theory of Thomson scattering can be developed by treating the photons as making up a classical electromagnetic wave (as discussed in §2.6.1), the theory of the Compton effect requires a treatment of photons as particles. The Compton effect becomes important when the photon energy is not negligible compared to the rest mass energy of the electron (as in the case of X-ray photons) and some energy can be transferred from the photon to the electron (see, for example, Yarwood, 1958, §7.20). On the other hand, when an electron with high kinetic energy interacts with a low-energy photon, we can have the *inverse Compton effect* in which energy is transferred from the electron to the photon. This happens when CMBR photons interact with the free electrons in a galaxy cluster, which are highly energetic because of the high temperature of the cluster gas. This transfer of energy from the electrons in the hot cluster gas to the CMBR photons is known as the *Sunyaev–Zeldovich effect* (Sunyaev and Zeldovich, 1972). As a result of this, some of the radio photons in the CMBR get scattered to become X-ray photons, leading to a depletion of CMBR intensity in radio frequencies in the directions of galaxy clusters. From this depletion in intensity, one can estimate the optical depth (usually $\ll 1$) of CMBR photons through the cluster gas. For a spherical cluster of radius R_c and internal electron density n_e, the maximum optical depth at the centre would be of order $2\sigma_T n_e R_c$, where σ_T is the Thomson scattering cross-section.

One important application of the Sunyaev–Zeldovich effect is that it can be used to estimate the distances of galaxy clusters, thereby leading to a determination of the Hubble constant. We get $n_e R_c$ from the depletion in the CMBR intensity. The angular size of the cluster is equal to $2R_c$ divided by its distance. The X-ray emission from the cluster gas by bremsstrahlung is governed by (8.70), which gives the emissivity per unit volume. On measuring the X-ray flux from the cluster received by us and combining it with the other measured quantities such as the angular size of the cluster and the depletion in the CMBR intensity, we can find the distance to the cluster (see Exercise 11.5). The Hubble constant derived from the Sunyaev–Zeldovich effect somehow turns out to be slightly lower than its value measured by the other methods.

11.8   Evidence for evolution during redshifts $z \sim 1$–6

We can get direct information about some material object in the astronomical Universe if it either emits radiation or absorbs radiation passing through it. We can study the distribution of primordial matter at the moment of its decoupling from radiation by analysing the CMBR which was emitted by this matter. After the matter-radiation decoupling, however, the matter in the Universe became transparent and did not emit any more radiation until stars and galaxies formed long afterwards. The era between the matter-radiation decoupling (around $z \sim 1100$) and the era when the first stars formed is often called the 'dark age' in cosmology. During this dark age, matter did not emit any radiation that we can detect today, although the CMBR that had got decoupled from matter remained present and kept on being redshifted to lower temperatures as the Universe expanded. We now discuss observations which give us an indication how the Universe might have looked like at the end of the dark age, when there were again radiation-emitting sources which we can try to discover today.

This new field of studying astronomical objects at redshifts lying in the range $z \sim$ 1–6 has blossomed only in the last few years when telescopes like the Hubble Space Telescope (HST) allowed astronomers to study faraway faint sources which could not be studied earlier. We do not plan to provide a full coverage of this newly emerging field here. We shall restrict our discussion here only to the question of what the Universe looked like at these high redshifts – especially to the question whether the Universe looked substantially different from the present Universe and whether we see clear indications of evolution. Another important issue is whether we can determine important cosmological parameters (such as $\Omega_{M,0}$) by using high-redshift observations. Since this topic requires a knowledge of relativistic cosmology, we postpone the discussion of this important topic to Chapter 14.

11.8.1   Quasars and galaxies at high redshift

Since quasars are intrinsically brighter than normal galaxies, they are much more likely to be discovered at large redshifts compared to normal galaxies. Quasars were therefore amongst the first objects at high redshifts to be studied systematically by astronomers. Figure 11.4 shows a plot of quasar number density in co-moving volume as a function of redshift. Keeping in mind that

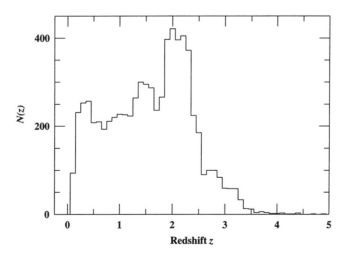

Fig. 11.4 The plot of quasar number density in co-moving volume as a function of redshift. From Peterson (1997, p. 17), based on the data catalogued by Hewitt and Burbidge (1993). (©Cambridge University Press.)

higher redshifts mean earlier times, it is very clear from this plot that the quasar number density has changed with the age of the Universe. It was highest around redshift $z \sim 2$. The evolution of quasar number density was one of the first pieces of evidence of evolution in the world of galaxies discovered by astronomers. As discussed in §9.4.3, we believe that quasar energy emission is caused by gas falling into a supermassive black hole. After the formation of a galaxy, it presumably takes some time for a supermassive black hole to develop at its centre. It is likely that the majority of galaxies formed well before the redshift of $z \sim 2$ when quasars were most abundant, but central black holes took time to form. Since the black hole has to be fed with infalling gas in order to produce quasar activity, we expect such activity to be more prevalent when more gas is available. It is possible that the availability of gas decreases with the age of a galaxy as more gas is used up for star formation. This scenario gives a qualitative explanation of why the quasar number density was maximum around $z \sim 2$. At earlier times, not many galaxies had supermassive black holes at their centres. At later times, the availability of gas for feeding the black holes decreased. This suggests that there must be many dead quasars in our present Universe. These are galaxies with supermassive black holes at their centres, which had acted as quasars once upon a time, but have now run out of gas to feed the central engine.

Since a powerful telescope can detect many millions of galaxies, it is notoriously difficult to isolate high-redshift galaxies in this vast sample. Also, a high-redshift normal galaxy would appear very faint to us (as will be shown §14.4.1) and can be imaged only after a very long exposure. In December 1995 HST imaged a very small portion of the sky without any special characteristics for about 10 days (Williams *et al.*, 1996). The resulting image, known as the *Hubble Deep Field*, is shown in Figure 11.5. It shows about 1500 galaxies at various stages of evolution, some of the galaxies being much fainter than any galaxies imaged before. From a detailed analysis of the Hubble Deep Field, it is concluded that the star formation rate was maximum during redshifts $z \sim 1$–1.5 (Hughes *et al.*, 1998).

To sum up, although some galaxies and quasars might have formed even before $z \sim 6$, such phenomena as quasar activity and star formation reached their maxima much later. It is, however, clear that the Universe revealed by the furthest quasars and furthest normal galaxies is quite different from the present Universe. There is unmistakable evidence for evolution. We now consider the material in the space between galaxies, to find out if this material gives any more clues in completing the story of the earliest galaxies.

11.8.2 The intergalactic medium

Apart from the gas in clusters of galaxies, is there matter in regions of space between clusters and outside of galaxies? Even if there is matter in the

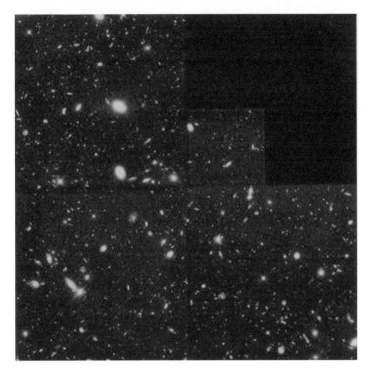

Fig. 11.5 The Hubble Deep Field, photographed with the Hubble Space Telescope. From Williams *et al.* (1996). (©American Astronomical Society. Reproduced with permission from *Astronomical Journal*.)

intergalactic space, the question is how we can detect it. The emission from the intergalactic medium lying outside galaxy clusters has not been detected in any band of the electromagnetic spectrum. The only other way of checking the existence of the intergalactic medium is to look for absorption lines in the spectra of objects lying very faraway. Since quasars are the most faraway objects which are bright enough to obtain spectra from, looking for absorption lines in the spectra of quasars is the best way of searching for the intergalactic medium.

Let us consider the Lyman-α absorption line caused by the transition $1s \rightarrow 2p$ in a hydrogen atom. If an absorbing system is mainly made up of neutral hydrogen atoms, then we expect this line to be one of the strongest absorption lines. The rest wavelength of this line is $\lambda_{L\alpha} = 1216$ Å. Suppose a quasar is at redshift z_{em}. Since quasars typically have broad emission lines, we expect a broad emission line at the redshifted wavelength $(1 + z_{em})\lambda_{L\alpha}$ of the Lyman-α line. If there is some absorbing material on the line of sight lying at some intermediate redshift z_{abs} (obviously we expect $0 < z_{abs} < z_{em}$), then we expect absorption at wavelength $(1 + z_{abs})\lambda_{L\alpha}$. If there is neutral hydrogen gas all along the line of sight, then we would expect to see an absorption trough from

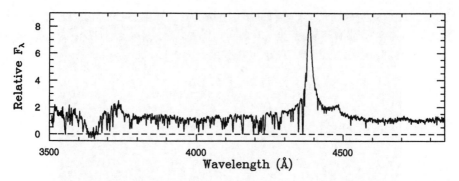

Fig. 11.6 The spectrum of a quasar at redshift $z_{em} = 2.6$. From Wolfe *et al.* (1993). (©American Astronomical Society. Reproduced with permission from *Astrophysical Journal*.)

$\lambda_{L\alpha}$ to $(1 + z_{em})\lambda_{L\alpha}$ in the spectrum of the quasar corresponding to the full run of possible values of z_{abs}. The presence or absence of such an absorption trough in the spectrum of a distant quasar would give us an estimate of the amount of neutral hydrogen gas over the line of sight (Gunn and Peterson, 1965).

Figure 11.6 shows the spectrum of a quasar at redshift $z_{em} = 2.60$, for which the Lyman-α emission line is at 4380 Å. We, however, do not see a continuous absorption trough from 1216 Å to 4380 Å. Instead of a trough, we find a large number of narrowly spaced absorption lines. These absorption lines are collectively referred to as the *Lyman-α forest*. This implies that we do not have a uniform distribution of neutral hydrogen gas along the line of sight. There must be many clouds of neutral hydrogen lying on the path at different redshifts, which are causing the absorption lines. In the particular spectrum shown in Figure 11.6, there is a prominent absorption feature at 3650 Å (corresponding to redshift $z_{abs} = 2.0$) where the radiation seems to fall almost to zero intensity. There must be a very large cloud at this redshift $z_{abs} = 2.0$. Such large dips in the spectra indicating the presence of large hydrogen clouds are found very often in the spectra of many distant quasars.

Readers wishing to know how to analyse these features in quasar spectra quantitatively may consult Peebles (1993, §23). Here we summarize the main conclusions qualitatively. The absence of an absorption trough, which is often referred to as the *Gunn–Peterson test* (Gunn and Peterson, 1965), shows that there is very little neutral hydrogen gas outside the clouds and one quantitatively finds that the number density of hydrogen atoms has to be less than about $10^{-6}\,m^{-3}$. For the sake of comparison, remember that the density of X-ray emitting gas in galaxy clusters is of order 10^3 particles m^{-3}. The large hydrogen clouds producing prominent dips in the quasar spectra (like the dip at 3650 Å in Figure 11.6) are estimated to have masses comparable to the mass of a typical galaxy. The most obvious possibility is that these are galaxies in the making. The smaller clouds producing the absorption lines of the Lyman-α

forest, however, have much smaller masses of the order of a few hundred $M_\odot$. Careful analysis of observational data shows that these smaller clouds are most abundant at redshifts $z \approx 2$–3 and become much less abundant at lower redshifts.

The spectra of distant quasars like the one shown in Figure 11.6 make it clear that neutral hydrogen is mainly found inside isolated clouds. There is very little neutral hydrogen outside these clouds. But does this mean that there is no material outside the clouds and space is really empty in those regions? A more plausible assumption is that there is hydrogen outside the clouds, but it has been ionized and hence is not producing the Lyman-α absorption line. As pointed out in §11.7, matter was ionized before $z \approx 1100$. Then neutral atoms formed, leading to the matter-radiation decoupling. When the first stars, galaxies and quasars started forming, the ionizing radiation from these objects presumably ionized the intergalactic medium again. This is called the *reionization*. The absence of neutral hydrogen atoms between the distant quasars and us (apart from the Lyman-α clouds) is believed to be a consequence of this reionization. However, if light started from a very distant quasar before the reionization, then the light path would initially pass through space filled with neutral hydrogen and we would expect to see a Gunn–Peterson trough in the spectrum at the lower-wavelength side of the redshifted Lyman-α line. There are indications that quasars with redshifts larger than $z \approx 6$ show such troughs in their spectra (Becker *et al.*, 2001).

Altogether, we get a picture of the Universe at redshifts $z \sim 2$–6 which is very different from the present-day Universe. Already some quasars have formed and ionized the intergalactic medium. Embedded in this ionized medium, there are clouds of neutral hydrogen (presumably their interiors are shielded from ionizing photons due to higher densities) with masses of order a few hundred $M_\odot$. There are also more massive clouds which appear like galaxies in the making.

11.9 Structure formation

As discussed in §11.7.1, matter was distributed fairly uniformly at the era $z \approx 1100$ when matter-radiation decoupling took place, the density perturbations at that era being of the order of 10^{-5}. On the other hand, the observations discussed in §11.8 suggest that the first stars, galaxies and quasars should have formed some time before $z \approx 6$. How do we connect the two? Presumably the very small density perturbations present in era $z \approx 1100$ grew by gravitational instability to become the first stars and galaxies before $z \approx 6$. Understanding the details of how this happened is the subject of *structure formation*. This is an enormously complex subject on which quite a lot of research is being done at the present time. Here we shall touch upon only some of the key issues.

In §8.3 we discussed the gravitational instability first studied by Jeans (1902), which showed that density enhancements having masses larger than the Jeans mass keep growing due to the stronger gravitational forces in the regions of density enhancements. The analysis in §8.3 was done for gas in a non-expanding region. To understand the growth of density perturbations in the expanding Universe, we have to carry out the Jeans analysis against an expanding background. This analysis is somewhat more complicated and is given in many standard textbooks (Kolb and Turner, 1990, §9.2; Narlikar, 1993, §7.2). Here we shall not reproduce that derivation (but see Exercise 11.6). Let us summarize the main conclusions only.

1. It is found that the perturbations remained frozen and could not grow as long as the Universe was radiation-dominated.
2. Only after the Universe becomes matter-dominated, can those perturbations grow for which the wavenumber k is less than k_J given by (8.21).
3. In contrast to the result in §8.3 that growing perturbations grow exponentially in time, a growing density perturbation in the matter-dominated expanding Universe is found to grow as

$$\frac{\delta\rho}{\rho} \propto t^{2/3}. \tag{11.35}$$

According to (10.60), the scale factor a in the matter-dominated Universe also grows as the 2/3 power of t. On the basis of (10.60) and (11.35), we can write

$$\frac{\delta\rho}{\rho} \propto a. \tag{11.36}$$

It should be remembered that this result is based on a linear analysis, like the linear analysis presented in §8.3. When the perturbation grows to be of the order of 1, the nonlinear effects become important and thereafter the perturbation can grow much faster than what is suggested by (11.36).

The result (11.36) at once leads us to a difficulty. The density perturbations were of order 10^{-5} at the era $z \approx 1100$. Since the scale factor has grown by a factor of 10^3 between that era and the present time, a straightforward application of (11.36) suggests that $\delta\rho/\rho$ at the present time should be of the order of only 10^{-2}. This certainly contradicts the existence of various structures in the Universe that we see at the present time. Since 10^{-2} is quite small compared to 1, we cannot hope to get around this difficulty by invoking nonlinear effects. Where could our arguments go wrong?

To find a clue for solving this puzzle, let us look at the expression of the critical wavenumber k_J as given by (8.21). For the early matter-dominated era, if we take $a \propto t^{2/3}$, the Friedmann equation (10.27) gives

$$\rho \approx \frac{1}{6\pi G t^2} \tag{11.37}$$

on neglecting the curvature term kc^2/a^2, which was insignificant at early times. On substituting this for ρ_0 in (8.21), we get

$$k_J^2 = \frac{2}{3c_s^2 t^2}.$$ (11.38)

The corresponding Jeans wavelength is

$$\lambda_J = \frac{2\pi}{k_J} = \sqrt{6\pi}\, c_s t.$$ (11.39)

A perturbation would grow only if its wavelength is larger than λ_J. We now consider the sound speed c_s appearing in (11.39). After the matter-radiation decoupling, it is given by (8.15). Before the formation of atoms, however, a perturbation in matter density would be accompanied by a perturbation in the radiation field which was coupled to matter. Since the pressure of the radiation field is $P = (1/3)\rho c^2$, the sound speed in the radiation field can be as large as

$$c_s = \frac{1}{\sqrt{3}}c.$$

On substituting this in (11.39), we find

$$\lambda_J = \sqrt{2\pi}\, ct.$$ (11.40)

This means that the Jeans length was even somewhat larger than the horizon size (of order ct) before matter-radiation decoupling and then suddenly fell to a much smaller value given by (11.39) after the decoupling when c_s becomes equal to the ordinary sound speed in the gas. Although perturbations can, in principle, grow after the Universe became matter-dominated, most of the perturbations would have wavelengths smaller than the Jeans length (11.40) and would not grow as long as matter and radiation remained coupled. Only after the radiation becomes decoupled, does the Jeans length become small and perturbations larger than it start growing.

We have discussed in §11.5 the possibility that much of the matter in the Universe is non-baryonic cold dark matter. If this is true, then the situation can be somewhat tricky. We expect only the baryonic matter to interact with radiation and to be coupled with it till the formation of atoms. Since the cold dark matter can have only the weak interaction, it must have become decoupled from the other components of the Universe when the temperature fell below MeV values. By the time the Universe became matter-dominated, the cold dark matter was totally decoupled and the Jeans length for the cold dark matter would be given by (11.39), with c_s representing the sound speed in the cold dark matter. This Jeans length would be much smaller than the Jeans length of baryonic matter given by (11.40) before the formation of atoms. So, as soon as the Universe became matter-dominated and it became possible for perturbations to grow, the perturbations in cold dark matter larger than its Jeans length would start

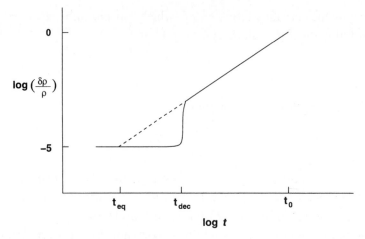

Fig. 11.7 Sketch indicating how perturbations in baryonic matter (solid line) and in cold dark matter (dashed line) would have grown.

growing. Eventually, when the atoms form, the baryonic matter also becomes decoupled and its Jeans length drops drastically, allowing perturbations larger than its Jeans length to grow. By that time, the perturbations in the cold dark matter would have grown considerably and would have produced gravitational potential wells in the regions where the cold dark matter got clumped. Once the baryonic perturbations are allowed to grow after the formation of atoms, the baryonic matter would quickly fall in the gravitational potential wells created by the cold dark matter.

Figure 11.7 gives a sketch of how the perturbations must have grown. When the Universe was radiation-dominated till $t = t_{eq}$, the perturbations in the baryonic matter and the cold dark matter must have had similar amplitudes and could not grow. Since perturbations in the baryonic matter remained frozen till the decoupling time $t = t_{dec}$ and CMBR observations tell us that the perturbations at the time t_{dec} were of amplitude 10^{-5}, we expect the primordial perturbations also to have this amplitude. Since cold dark matter perturbations started growing from $t = t_{eq}$ when $a = a_{eq}$ given by (10.52) and the perturbation growth rate is given by (11.36), we expect the dark matter perturbations to become close to 1 by the present time. After $t = t_{dec}$, the baryonic perturbations fell in the potential wells of the cold dark matter and started following the cold dark matter perturbations, as indicated by the solid line in Figure 11.7. Hence the baryonic perturbations also should become of order 1 by the present time. When the perturbations are no longer very small compared to 1, nonlinear effects start becoming important and the clustering of matter to produce various structures proceeds at a much faster rate. Evolution due to these nonlinear effects can be studied best by carrying on detailed numerical simulations.

Many ambitious numerical simulations of structure formation are being carried out by different groups, following the pioneering early work by Davis *et al.* (1985). These simulations are still not realistic enough to show matter eventually forming stars. First of all, the grid spacing in most of the simulations is much larger than the size of a star. Secondly, star formation involves several complicated physical processes and it is not easy to include them in a simulation which has to follow the large-scale perturbations as well. However, these simulations indicate that structures like what we see today may indeed form if the Universe contains a significant amount of cold dark matter in addition to baryonic matter. This is taken as further compelling evidence that the dark matter in the Universe is cold and not hot. Only if the Universe has a large amount of cold dark matter, would baryonic perturbations having amplitude of order 10^{-5} at t_{dec} be able to grow enough by the present time, by falling in the potential wells created by the cold dark matter, so as to explain the observed structures of today.

Exercises

11.1 (a) Using the definition of the Riemann zeta function

$$\zeta(n) = \sum_{k=1}^{\infty} k^{-n}$$

derive (11.9) for bosons from (11.7). Express (11.8) for bosons in terms of $\zeta(4)$ and then obtain (11.10) for bosons by using the fact that $\zeta(4) = \pi^4/90$. [Hint: You can use

$$\frac{1}{e^x - 1} = e^{-x} + e^{-2x} + e^{-3x} + e^{-4x} + \cdots$$

to evaluate the integrals in (11.7) and (11.8).]
    (b) Defining

$$I_n^{\pm} = \int_0^{\infty} \frac{x^n \, dx}{e^x \pm 1},$$

show that $I_n^- - I_n^+ = 2^{-n} I_n^-$. Use this relation to derive expressions for fermions in (11.9) and (11.10) from the expressions for bosons.

11.2 If the reaction rate for the reaction $n + \nu_e \longleftrightarrow p + e^-$ were to fall below the expansion rate of the Universe at a temperature $T \approx 0.4\,\mathrm{MeV}$, what would have been the helium abundance in the Universe?

11.3 In the early Universe before electron-positron annihilation, the electron number density must have been slightly higher than the positron number density. Show that this implies that the chemical potentials for electrons and positrons have to be non-zero. How can one calculate these chemical potentials theoretically?

11.4 Estimate the gas pressure in the Universe at the epoch corresponding to $z \approx 1100$ and consider hydrogen gas kept at that pressure. Apply the Saha equation (2.29) to calculate numerically the ionization fraction x for different temperatures T and plot x as a function of T. You will find that x changes from values close to 0 to values close to 1 within a not very wide temperature range. What is the approximate temperature at which this transition takes place?

11.5 Suppose we find the optical depth τ of CMBR photons passing through the middle of a galaxy cluster from the Sunyaev–Zeldovich effect and we also measure the X-ray flux f_ν at frequency ν from the hot gas in the cluster. Assuming that the hot gas makes up a sphere of radius R_c with uniform electron density n_e inside, we obviously have $\tau = 2\sigma_T R_c n_e$ and

$$f_\nu = \frac{\frac{4}{3}\pi R_c^3 \epsilon_\nu}{4\pi D^2},$$

where ϵ_ν is the emissivity per unit volume of the hot gas and D is the distance of the galaxy cluster. Following (8.70), we can write

$$\epsilon_\nu = \frac{An_e^2}{\sqrt{T}}e^{-h\nu/\kappa_B T}.$$

Now show that

$$D = \frac{A\,\Delta\theta}{24\sigma_T^2\sqrt{T}}e^{-h\nu/\kappa_B T}\frac{\tau^2}{f_\nu},$$

where $\Delta\theta = 2R_c/D$ is the observed angular size of the X-ray emitting gas sphere. This expression for D is used to determine the distances of galaxy clusters. [Note: If the galaxy cluster is at a large redshift, then some corrections have to be applied to the above expression for D.]

11.6 We have discussed in §10.4 how the Friedmann equation can be obtained classically by considering the radial expansion of a spherical shell located at $a(t)$ at time t. Suppose there is a perturbation in the central region such that the radius $a(t)$ has become $a(t) + l(t)$. For small $l(t)$, show that

$$\frac{d^2l}{dt^2} = \frac{2GM}{a^3}l = \frac{8\pi G}{3}\rho(t)l,$$

where M is the mass inside the spherical shell and

$$\rho = \frac{3}{4\pi}\frac{M}{a^3}.$$

If the material within the shell $a(t)$ evolves slightly differently due to this perturbation compared to what we would expect from the Friedmann equation, the density difference compared to the unperturbed density leads to the parameter

$$\delta = \frac{\delta\rho}{\rho} = -3\frac{l}{a}.$$

For a marginally critical situation (i.e. for a situation in which the total mechanical energy of the unperturbed shell is zero), derive the equation for δ:

$$\ddot{\delta} + \frac{4}{3t}\dot{\delta} - \frac{2}{3t^2}\delta = 0.$$

This equation tells us how a density perturbation in the central region of expansion would evolve. It turns out that the evolution equation of density perturbations in the expanding matter-dominated Universe is exactly this if $k \ll k_J$. Show that

$$\delta \propto t^{2/3}$$

is a solution.

12

Elements of tensors and general relativity

12.1 Introduction

We have pointed out in §10.2 that in general relativity we have to deal with the curvature of spacetime and that tensors provide a natural mathematical language for describing such curvature. We now plan to give an introduction to tensor analysis and then an introduction to general relativity at a technical level. It will be useful for readers to be familiar with the qualitative concepts introduced in §10.2 before studying this chapter.

Since general relativity is a challenging subject, it is helpful to clearly distinguish the purely mathematical topics from the physical concepts of general relativity. So, when we develop tensor analysis in the next section, we shall develop it as a purely mathematical subject without bringing in general relativistic concerns at all. The two-dimensional metrics (10.7), (10.8) and (10.9) introduced in §10.2 will be used as illustrative examples repeatedly to clarify various points. When various formulae of tensor analysis are applied to metrics of dimensions higher than two, the algebra can be horrendous. It is, therefore, advisable to develop a familiarity with tensors by first applying the important results to two-dimensional surfaces.

After introducing the basics of tensor analysis in the next section, we shall start developing the basic concepts of general relativity from §12.3. Since general relativity is now used in many areas of astrophysics, a minimum knowledge of general relativity is expected nowadays from a student of astrophysics, irrespective of the area of specialization. What will be presented in this chapter is somewhat like that bare minimum. There are many excellent textbooks on this subject, to which the reader can turn in order to learn the subject at a greater depth.

12.2 The world of tensors

It requires a little bit of practice to feel comfortable with tensors. While we shall introduce all the basic notions, we shall be rather brief, and some previous acquaintance with tensor algebra at least may be helpful in following our discussion.

12.2.1 What is a tensor?

In elementary textbooks of physics, usually two kinds of physical quantities are introduced: scalars and vectors. A vector is defined as a quantity which has both magnitude and direction. There is, however, an alternative way of defining a vector. A vector has components of which each one is associated with one coordinate axis (i.e. components A_x, A_y and A_z of the vector $\mathbf{A}$ are associated with x, y and z axes respectively) and they transform in some particular way when we change from one coordinate system to another. Since this alternative definition of vectors provides a natural entry point into the world of tensors, let us consider in some detail how the components of a vector transform on changing the coordinate system.

In order to find out how the components of a vector transform, we first need to know how the components of a vector are defined in a coordinate system. Let us take the concrete examples of the generalized velocity and the generalized force which we encounter in Lagrangian mechanics (see, for example, Goldstein, 1980, §1–4; Landau and Lifshitz, 1976, §1). Suppose we denote the generalized coordinates in a system by x^i, where i can have values $i = 1, 2, \ldots, N$. The rationale of writing i as a superscript rather than a subscript will be clear as we proceed. A component of the generalized velocity is given by dx^i/dt, whereas a component of the generalized force is given by $-\partial V/\partial x^i$ where V is the potential. If components of generalized velocity and generalized force are defined in another coordinate system $\bar{x}^i$ in exactly the same way, then the chain rule of partial differentiation implies

$$\frac{d\bar{x}^i}{dt} = \sum_{k=1}^{N} \frac{dx^k}{dt} \frac{\partial \bar{x}^i}{\partial x^k}, \tag{12.1}$$

$$\frac{\partial V}{\partial \bar{x}^i} = \sum_{k=1}^{N} \frac{\partial V}{\partial x^k} \frac{\partial x^k}{\partial \bar{x}^i}. \tag{12.2}$$

It should be noted that these two transformation laws are slightly different. Vectors transforming like dx^i/dt are called *contravariant vectors* and are indicated by superscripts, whereas vectors transforming like $\partial V/\partial x^i$ are called *covariant vectors* and are indicated by subscripts (i in x^i or $\bar{x}^i$ appearing at the bottom of a derivative should be treated as subscript). We also introduce the well-known *summation convention* that if an index is repeated twice in a term, once as a sub-script and once as a superscript, then it automatically implies summation over the possible values of that index and it is not necessary to put the summation sign explicitly. Using this summation convention, the transformation laws for a contravariant vector A^i and a covariant vector A_i would be

$$\overline{A}^i = A^k \frac{\partial \overline{x}^i}{\partial x^k}, \tag{12.3}$$

$$\overline{A}_i = A_k \frac{\partial x^k}{\partial \overline{x}^i}, \tag{12.4}$$

where $\overline{A}^i$ and $\overline{A}_i$ are components of these contravariant and covariant vectors in the coordinate system $\overline{x}^i$. By comparing (12.3) with (12.1) and (12.4) with (12.2), it is clear that the generalized velocity dx^i/dt transforms as a contravariant vector and the generalized force $-\partial V/\partial x^i$ transforms as a covariant vector. If we consider the transformation from one Cartesian frame to another (for example, due to a rotation from one frame to the other in two dimensions), it is easy to show that

$$\frac{\partial \overline{x}^i}{\partial x^k} = \frac{\partial x^k}{\partial \overline{x}^i}.$$

This implies that the distinction between contravariant and covariant vectors disappears if we consider only transformations between Cartesian frames.

A component of a vector is associated with only one coordinate axis. In the case of a general tensor, a component can be associated with several coordinate axes. So a component will generally have several indices. The transformation law of a general tensor will be the following:

$$\overline{T}^{ab..d}_{l..n} = T^{\alpha\beta..\delta}_{\lambda..\nu} \frac{\partial \overline{x}^a}{\partial x^\alpha} \frac{\partial \overline{x}^b}{\partial x^\beta} .. \frac{\partial \overline{x}^d}{\partial x^\delta} \frac{\partial x^\lambda}{\partial \overline{x}^l} .. \frac{\partial x^\nu}{\partial \overline{x}^n}. \tag{12.5}$$

Note that some indices are put as superscripts and some as subscripts depending on whether the corresponding parts of the transformation are like contravariant vectors or covariant vectors.

From the transformation law (12.5) of tensors, it is very easy to show that the product $A_i B_k$ of two vectors A_i and B_k should transform like a tensor with two covariant components i and k. This can be generalized to the result that the product of two tensors gives a tensor of higher rank. One very important operation is the *contraction* of tensors. Suppose we write $n = d$ in the tensor $\overline{T}^{ab..d}_{l..n}$. This, by the summation convention, implies that we are summing over all possible values of $n = d$. It follows from (12.5) that

$$\overline{T}^{ab..d}_{l..d} = T^{\alpha\beta..\delta}_{\lambda..\nu} \frac{\partial \overline{x}^a}{\partial x^\alpha} \frac{\partial \overline{x}^b}{\partial x^\beta} .. \frac{\partial \overline{x}^d}{\partial x^\delta} \frac{\partial x^\lambda}{\partial \overline{x}^l} .. \frac{\partial x^\nu}{\partial \overline{x}^d}.$$

Using the fact that

$$\frac{\partial \overline{x}^d}{\partial x^\delta}\frac{\partial x^\nu}{\partial \overline{x}^d} = \delta_\delta^\nu,$$

we easily get

$$\overline{T}_{l..d}^{ab..d} = T_{\lambda..\delta}^{\alpha\beta..\delta}\frac{\partial \overline{x}^a}{\partial x^\alpha}\frac{\partial \overline{x}^b}{\partial x^\beta}..\frac{\partial x^\lambda}{\partial \overline{x}^l}.. \tag{12.6}$$

on making use of the obvious result $T_{\lambda..\nu}^{\alpha\beta..\delta}\delta_\delta^\nu = T_{\lambda..\delta}^{\alpha\beta..\delta}$ following from the properties of the Kronecker δ. We leave it as an exercise for the reader to verify that δ_k^i transforms like a tensor, but not δ_{ik} or δ^{ik} (the Kronecker δ in any form is assumed to have the value 1 if $i = k$ and 0 otherwise). It is clear from (12.6) that $T_{\lambda..\delta}^{\alpha\beta..\delta}$ transforms like a tensor with one contravariant index and one covariant index less compared to $T_{\lambda..\nu}^{\alpha\beta..\delta}$. Thus the operation of contraction reduces the rank of the tensor (by reducing one contravariant rank and one covariant rank).

12.2.2   The metric tensor

We have pointed out in §10.2 that the distance between two nearby points in a space is given by (10.5). Using our present notation and the summation convention, we can write

$$ds^2 = g_{ik}\,dx^i dx^k. \tag{12.7}$$

Although x^i does not transform as a vector, dx^i is clearly a contravariant vector. Hence $g_{ik}dx^l ds^m$ must be a tensor which, after two contractions, would give a scalar, implying that ds^2 is a scalar.

We pointed out in §10.2 that it is the metric tensor g_{ik} which determines whether a space is curved or not. We shall now develop the mathematical machinery to calculate curvature. As we develop the techniques, we shall illustrate them by applying them to the metrics

$$ds^2 = dr^2 + r^2\,d\theta^2, \tag{12.8}$$

$$ds^2 = a^2(d\theta^2 + \sin^2\theta\,d\phi^2), \tag{12.9}$$

where (12.8) corresponds to a plane using polar coordinates and (12.9) corresponds to the surface of a sphere. We shall also sometimes consider the metric

$$ds^2 = a^2(dx_1^2 + \sinh^2 x_1\,dx_2^2), \tag{12.10}$$

which corresponds to the surface having the property of a saddle point at every point. We pointed out in §10.2 that these three metrics can be written in the forms (10.7)–(10.9) by using very similar notation and also discussed that these correspond to the only three possible uniform two-dimensional surfaces (i.e. surfaces in which every point is equivalent). It may be noted that (12.7) allows for cross-terms of the form $g_{12}dx^1dx^2$. The metrics (12.8)–(12.10), however, have only pure quadratic terms and no cross-terms. If the coordinate system is orthogonal, then we do not have cross-terms in the expression of ds^2. In this book, we shall restrict our discussions to only simple metrics without cross-terms corresponding to orthogonal coordinates. In other words, we shall only be concerned with metric tensors g_{ik} which have non-zero terms on the diagonal alone when represented in the form of a matrix. For the three metrics (12.8)–(12.10), the components of the metric tensor respectively are

$$g_{rr} = 1, \ g_{\theta\theta} = r^2, \ g_{r\theta} = g_{\theta r} = 0, \tag{12.11}$$

$$g_{\theta\theta} = a^2, \ g_{\phi\phi} = a^2 \sin^2\theta, \ g_{\theta\phi} = g_{\phi\theta} = 0, \tag{12.12}$$

$$g_{11} = a^2, \ g_{22} = a^2 \sinh^2 x_1, \ g_{12} = g_{21} = 0. \tag{12.13}$$

There are astrophysical situations where the cross-terms become important, such as the Kerr metric of a rotating black hole, which is beyond this elementary book.

From a contravariant vector A^i, it is possible to construct the corresponding covariant vector in the following way

$$A_i = g_{ik}A^k. \tag{12.14}$$

This is often called the *lowering of an index*. It is very easy to do it if the metric is diagonal. For example, if (A^r, A^θ) are the contravariant components of a vector in the plane with the metric tensor (12.11), then the corresponding covariant components are $(A_r = A^r, A_\theta = r^2A^\theta)$. Suppose dx^i and dx^k have corresponding covariant vectors dx_i and dx_k obtained according to (12.14). It should be possible to write the metric in the form

$$ds^2 = g^{ik}dx_idx_k. \tag{12.15}$$

By requiring that ds^2 given by (12.7) and (12.15) should be equal, the reader is asked to show that

$$g^{ik}g_{kl} = \delta^i_l. \tag{12.16}$$

For a diagonal metric tensor g_{ik}, it is particularly easy to obtain the corresponding g^{ik}. One merely has to take the inverse of the diagonal elements while leaving the off-diagonal elements zero in order to satisfy (12.16). For example, for the metric tensor (12.12) corresponding to the surface of a sphere, it is easily seen that

$$g^{\theta\theta} = \frac{1}{a^2}, \ g^{\phi\phi} = \frac{1}{a^2 \sin^2\theta}, \ g^{\theta\phi} = g^{\phi\theta} = 0 \tag{12.17}$$

would satisfy (12.16). Once the contravariant metric tensor g^{ik} has been introduced, we can use it to raise an index and obtain a contravariant vector from a

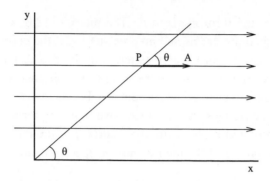

Fig. 12.1 A constant vector field $\mathbf{A}$, shown with the point P where we consider the components of this vector field.

covariant vector in the following way

$$A^i = g^{ik} A_k. \tag{12.18}$$

Starting from a contravariant vector A^i, if we once lower the index by (12.14) and then raise it again by (12.18), then it is easy to use (12.16) to show that we get back the same vector A^i. It may be noticed that we are repeatedly leaving various simple steps to the readers so that they can carry out these steps themselves and become conversant with tensor operations.

We often have to deal with components of a vector along the coordinate directions. For a particle moving in a plane, for example, the components of velocity in polar coordinates are $(\dot{r}, r\dot{\theta})$. We can call these *vectorial components*. It is obvious that the vectorial components in different orthogonal coordinate systems will not transform amongst each other according to either (12.3) or (12.4). However, if we divide the two vectorial components of velocity by $\sqrt{g_{rr}} = 1$ and $\sqrt{g_{\theta\theta}} = r$ respectively, then we get the contravariant velocity vector $(\dot{r}, \dot{\theta})$ in polar coordinates. In general, if we divide the i-th vectorial component of a vector in an orthogonal coordinate system by $\sqrt{g_{ii}}$ and do this for each component, then we get a set of components of the vector which would transform between orthogonal coordinates in accordance with the rule (12.3) for contravariant vectors. Similarly, if we multiply the vectorial components by $\sqrt{g_{ii}}$, then we would get the components of the corresponding covariant vector. As an example, let us consider a constant vector field $\mathbf{A}$ in a plane and find its components in polar coordinates if we assume $\mathbf{A}$ to transform like a contravariant vector. Let us measure θ from the direction of $\mathbf{A}$ as shown in Figure 12.1. At a point P with coordinates (r, θ), the vectorial components of $\mathbf{A}$ in polar coordinates are $(A \cos\theta, -A \sin\theta)$. Making use of (12.11), we easily see that the contravariant form of the constant vector in polar coordinates will be

$$A^r = A \cos\theta, \quad A^\theta = -A \frac{\sin\theta}{r}, \tag{12.19}$$

whereas the covariant form will be

$$A_r = A \cos \theta, \quad A_\theta = -Ar \sin \theta. \tag{12.20}$$

One can easily check that these expressions for different components will follow on applying (12.3) and (12.4) to the components in Cartesian coordinates: $A^x = A_x = A$, $A^y = A_y = 0$. It should be apparent that the question of whether a vector is contravariant or covariant does not make sense until we are told how its components in an arbitrary coordinate system are to be taken. For example, from a knowledge of how the generalized velocity or the generalized force is to be defined in different coordinates, we could use (12.1) and (12.2) to conclude that they transform as contravariant and covariant vectors respectively.

12.2.3   Differentiation of tensors

Let us consider a contravariant vector field $A^k(x^m)$. In another coordinate system $\overline{x}^l$, this same vector field will be denoted by $\overline{A}^i(\overline{x}^l)$, the transformation law being given by (12.3). It readily follows from (12.3) that

$$\frac{\partial \overline{A}^i}{\partial \overline{x}^l} = \frac{\partial A^k}{\partial x^m} \frac{\partial x^m}{\partial \overline{x}^l} \frac{\partial \overline{x}^i}{\partial x^k} + A^k \frac{\partial x^m}{\partial \overline{x}^l} \frac{\partial^2 \overline{x}^i}{\partial x^m \partial x^k}.$$

Due to the presence of the second term on the right-hand side, it is clear that the derivative $\partial A^k / \partial x^m$ does not transform like a tensor. To understand the physical reason behind it, we consider the contravariant form of a constant vector field given by (12.19). It follows from (12.19) that

$$\frac{\partial A^r}{\partial \theta} = -A \sin \theta.$$

Even though we may expect the derivative of a constant vector field to be zero, that is not the case. This is presumably due to the fact that we are using curvilinear coordinates. From the derivative, we have to remove the part coming due to the curvature of the coordinate system in order to get a more physically meaningful expression of the derivative. We now discuss how to do it.

In order to differentiate a vector field $\mathbf{A}$, we need to subtract $\mathbf{A}(\mathbf{x})$ from $\mathbf{A}(\mathbf{x} + d\mathbf{x})$. Now we can sensibly talk of adding or subtracting vectors only if they are at the same point. So we need to do what is called a *parallel transport* of $\mathbf{A}(\mathbf{x})$ to $\mathbf{x} + d\mathbf{x}$ before we subtract it from $\mathbf{A}(\mathbf{x} + d\mathbf{x})$. We expect that $\mathbf{A}$ will change to $\mathbf{A} + \delta\mathbf{A}$ under such a parallel transport. For example, when we transport a vector $\mathbf{A}$ even on a plane surface, its components in polar coordinates (A^r, A^θ) will in general change. On physical grounds, we expect that the change δA^i in A^i under parallel transport from x^l to $x^l + dx^l$ will be proportional to the displacement and to the vector itself. Hence we can write

$$\delta A^i = -\Gamma^i_{kl} A^k dx^l, \tag{12.21}$$

where Γ^i_{kl} is known as the *Christoffel symbol*, named after Christoffel (1869) who introduced it. The Christoffel symbol is not a tensor because δA^i is not a tensor. We now expect a proper derivative of A^i to be given by

$$\frac{DA^i}{Dx^l} = \lim_{dx^l \to 0} \frac{A^i(x^l + dx^l) - [A^i(x^l) + \delta A^i]}{dx^l}.$$

On substituting for δA^i from (12.21) and writing

$$A^i(x^l + dx^l) = A^i(x^l) + \frac{\partial A^i}{\partial x^l} dx^l,$$

we get

$$\frac{DA^i}{Dx^l} = \frac{\partial A^i}{\partial x^l} + \Gamma^i_{kl} A^k. \tag{12.22}$$

This is known as the *covariant derivative*. It may be noted that the ordinary derivative $\partial A^i / \partial x^l$ and the covariant derivative DA^i / Dx^l are sometimes denoted by symbols $A^i_{,l}$ and $A^i_{;l}$. Other popular symbols for these derivatives are $\partial_l A^i$ and $\nabla_l A^i$. We shall, however, use the longer notation in this book in order to avoid introducing too many new notations in our short discussion of general relativity.

In order to calculate the covariant derivate of a vector, we first need to figure out how to evaluate the Christoffel symbols appearing in the expression (12.22) of the covariant derivative. It is possible to find the Christoffel symbols from the metric tensor. For deriving the relation of Christoffel symbols to the metric tensor, we first have to figure out the expressions of covariant derivatives for a covariant vector A_i and tensors of higher rank. By noting that $A_i B^i$ is a scalar for which we must have $\delta(A_i B^i) = 0$, we obtain

$$B^i \, \delta A_i = -A_i \, \delta B^i = A_i \, \Gamma^i_{kl} B^k \, dx^l.$$

For indices which are summed (i.e. which are repeated twice above and below), we can change the symbols without affecting anything else. So we can write the above relation as

$$B^i \, \delta A_i = A_k \Gamma^k_{il} B^i \, dx^l,$$

from which it follows that

$$\delta A_i = \Gamma^k_{il} A_k \, dx^l. \tag{12.23}$$

It is now easy to show that

$$\frac{DA_i}{Dx^l} = \frac{\partial A_i}{\partial x^l} - \Gamma^k_{il} A_k. \tag{12.24}$$

We leave it for the reader to argue that the covariant derivative of the tensor A_{ik} must be given by

$$\frac{DA_{ik}}{Dx^l} = \frac{\partial A_{ik}}{\partial x^l} - \Gamma^m_{kl} A_{im} - \Gamma^m_{il} A_{mk}. \tag{12.25}$$

We now show that the Christoffel symbol is symmetric in its bottom two indices. Let us consider a vector

$$A_i = \frac{\partial V}{\partial x^i},$$

where V is a scalar field. We readily find that

$$\frac{DA_i}{Dx^k} - \frac{DA_k}{Dx^i} = (\Gamma^l_{ki} - \Gamma^l_{ik})\frac{\partial V}{\partial x^l}.$$

We note that the left-hand side is a tensor, which should transform between frames by obeying tensor transformation formulae. It is obvious that the left-hand side in a Cartesian frame is zero and hence it must be zero in all frames. Then the right-hand side also should be zero, implying

$$\Gamma^l_{ik} = \Gamma^l_{ki}. \tag{12.26}$$

Now, if A_i is the covariant vector associated with A^i, then

$$\frac{DA_i}{Dx^l} = \frac{D}{Dx^l}(g_{ik}A^k) = g_{ik}\frac{DA^k}{Dx^l} + A^k\frac{Dg_{ik}}{Dx^l}.$$

Since A_i and A^i essentially correspond to the same physical entity, their covariant derivatives also must be the same physical entity and should be related to each other as

$$\frac{DA_i}{Dx^l} = g_{ik}\frac{DA^k}{Dx^l}.$$

It then follows that we must have

$$\frac{Dg_{ik}}{Dx^l} = 0. \tag{12.27}$$

From (12.25) and (12.27), we have

$$\frac{\partial g_{ik}}{\partial x^l} = \Gamma^m_{kl} g_{im} + \Gamma^m_{il} g_{mk}. \tag{12.28}$$

On permuting the symbols, we also have

$$\frac{\partial g_{li}}{\partial x^k} = \Gamma^m_{ik} g_{lm} + \Gamma^m_{lk} g_{mi}, \tag{12.29}$$

$$\frac{\partial g_{kl}}{\partial x^i} = \Gamma^m_{li} g_{km} + \Gamma^m_{ki} g_{ml}. \tag{12.30}$$

On subtracting (12.28) from the sum of (12.29) and (12.30), we get

$$2\Gamma^n_{ik}g_{ln} = \frac{\partial g_{li}}{\partial x^k} + \frac{\partial g_{kl}}{\partial x^i} - \frac{\partial g_{ik}}{\partial x^l}$$

on keeping in mind the symmetry property (12.26) of the Christoffel symbol (as well as the symmetry property of the metric tensor). On multiplying this equation by g^{ml} and making use of (12.16), we finally have

$$\Gamma^m_{ik} = \frac{1}{2}g^{ml}\left(\frac{\partial g_{li}}{\partial x^k} + \frac{\partial g_{lk}}{\partial x^i} - \frac{\partial g_{ik}}{\partial x^l}\right). \tag{12.31}$$

This is the final expression of the Christoffel symbol in terms of the metric tensor. If we have the metric tensor for a space, we can calculate the Christoffel symbols by using (12.31) and then work out any covariant derivatives. It is obvious that the Christoffel symbols are zero in a Cartesian coordinate system. Even when the space is curved, it can be shown that it is possible to introduce Cartesian coordinates in a local region in such a way that the spatial derivatives of the metric tensor are zero and the Christoffel symbols vanish. It is, however, not possible to make the higher derivatives of the metric tensor zero in a general situation.

Calculating Christoffel symbols for spaces of several dimensions can involve quite a bit of algebra, even though the algebra is usually straightforward if the metric tensor is not too complicated. For four-dimensional spacetime, for example, ik in Γ^m_{ik} can have 10 independent combinations in view of the symmetry. Since m can have four possible values, it is clear that the Christoffel symbol will have 40 components in four-dimensional spacetime. We shall see in §12.2.4 that one has to do further computations to obtain the curvature of space-time from the Christoffel symbols. In general relativistic applications, often we have to do this very long algebra involving quantities with many components. We shall only consider the two-dimensional metrics (12.8) and (12.9) here for illustrative purposes. The metric tensors for them are explicitly written down in (12.11) and (12.12) respectively. On substituting (12.11) into (12.31), we find that the Christoffel symbols for the plane in polar coordinates are

$$\Gamma^r_{\theta\theta} = -r, \ \Gamma^\theta_{r\theta} = \frac{1}{r}, \tag{12.32}$$

whereas the other four components (since the Christoffel symbol has six independent components in two dimensions) turn out to be zero. For the surface of the sphere with the metric tensor given by (12.12), the only non-zero components are

$$\Gamma^\theta_{\phi\phi} = -\sin\theta\cos\theta, \ \Gamma^\phi_{\theta\phi} = \cot\theta. \tag{12.33}$$

We have seen that the ordinary derivatives of the constant vector field given by (12.19) are non-zero. Now we are ready to show that the covariant derivatives are zero, as we expect in the case of a constant vector field. Using (12.22), we can write down

$$\frac{DA^r}{D\theta} = \frac{\partial A^r}{\partial \theta} + \Gamma^r_{k\theta} A^k = \frac{\partial A^r}{\partial \theta} + \Gamma^r_{\theta\theta} A^\theta,$$

since the other term is zero due to the fact that $\Gamma^r_{r\theta} = 0$. On substituting for (A^r, A^θ) from (12.19) and using $\Gamma^r_{\theta\theta} = -r$, we find that

$$\frac{DA^r}{D\theta} = 0.$$

The other components of the covariant derivative can also similarly be shown to be zero. Since the covariant derivative is a proper tensor and we know that the covariant derivative of a constant vector field in Cartesian coordinates (where the derivative reduces to an ordinary derivative) is zero, we would expect it to be zero in the other coordinates as well. The small demonstration of this by explicit calculations has hopefully given the reader some idea of how such calculations are done.

12.2.4 Curvature

Suppose a vector A^i lying on a flat surface is made to undergo parallel transport along a closed path and eventually brought back to its original location. We expect the final vector to be identical with the initial vector. On the other hand, if we have a curved surface, then the parallel transport of a vector along some arbitrary closed path may not bring it back to its original self. From (12.23) we conclude that the change in the A_k after parallel transport along a closed path C will be

$$\Delta A_k = \oint_C \Gamma^i_{kl} A_i \, dx^l. \tag{12.34}$$

Whether a surface is plane or curved can be inferred by finding out if the right-hand side of (12.34) is zero or non-zero for arbitrary closed paths. The same considerations should apply to higher dimensions as well. We can conclude a space to have zero curvature if $\oint_C \Gamma^i_{kl} A_i dx^l$ around any arbitrary closed path is always zero. On the other hand, if we can find some closed path such that this line integral over it is non-zero, then the space must be curved.

To proceed further, we have to convert the line integral of (12.34) into a surface integral. Let us try to write down Stokes's theorem of ordinary vector analysis in tensorial notation. We consider ordinary three-dimensional Cartesian space. An element of area $d\mathbf{s}$ is a pseudovector from which we can construct a tensor

$$df^{ik} = \begin{pmatrix} 0 & ds_z & -ds_y \\ -ds_z & 0 & ds_x \\ ds_y & -ds_x & 0 \end{pmatrix}.$$

On using this notation, Stokes's theorem can be written as

$$\oint_C A_i \, dx^i = \frac{1}{2} \int df^{ik} \left(\frac{\partial A_k}{\partial x^i} - \frac{\partial A_i}{\partial x^k} \right), \tag{12.35}$$

where the right-hand side is a surface integral over a surface of which C is the boundary. This tensorial expression of Stokes's theorem holds for curved space and for higher dimensions also, though we shall not try to justify it here. On converting the right-hand side of (12.34) into a surface integral with the help of Stokes's theorem, we get

$$\Delta A_k = \frac{1}{2} \int \left[\frac{\partial}{\partial x^l} (\Gamma^i_{km} A_i) - \frac{\partial}{\partial x^m} (\Gamma^i_{kl} A_i) \right] df^{lm}$$

$$= \frac{1}{2} \int \left[\frac{\partial \Gamma^i_{km}}{\partial x^l} A_i - \frac{\partial \Gamma^i_{kl}}{\partial x^m} A_i + \Gamma^i_{km} \frac{\partial A_i}{\partial x^l} - \Gamma^i_{kl} \frac{\partial A_i}{\partial x^m} \right] df^{lm}.$$

Now the change in A_i in the present situation is caused by parallel transport from its original position. So the change in A_i must be given by (12.23) from which it follows that

$$\frac{\partial A_i}{\partial x^l} = \Gamma^n_{il} A_n.$$

Using this we get

$$\Delta A_k = \frac{1}{2} \int \left[\frac{\partial \Gamma^i_{km}}{\partial x^l} - \frac{\partial \Gamma^i_{kl}}{\partial x^m} + \Gamma^n_{km} \Gamma^i_{nl} - \Gamma^n_{kl} \Gamma^i_{nm} \right] A_i \, df^{lm}.$$

This can be written as

$$\Delta A_k = \frac{1}{2} \int R^i_{klm} A_i \, df^{lm}, \tag{12.36}$$

where

$$R^i_{klm} = \frac{\partial \Gamma^i_{km}}{\partial x^l} - \frac{\partial \Gamma^i_{kl}}{\partial x^m} + \Gamma^n_{km} \Gamma^i_{nl} - \Gamma^n_{kl} \Gamma^i_{nm}. \tag{12.37}$$

As we have already pointed out, whether a space is flat or curved can be determined by finding out if ΔA_k is always zero or not. This in turn depends of whether R^i_{kml} is zero or non-zero, as should be clear from (12.36). We thus conclude that R^i_{klm} given by (12.37) is a measure of the curvature of space. It is called the *Riemann curvature tensor*, after Riemann who carried out pathbreaking studies on the curvature of space (Riemann, 1868).

Just from the definition of R^i_{klm} given by (12.37), the following symmetry properties follow

$$R^i_{klm} = -R^i_{kml},$$
(12.38)

$$R^i_{klm} + R^i_{mkl} + R^i_{lmk} = 0.$$
(12.39)

Another important result, known as the *Bianchi identity*, is

$$\frac{DR^n_{ikl}}{Dx^m} + \frac{DR^n_{imk}}{Dx^l} + \frac{DR^n_{ilm}}{Dx^k} = 0.$$
(12.40)

The best way of proving this identity is to go to a Cartesian frame and use the result following from (12.5) that, if a tensor has all components zero in a Cartesian frame, it must be identically zero in all frames. Now, if the space under consideration is curved, it is possible to introduce a Cartesian coordinate only in a local region. The Christoffel symbol Γ^i_{km} can be made zero in the local Cartesian frame, but its derivatives may not be zero. It follows from (12.37) that at this local point we should have

$$\frac{DR^n_{ikl}}{Dx^m} = \frac{\partial^2 \Gamma^n_{il}}{\partial x^m \partial x^k} - \frac{\partial^2 \Gamma^n_{ik}}{\partial x^m \partial x^l},$$

since the covariant derivative reduces to the ordinary derivative in the local region of the Cartesian frame. If we write similar expressions for the other terms in (12.40) and add them up, we establish the identity in the local Cartesian frame. Due to its tensorial nature, it then follows that the Bianchi identity must be a general identity true in all frames.

From the curvature tensor R^i_{klm}, we can obtain a tensor R_{km} of lower rank by contracting i with l:

$$R_{km} = R^i_{kim} = \frac{\partial \Gamma^i_{km}}{\partial x^i} - \frac{\partial \Gamma^i_{ki}}{\partial x^m} + \Gamma^n_{km}\Gamma^i_{ni} - \Gamma^n_{ki}\Gamma^i_{nm}.$$
(12.41)

This tensor R_{km} is known as the *Ricci tensor*. It is straightforward to show that R_{km} is symmetric, i.e.

$$R_{km} = R_{mk}.$$

We can finally obtain a scalar

$$R = g^{mk} R_{mk}$$
(12.42)

known as the *scalar curvature*.

So far our discussion of curvature has been completely formal. Readers may now develop some appreciation of this tensorial machinery by applying it to the metric (12.9) pertaining to the surface of a sphere, of which the non-zero Christoffel symbols are given in (12.33). Let us try to calculate the scalar curvature, for which we need $R_{\theta\theta}$ and $R_{\phi\phi}$. Now

$$R_{\theta\theta} = R^{\theta}_{\theta\theta\theta} + R^{\phi}_{\theta\phi\theta} = R^{\phi}_{\theta\phi\theta},$$

since $R^{\theta}_{\theta\theta\theta} = 0$ due to the antisymmetry property (12.38). We thus need to calculate only one component of the curvature tensor to obtain $R_{\theta\theta}$. On using the expressions of Christoffel symbols given by (12.33), it easily follows from the definition (12.37) of the curvature tensor that

$$R_{\theta\theta} = R^{\phi}_{\theta\phi\theta} = 1.$$

A similar calculation gives

$$R_{\phi\phi} = R^{\theta}_{\phi\theta\phi} = \sin^2\theta.$$

Finally the scalar curvature is given by

$$R = g^{\theta\theta} R_{\theta\theta} + g^{\phi\phi} R_{\phi\phi} = \frac{1}{a^2}.1 + \frac{1}{a^2 \sin^2\theta}.\sin^2\theta = \frac{2}{a^2} \qquad (12.43)$$

on making use of (12.17). Although the Riemann tensor has many components, we did not have to do too much algebra to obtain the scalar curvature. Thus calculating the curvature of a simple two-dimensional metric is not too complicated. The curvature calculation for any four-dimensional spacetime metric usually involves a horrendous amount of straightforward algebra. We also note from (12.43) that the curvature scalar is a constant over the surface of the sphere, which is expected from the fact that this surface is uniform. We leave it as an exercise for the reader to show that the scalar curvature for the metric (12.10) is $-2/a^2$. The metrics (12.9) and (12.10) give the only two uniformly curved surfaces possible in two dimensions, one with uniform positive curvature and the other with uniform negative curvature. It is easy to show that all components of the curvature tensor are zero for the metric (12.8) corresponding to a plane. Given the metric of a space (or a spacetime), we now have the machinery to find out whether it is curved or flat.

We end our discussion of curvature by introducing another tensor which turns out to be very important in the formulation of general relativity, as we shall see later. It is the *Einstein tensor* defined as

$$G_{ik} = R_{ik} - \frac{1}{2} g_{ik} R. \qquad (12.44)$$

One of its very important properties is that it is a divergenceless tensor satisfying

$$\frac{DG_{ik}}{Dx_k} = 0. \qquad (12.45)$$

From the definition (12.44) of the Einstein tensor, this implies

$$\frac{DR_{ik}}{Dx_k} - \frac{1}{2} g_{ik} \frac{\partial R}{\partial x_k} = 0. \qquad (12.46)$$

To prove this, we need to begin from the Bianchi identity (12.40). Writing $R^n_{imk} = -R^n_{ikm}$ in the second term of (12.40), we (i) contract n and k, and (ii) multiply by g^{im}. This gives

$$\frac{DR^m_l}{Dx^m} - \frac{\partial R}{\partial x^l} + \frac{D}{Dx^n}(g^{im}R^n_{ilm}) = 0 \tag{12.47}$$

on remembering that the covariant derivative of the metric tensor is zero, as indicated by (12.27), and the covariant derivative of a scalar reduces to its ordinary derivative. In Exercise 12.3 the reader is given some hints to show that $g^{im}R^n_{ilm} = R^n_l$. This implies that the last term in (12.47) is the same as the first term, giving an equation which is easily seen to be equivalent with (12.46). This essentially establishes that the Einstein tensor is divergenceless as encapsulated in (12.45) – a result which is going to be crucially important when we formulate general relativity.

12.2.5 Geodesics

The shortest path between two points in a plane surface or in a flat space is a straight line. If the surface or the space is curved, then the shortest path between two points is called a *geodesic*. We have pointed out in §10.2 that one of the central ideas of general relativity is that a particle moves along a geodesic in the four-dimensional spacetime. Before we get into the physics of general relativity, the last mathematical question we have to address is to show how we obtain geodesics in a particular space of which the metric tensor is known.

Let us consider an arbitrary path between two points A and B. The length ds of a small segment of this path is given by (12.7). Hence the length of the whole path must be

$$s = \int_A^B \sqrt{g_{ik}\frac{dx^i}{d\lambda}\frac{dx^k}{d\lambda}}\, d\lambda, \tag{12.48}$$

where λ is a parameter measured along the path. On writing

$$L = \sqrt{g_{ik}\frac{dx^i}{d\lambda}\frac{dx^k}{d\lambda}} \tag{12.49}$$

the length of the path is

$$s = \int_A^B L\, d\lambda$$

and the condition for the path to be an extremum is given by the Lagrange equation

$$\frac{d}{d\lambda}\left(\frac{\partial L}{\partial(dx^i/d\lambda)}\right) - \frac{\partial L}{\partial x^i} = 0.$$

(See, for example, Mathews and Walker, 1979, §12–1.) On substituting the expression of L given by (12.49) into this Lagrange equation and remembering that

$$\sqrt{g_{ik}\frac{dx^i}{d\lambda}\frac{dx^k}{d\lambda}} = \frac{ds}{d\lambda},$$

a few steps of easy algebra give

$$\frac{d}{ds}\left(g_{ik}\frac{dx^k}{ds}\right) - \frac{1}{2}\frac{\partial g_{kl}}{\partial x^i}\frac{dx^k}{ds}\frac{dx^l}{ds} = 0. \tag{12.50}$$

Now the first term of this equation is

$$\frac{d}{ds}\left(g_{ik}\frac{dx^k}{ds}\right) = g_{ik}\frac{d^2x^k}{ds^2} + \frac{\partial g_{ik}}{\partial x^l}\frac{dx^l}{ds}\frac{dx^k}{ds}$$

$$= g_{ik}\frac{d^2x^k}{ds^2} + \frac{1}{2}\left(\frac{\partial g_{ik}}{\partial x^l} + \frac{\partial g_{il}}{\partial x^k}\right)\frac{dx^l}{ds}\frac{dx^k}{ds}.$$

On substituting this in (12.50), we have

$$g_{ik}\frac{d^2x^k}{ds^2} = -\frac{1}{2}\left(\frac{\partial g_{ik}}{\partial x^l} + \frac{\partial g_{il}}{\partial x^k} - \frac{\partial g_{kl}}{\partial x^i}\right)\frac{dx^k}{ds}\frac{dx^l}{ds}.$$

Multiplying this by g^{mi}, we finally get

$$\frac{d^2x^m}{ds^2} = -\Gamma^m_{kl}\frac{dx^k}{ds}\frac{dx^l}{ds}, \tag{12.51}$$

where Γ^m_{kl} is the Christoffel symbol defined in (12.31). This equation (12.51) is finally the *geodesic equation* which has to be satisfied by a curve in space if it happens to be a geodesic in that space.

Whenever we obtain any important tensorial relation, we have been illustrating it by applying it to one of the two-dimensional metrics. We follow the same approach now. Since the metric (12.8) corresponds to a plane surface, the geodesic for this metric must be a straight line. Hence we expect that the geodesic equation (12.51) applied to this metric should give us a straight line. We now explicitly show this. The non-zero Christoffel symbols of this metric (12.8) are given in (12.32). On substituting these in (12.51), we get the following two equations

$$\frac{d^2r}{ds^2} = r\left(\frac{d\theta}{ds}\right)^2, \tag{12.52}$$

$$\frac{d^2\theta}{ds^2} = -\frac{2}{r}\frac{dr}{ds}\frac{d\theta}{ds}. \tag{12.53}$$

The factor 2 in (12.53) comes from the two equal terms arising out of the combinations $r\theta$ and θr. Now (12.53) is equivalent to

$$\frac{1}{r^2}\frac{d}{ds}\left(r^2\frac{d\theta}{ds}\right)=0.$$

It follows from this that $r^2(d\theta/ds)$ must be a constant along the geodesic. We therefore write

$$\frac{d\theta}{ds}=\frac{l}{r^2},\tag{12.54}$$

where l is a constant. On dividing the metric (12.8) by ds^2, we obtain

$$\left(\frac{dr}{ds}\right)^2=1-r^2\left(\frac{d\theta}{ds}\right)^2=1-\frac{l^2}{r^2}$$

so that

$$\frac{dr}{ds}=\pm\sqrt{1-\frac{l^2}{r^2}}.\tag{12.55}$$

Dividing (12.54) by (12.55), we get

$$\frac{d\theta}{dr}=\pm\frac{l/r^2}{\sqrt{1-\frac{l^2}{r^2}}},$$

of which a solution is

$$\theta=\theta_0\pm\cos^{-1}\left(\frac{l}{r}\right),$$

where θ_0 is the constant of integration. This solution can be put in the form

$$r\cos(\theta-\theta_0)=l,\tag{12.56}$$

which is clearly the equation of a straight line. This completes our proof that the geodesic in a plane is a straight line, even though it is mathematically not so apparent when we use polar coordinates.

12.3 Metric for weak gravitational field

We have finished developing the mathematical machinery necessary for formulating general relativity. Now we are ready to get into a discussion of the physics of general relativity. Since it is a complex and difficult subject, we shall proceed cautiously. Suppose we consider the motion of a non-relativistic particle in a weak gravitational field. It will become clear soon what we quantitatively mean by the adjective 'weak' applied to the gravitational field. If the gravitational field is sufficiently weak, we can certainly treat the motion of the particle by using the Newtonian theory of gravity. We can, however, use general relativity also to solve this problem, even though that might be like using a sledgehammer to crack a nut. Both the Newtonian theory and general relativity should give the same result in the regime where both are valid. We introduce some first concepts of general relativity by discussing how it can be applied to study the motion of a non-relativistic particle in a weak gravitational field for which the Newtonian theory is adequate.

We have pointed out in §10.2 (see Table 10.1) that the motion of a particle in a gravitational field is obtained in general relativity by using the fact that the motion has to be along a geodesic. We also saw in §12.2.5 that the geodesic between two spacetime points A and B is the path along which the path length $s = \int_A^B ds$ is an extremum, ds being given by (12.7). So the path of a particle in the four-dimensional spacetime between the two points A and B can be obtained in general relativity by making s an extremum. On the other hand, classical mechanics tells us that we can solve the motion by applying Hamilton's principle that the action $I = \int_A^B L \, dt$ has to be an extremum between A and B, where L is the Lagrangian (see, for example, Goldstein, 1980, Chapter 2; Landau and Lifshitz, 1976, §2). For the motion of a non-relativistic particle in a weak gravitational field, both classical mechanics and general relativity should give correct results, and the results should be identical. In other words, the extremum of $s = \int_A^B ds$ and the extremum of $I = \int_A^B L \, dt$ should give the same path. This is possible only if ds for a weak gravitational field in general relativity is essentially the same thing as $L \, dt$ in classical mechanics (except for any additive or multiplicative constants). We know that the Lagrangian L for a particle moving in a gravitational potential Φ is given by

$$L = \frac{1}{2}mv^2 - m\Phi$$

(see, for example, Goldstein, 1980, §1–4; Landau and Lifshitz, 1976, §5). Hence the non-relativistic action can be taken to be

$$I_{NR} = \int_A^B \left[\frac{1}{2}mv^2 - m\Phi \right] dt - mc^2 \int_A^B dt. \tag{12.57}$$

The last term has a constant value $-mc^2(t_B - t_A)$ and would not contribute in the calculation of the extremum. As we proceed further, it will be clear why we are including this term. We shall now try to figure out how to write ds in this situation such that the extremum of $\int_A^B ds$ gives the same result as the extremum of I_{NR} given by (12.57).

Before considering the case for a weak gravitational field, let us first look at the case of zero gravitational field. In the absence of a gravitational field, general relativity reduces to special relativity. We write the spacetime coordinates as $x^0 = ct, x^1, x^2, x^3$ in view of the fact that their differentials make up a contravariant vector and we should use superscripts. If ds is the separation between two nearby spacetime events, we know that ds^2, which is often written as $-c^2d\tau^2$, can be written in the special relativistic situation in the form

$$ds^2 = -c^2d\tau^2 = -(dx^0)^2 + (dx^1)^2 + (dx^2)^2 + (dx^3)^2. \qquad (12.58)$$

We can write this as

$$ds^2 = -c^2d\tau^2 = \eta_{ik} \, dx^i \, dx^k, \qquad (12.59)$$

where the special relativistic metric η_{ik} is given by

$$\eta_{ik} = \begin{pmatrix} -1 & 0 & 0 & 0 \\ 0 & 1 & 0 & 0 \\ 0 & 0 & 1 & 0 \\ 0 & 0 & 0 & 1 \end{pmatrix}. \qquad (12.60)$$

One can easily check that a geodesic has to be a straight line in this spacetime (which is flat). A straight line in the special relativistic spacetime corresponds to the world-line of a uniformly moving particle. Thus, if the geodesics give the paths which particles follow, we come to the conclusion that particles move uniformly in the absence of a gravitational field.

It is clear that the extremum of $\int_A^B d\tau$, with $d\tau$ given by (12.58), would give the geodesic in the special relativistic case (which is a straight line). Let us now multiply $\int_A^B d\tau$ by $-mc^2$ and argue that the resulting quantity

$$I_{\Phi=0} = -mc^2 \int_A^B d\tau \qquad (12.61)$$

should be the action for a free particle (i.e. a particle in zero gravitational field) in classical mechanics. If (12.61) indeed gives the classical action, then it would follow that a geodesic is a path along which the action is also an extremum and the path followed by the particle according to classical mechanics would then be a geodesic in spacetime. Suppose we consider a moving particle which is at spacetime points (x^0, x^1, x^2, x^3) and $(x^0 + dx^0, x^1 + dx^1, x^2 + dx^2, x^3 + dx^3)$ at the beginning and the end of a short interval. The velocity of the particle is given by

$$v^2 = \frac{(dx^1)^2 + (dx^2)^2 + (dx^3)^2}{dt^2} = c^2\frac{(dx^1)^2 + (dx^2)^2 + (dx^3)^2}{(dx^0)^2}, \qquad (12.62)$$

since $dx^0 = cdt$. It then follows from (12.58) that

$$d\tau = \frac{dx^0}{c}\sqrt{1 - \frac{v^2}{c^2}} = dt\sqrt{1 - \frac{v^2}{c^2}}. \qquad (12.63)$$

On substituting this in (12.61), we have

$$I_{\Phi=0} = -mc^2 \int_A^B dt\sqrt{1 - \frac{v^2}{c^2}}. \qquad (12.64)$$

If $v^2 \ll c^2$, then we can make a Taylor series expansion of the square root, which would give

$$I_{\Phi=0} \approx -mc^2 \int_A^B dt + \int_A^B \frac{1}{2} mv^2 \, dt.$$

This is the same as I_{NR} given by (12.57) after setting $\Phi = 0$. Thus the non-relativistic limit of (12.61) gives the usual non-relativistic action for a free particle. So, if we take (12.61) as our action, we shall get the correct path in the non-relativistic situation, which would be a geodesic in spacetime. Although we are here discussing the motion of a non-relativistic particle, it may be mentioned that (12.61) is the correct expression for the action of a free particle even when the particle moves relativistically (see Landau and Lifshitz, 1975, §8).

When a weak gravitational field is present, we need to add a part due to gravitational interaction in (12.61). Let us take

$$I = -mc^2 \int_A^B d\tau - \int_A^B m\Phi \, dt$$

which, by virtue of (12.63), becomes

$$I = -\int_A^B dt \left[mc^2 \sqrt{1 - \frac{v^2}{c^2}} + m\Phi \right]. \tag{12.65}$$

It is trivial to check that the non-relativistic limit ($v^2 \ll c^2$) of (12.65) is (12.57). The form of (12.65) clarifies the quantitative meaning of a weak gravitational field. The extra term in the action due to the gravitational field should be small compared to the rest of the action. It follows from (12.65) that the condition for a weak gravitational field is

$$\Phi \ll c^2. \tag{12.66}$$

In §1.5 we pointed out that the condition for the Newtonian theory of gravity to be adequate is that f defined by (1.11) is small compared to 1. It is easy to see that this is effectively the same condition as (12.66). When (12.66) is satisfied, (12.65) must be approximately equivalent to

$$I \approx -mc^2 \int_A^B dt \sqrt{1 - \frac{v^2}{c^2} + \frac{2\Phi}{c^2}}.$$

This can be written as

$$I \approx -mc \int_A^B \sqrt{\left(1 + \frac{2\Phi}{c^2}\right) c^2 \, dt^2 - v^2 \, dt^2}.$$

On making use of (12.62), this becomes

$$I \approx -mc \int_A^B \sqrt{\left(1 + \frac{2\Phi}{c^2}\right)(dx^0)^2 - (dx^1)^2 - (dx^2)^2 - (dx^3)^2}. \quad (12.67)$$

If we take the metric of the four-dimensional spacetime to be

$$ds^2 = -c^2 d\tau^2 = -\left(1 + \frac{2\Phi}{c^2}\right)(dx^0)^2 + (dx^1)^2 + (dx^2)^2 + (dx^3)^2,$$

$$(12.68)$$

then the geodesics in this spacetime will coincide with the paths we would get by making the action given by (12.67) an extremum. Since (12.67) in turn reduces to (12.57) in the non-relativistic situation with a weak gravitational field, we conclude that a particle moving non-relativistically in a weak gravitational field, of which the motion can be found by making (12.57) an extremum, should follow a geodesic in the spacetime described by (12.68). This suggests that (12.68) is the metric of spacetime with a weak gravitational field, where Φ is the classical gravitational potential. We can write (12.68) as

$$ds^2 = g_{ik} dx^i dx^k, \quad (12.69)$$

where the metric tensor g_{ik} in the presence of a weak gravitational field is given by

$$g_{ik} = \begin{pmatrix} -\left[1 + \frac{2\Phi}{c^2}\right] & 0 & 0 & 0 \\ 0 & 1 & 0 & 0 \\ 0 & 0 & 1 & 0 \\ 0 & 0 & 0 & 1 \end{pmatrix}. \quad (12.70)$$

It is obvious that the general relativistic metric tensor (12.70) reduces to the special relativistic metric tensor (12.60) in the absence of the gravitational field.

It is instructive to work out the geodesic equation (12.51) for the metric tensor given by (12.70) and to show that it is the same as the classical equation for non-relativistic motion in a weak gravitational field. In the following discussion, we shall use the superscript α to indicate indices 1, 2 or 3, but not 0. For non-relativistic motion, the change dx^α in the particle's position during an interval dt has to be much smaller than $dx^0 = cdt$. Hence the dominant terms in (12.51) are

$$\frac{d^2 x^m}{ds^2} = -\Gamma_{00}^m \frac{dx^0}{ds} \frac{dx^0}{ds}, \quad (12.71)$$

where there is no summation now. To proceed further, we have to calculate the Christoffel symbol Γ_{00}^m, which involves derivatives of the metric tensor as seen in (12.31). If the gravitational potential Φ is independent of time, then the only non-zero derivative of the metric tensor (12.70) is

$$\frac{\partial g_{00}}{\partial x^\alpha} = -\frac{2}{c^2}\frac{\partial \Phi}{\partial x^\alpha}.$$

On using this, we find from (12.31) that

$$\Gamma^\alpha_{00} = \frac{1}{c^2}\frac{\partial \Phi}{\partial x_\alpha}, \tag{12.72}$$

if we neglect terms quadratic in the small quantity Φ. Substituting this in (12.71) and making use of the fact $ds = i\,c\,d\tau$, we get

$$\frac{d^2 x^\alpha}{d\tau^2} = -\frac{1}{c^2}\frac{\partial \Phi}{\partial x_\alpha}\frac{dx^0}{d\tau}\frac{dx^0}{d\tau}. \tag{12.73}$$

It follows from (12.62) and (12.68) that

$$c\,d\tau = dx^0\sqrt{1 + \frac{2\Phi}{c^2} - \frac{v^2}{c^2}}. \tag{12.74}$$

For non-relativistic motions in a weak gravitational field for which we neglect quadratics of Φ, we need to substitute

$$\frac{dx^0}{d\tau} = c$$

in (12.73) and take $d\tau = dt$. Then (12.73) gives

$$\frac{d^2 x^\alpha}{dt^2} = -\frac{\partial \Phi}{\partial x_\alpha},$$

which is the classical equation of motion. This completes our proof that general relativity and ordinary classical mechanics would give the same result if (12.68) is the metric for a weak gravitational field.

12.4 Formulation of general relativity

After discussing how general relativity can be used to study a weak gravitational field, we are now ready to present the complete formulation of general relativity. As pointed out in §10.2, the central equation of general relativity is Einstein's equation telling us how the curvature of spacetime is related to the density of mass-energy present in spacetime. In §12.2.4 we have introduced several tensors associated with the curvature of space. It is the Einstein tensor G_{ik} defined in (12.44) which turns out to be a particularly convenient tensor in the basic formulation of the theory. Suppose we are able to find a suitable second-rank tensor describing the mass-energy density. Then making G_{ik} proportional to this tensor would give us an equation that would imply that the curvature of space-time is caused by mass-energy. Since the divergence of G_{ik} is zero according to (12.45), the divergence of the tensor giving the mass-energy density also has to be zero for the sake of consistency. The job before us now is to find a divergenceless second-rank tensor which describes the mass-energy density. This tensor is called the *energy-momentum tensor*. We shall now show that a convenient tensor of this kind exists.

12.4.1 The energy-momentum tensor

For the time being, let us forget about relativity and show that the classical hydrodynamic equations can be put in a form such that the divergence of a second-rank tensor is zero. Then we shall consider how to generalize this tensor to general relativity and thereby obtain the energy-momentum tensor. As in the previous section, let us write x^0 for ct and x^1, x^2, x^3 for the three spatial coordinates. The Roman indices i, j,... will run over the values 0, 1, 2, 3, whereas the Greek indices α, β, ... will run over only 1, 2, 3. We have pointed out in §12.2.1 that the generalized velocity transforms as a contravariant vector. So we shall write the velocity components with indices at the top indicative of contravariant tensors. It is easy to see that the continuity equation (8.3) can be written in the form

$$\frac{\partial S^i}{\partial x^i} = 0, \tag{12.75}$$

where S^i is a 4-vector with components $(\rho c, \rho v^1, \rho v^2, \rho v^3)$ and the index i repeated twice implies that we are summing over 0, 1, 2, 3. We have

$$\frac{\partial}{\partial t}(\rho v^\alpha) = v^\alpha \frac{\partial \rho}{\partial t} + \rho \frac{\partial v^\alpha}{\partial t}.$$

Let us now substitute for $\partial \rho / \partial t$ from the continuity equation (8.3) and for $\partial v^\alpha / \partial t$ from the Euler equation (8.9) after setting the external force $\mathbf{F}$ to zero. This gives

$$\frac{\partial}{\partial t}(\rho v^\alpha) = -v^\alpha \frac{\partial}{\partial x^\beta}(\rho v^\beta) - \rho v^\beta \frac{\partial v^\alpha}{\partial x^\beta} - \frac{\partial P}{\partial x_\alpha},$$

where a Greek index α or β repeated twice signifies summation over only the spatial components 1, 2, 3. It is easy to see that the above equation can be written in the form

$$\frac{\partial}{\partial t}(\rho v^\alpha) + \frac{\partial T^{\alpha\beta}}{\partial x^\beta} = 0, \tag{12.76}$$

where

$$T^{\alpha\beta} = P \delta^{\alpha\beta} + \rho v^\alpha v^\beta. \tag{12.77}$$

We can now combine (12.75) and (12.76) in the compact form

$$\frac{\partial (\mathcal{T}_{NR})^{ik}}{\partial x^k} = 0, \tag{12.78}$$

where $(\mathcal{T}_{NR})^{ik}$ is the non-relativistic four-dimensional energy-momentum tensor of which the various components are given by

$$(\mathcal{T}_{NR})^{00} = \rho c^2, \quad (\mathcal{T}_{NR})^{0\alpha} = (\mathcal{T}_{NR})^{\alpha 0} = \rho c v^\alpha, \quad (\mathcal{T}_{NR})^{\alpha\beta} = T^{\alpha\beta}. \tag{12.79}$$

It should be noted that we have not invoked relativity in obtaining (12.78). Writing x^0 for ct has been merely a matter of notation. The equation (12.78) combines the equations of continuity and motion of classical hydrodynamics, showing that we can have a divergenceless second-rank tensor $(\mathcal{T}_{NR})^{ik}$ in the non-relativistic situation. We now have to generalize $(\mathcal{T}_{NR})^{ik}$ to obtain the fully relativistic energy-momentum tensor.

Let us first consider how we generalize the concept of velocity in general relativity. Suppose a particle has positions x^i and $x^i + dx^i$ before and after an infinitesimal interval. The difference dx^i is a 4-vector and the quotient obtained by dividing it by a scalar will be a 4-vector as well. As in (12.58) and (12.68), we introduce the time-like interval $d\tau$ defined through

$$ds^2 = -c^2 d\tau^2 = g_{ik}\, dx^i\, dx^k. \tag{12.80}$$

From discussions in the previous section, it should be clear that $d\tau \rightarrow dt$ for non-relativistic motion in a weak gravitational field. Since $d\tau$ as introduced in (12.80) must be a scalar, dx^i divided by $d\tau$ should give us a 4-vector. We now define the relativistic velocity 4-vector as

$$u^i = \frac{1}{c}\frac{dx^i}{d\tau}. \tag{12.81}$$

In the non-relativistic limit, this clearly reduces to

$$u^i \rightarrow \left(1, \frac{v^1}{c}, \frac{v^2}{c}, \frac{v^3}{c}\right). \tag{12.82}$$

One interesting property of the velocity 4-vector is that

$$u^i u_i = -1. \tag{12.83}$$

This can be easily proved if we use (12.14) to obtain u_i from u^i and then use (12.80)–(12.81). We now define the energy-momentum tensor

$$T^{ik} = \rho c^2 u^i u^k + P(g^{ik} + u^i u^k). \tag{12.84}$$

We leave it as an exercise for the reader to verify that this reduces in the non-relativistic limit to the non-relativistic tensor $(\mathcal{T}_{NR})^{ik}$ as given by (12.79). You need to assume that $P \ll \rho c^2$ for a non-relativistic fluid. The quantities like ρ and P are defined with respect to the rest frame of the fluid.

Since T^{ik} is a proper relativistic second-rank tensor and reduces to the non-relativistic expression in the appropriate limit, we take it to be the relativistic generalization of the energy-momentum tensor. Now we can generalize the equation (12.78) also. We need to replace the ordinary derivative by the covariant derivative apart from replacing $(T_{NR})^{ik}$ by T^{ik}. This gives

$$\frac{DT^{ik}}{Dx^k} = 0. \tag{12.85}$$

This completes our task of finding a properly relativistic divergenceless second-rank tensor, which can be presumed to act as the source of curvature of spacetime.

Before leaving the discussion of the energy-momentum tensor, let us consider one important special case of this tensor. We consider a fluid at rest. Then the spatial components of the 4-velocity must be zero, i.e.

$$u^i = (u^0, 0, 0, 0). \tag{12.86}$$

It then follows from (12.83) that

$$u^0 u_0 = -1. \tag{12.87}$$

On lowering the index k in (12.84) by the usual procedure (12.14), we have

$$T^i_k = \rho c^2 u^i u_k + P \left(\delta^i_k + u^i u_k \right).$$

On making use of (12.86) and (12.87), this gives

$$T^i_k = \begin{pmatrix} -\rho c^2 & 0 & 0 & 0 \\ 0 & P & 0 & 0 \\ 0 & 0 & P & 0 \\ 0 & 0 & 0 & P \end{pmatrix}. \tag{12.88}$$

We pointed out in §10.3 that the Robertson–Walker metric (10.19) corresponds to a co-moving coordinate system in which the material of the Universe is assumed to be at rest. So the energy-momentum tensor of the Universe is given by (12.88) when we use the co-moving coordinate system. We shall use this result when we develop relativistic cosmology in §14.1.

12.4.2 Einstein's equation

Since G_{ik} defined in (12.44) and T^{ik} defined in (12.84) are both divergenceless tensors, one being a measure of the curvature of spacetime and the other being a measure of energy-momentum density, it is tempting to write

$$G_{ik} = \kappa T_{ik}, \tag{12.89}$$

where κ is a constant. This equation would imply that the curvature of spacetime is produced by the energy-momentum density, which is a basic requirement of general relativity. It should be emphasized that we have not 'derived' (12.89), but have given a string of arguments that such an equation may be expected. Since the divergences of both sides are zero, there would not be any mathematical inconsistency when we take a divergence of (12.89). If we want the curvature of spacetime to be produced by mass-energy, then an equation like (12.89) would be a natural possibility. Whether (12.89) is a really correct equation can be determined only by checking if results derived from (12.89) are confirmed by experiments. We shall discuss some experimental confirmations of general relativity in the next chapter.

To complete our discussion, we need to determine the constant κ. If we can determine the value of κ in a special case, then that would be the true value of κ in all situations if it is a universal constant. We have discussed in §12.3 how general relativity can be formulated in the case of a weak gravitational field. We shall now determine κ by applying (12.89) to a weak gravitational field and comparing it with the Newtonian theory of gravity.

Using (12.44) we can write (12.89) in the form

$$R^i_k - \frac{1}{2}\delta^i_k R = \kappa T^i_k. \tag{12.90}$$

We now carry on a contraction between the indices i and k, keeping in mind that $\delta^i_i = 4$ because of the four dimensions of spacetime, allowing i to have values 0, 1, 2, 3. This gives

$$-R = \kappa T,$$

where $T = T^i_i$. On substituting $-\kappa T$ for R in (12.90), we have

$$R^i_k = \kappa \left(T^i_k - \frac{1}{2}\delta^i_k T \right).$$

We shall now consider the following particular component of this equation

$$R^0_0 = \kappa \left(T^0_0 - \frac{1}{2}\delta^0_0 T \right). \tag{12.91}$$

Let us apply (12.91) to the case of a weak gravitational field produced by a distribution of matter at rest in our coordinate system. The expression for the energy-momentum tensor for matter at rest is given by (12.88). If $P \ll \rho c^2$, then

$$T \approx T^0_0 = -\rho c^2.$$

On substituting this in (12.91), we get

$$R^0_0 = -\frac{1}{2}\kappa \rho c^2. \tag{12.92}$$

Our job now is to find the expression for R_0^0 from the metric (12.68) for the weak gravitational field. It follows from (12.41) that

$$R_{00} = R_{0i0}^i = \frac{\partial \Gamma_{00}^i}{\partial x^i} - \frac{\partial \Gamma_{0i}^i}{\partial x^0},$$ (12.93)

if we neglect the quadratic terms in Christoffel symbols for a weak gravitational field. Since we are considering matter to be at rest, the field should be independent of time and any derivative with respect to $x^0 = ct$ should give zero. Then (12.93) reduces to

$$R_{00} = \frac{\partial \Gamma_{00}^\alpha}{\partial x^\alpha},$$

where α repeated twice implies summation over 1, 2, 3 as usual. The Christoffel symbol Γ_{00}^α for a weak gravitational field was already determined in (12.72). On substituting this,

$$R_{00} = \frac{\partial}{\partial x^\alpha} \left(\frac{1}{c^2} \frac{\partial \Phi}{\partial x_\alpha} \right) = \frac{1}{c^2} \nabla^2 \Phi.$$

To get R_0^0, we merely have to multiply R_{00} by $g^{00} \approx -1$, since terms quadratic in Φ are neglected. It then follows from (12.92) that

$$\nabla^2 \Phi = \frac{\kappa c^4}{2} \rho.$$ (12.94)

The Newtonian theory of gravity leads to the gravitational Poisson equation

$$\nabla^2 \Phi = 4\pi G \rho.$$

Comparing this with (12.94), we finally conclude that

$$\kappa = \frac{8\pi G}{c^4}.$$ (12.95)

On substituting the value of κ in (12.89), we have

$$G_{ik} = \frac{8\pi G}{c^4} T_{ik}.$$ (12.96)

This is the famous *Einstein equation* and tells how matter-energy acts as a source of the curvature of spacetime (Einstein, 1916).

The compact tensorial notation makes Einstein's equation (12.96) appear deceptively simple. Although it is one of the most beautiful equations of mathematical physics, it also happens to be one of the most difficult equations to handle. Since particles move along geodesics and we need a knowledge of the metric tensor to determine the geodesics, most of the practical problems in general relativity require the determination of the metric tensor for a given matter-energy distribution. The connection between the metric tensor g_{ik} and the Einstein tensor G_{ik} follows from (12.31), (12.41) and (12.44). If we know the energy-momentum tensor T_{ik} in a particular situation, Einstein's equation

(12.96) at once gives us the Einstein tensor G_{ik}. But determining the metric tensor g_{ik} after that is not an easy job. There are very few cases of practical importance where one can determine the metric tensor that would satisfy Einstein's equation. We shall consider some applications of general relativity in the next two chapters, where we shall present some solutions of Einstein's equation.

Exercises

12.1 Fully work out all the components of the Christoffel symbol, the Riemann tensor and the Ricci tensor as well as the scalar curvature for the following metrics

$$ds^2 = dr^2 + r^2\, d\theta^2,$$

$$ds^2 = a^2(d\theta^2 + \sin^2\theta\, d\phi^2),$$

$$ds^2 = a^2(d\chi^2 + \sinh^2\chi\, d\eta^2).$$

12.2 Show that the two covariant derivatives in general do not commute. For a contravariant vector A^i, show especially that

$$\left(\frac{D}{Dx^k}\frac{D}{Dx^l} - \frac{D}{Dx^l}\frac{D}{Dx^k}\right) A^i = -R^i_{mlk} A^m.$$

12.3 From the tensor R^i_{klm} defined in (12.37), construct the tensor

$$R_{iklm} = g_{in} R^n_{klm}$$

and show that

$$R_{iklm} = \frac{1}{2}\left(\frac{\partial^2 g_{im}}{\partial x^k\, \partial x^l} + \frac{\partial^2 g_{kl}}{\partial x^i\, \partial x^m} - \frac{\partial^2 g_{il}}{\partial x^k\, \partial x^m} - \frac{\partial^2 g_{km}}{\partial x^i\, \partial x^l}\right)$$

$$+ g_{np}(\Gamma^n_{kl}\Gamma^p_{im} - \Gamma^n_{km}\Gamma^p_{il})$$

from which you can demonstrate the following symmetry properties

$$R_{iklm} = -R_{kilm}, \quad R_{iklm} = -R_{ikml}, \quad R_{iklm} = R_{lmik}.$$

We need to show that $g^{im} R^n_{ilm}$ appearing in (12.47) has to be equal to R^n_l in order to complete the proof that the Einstein tensor is divergenceless. You are now asked to prove this by using the symmetry properties just obtained. [Hint: You have to write

$$g^{im} R^n_{ilm} = g^{im} g^{nk} R_{kilm}$$

and then use some symmetry properties of R_{kilm}.]

12.4 Suppose $x^i(s)$ and $x^i(s) + \delta x^i(s)$ are the points on two infinitesimally separated geodesics, where s is a parameter measured along either of these geodesics. Show that δx^i, which measures the separation between the geodesics, satisfies

$$\frac{D^2}{Ds^2}\delta x^i = -R^i_{klm}\,\delta x^l\,\frac{dx^k}{ds}\frac{dx^m}{ds},$$

where $D/Ds = (dx^i/ds)D/Dx^i$ is the covariant derivative along the arclength of either geodesic.

12.5 Prove that the equator on a spherical surface (which is a great circle) is a geodesic, but any other circle of constant latitude parallel to the equator is not a geodesic.

12.6 The special relativistic metric is

$$ds^2 = -c^2 dt^2 + dx^2 + dy^2 + dz^2.$$

Consider a frame rotating uniformly around the z axis with respect to this frame such that

$$t' = t,\ z' = z,\ x' = x\cos\Omega t + y\sin\Omega t,\ y' = -x\sin\Omega t + y\cos\Omega t.$$

Find out the metrics g_{ik} and g^{ik} in the rotating frame and explicitly verify that

$$g^{ik}g_{km} = \delta^i_m.$$

12.7 The metric

$$ds^2 = -\left(1 + \frac{2\Phi}{c^2}\right)c^2 dt^2 + dx^2 + dy^2 + dz^2$$

can describe the motion of a non-relativistic particle in a weak gravitational field (where Φ is the non-relativistic gravitational potential due to some density distribution ρ). Calculate the Einstein tensor G_{xx} for this metric and verify if the corresponding component of Einstein's equation is satisfied. If not, how do you account for it?

13

Some applications of general relativity

13.1 Time and length measurements

While the formulation of general relativity presented in §12.4 may appear somewhat formal, a physical theory ultimately has to make contact with the results of measurements. Before considering applications of general relativity, we need to understand how the results of time and length measurements can be expressed in terms of the quantities appearing in the mathematical theory.

We keep following the notation that Roman indices $i, j, \ldots$ will run over the values 0, 1, 2, 3, whereas the Greek indices $\alpha, \beta, \ldots$ will run over only 1, 2, 3. Writing the time part of the metric separately, we can write the spacetime metric in the form

$$ds^2 = g_{00}(dx^0)^2 + 2g_{0\alpha} \, dx^0 \, dx^\alpha + g_{\alpha\beta} \, dx^\alpha \, dx^\beta, \qquad (13.1)$$

where x^0 is the time-like coordinate. The metric tensor component $g_{0\alpha}$ gives rise to cross-terms between the time and space coordinates in (13.1). In our elementary treatment of general relativity, we shall restrict ourselves to examples in which $g_{0\alpha} = 0$. The mathematical theory becomes much more complicated if $g_{0\alpha}$ is not equal to zero, which happens when rotation is present in a system (see Exercise 12.6). For example, the metric around a rotating black hole (known as the Kerr metric) has non-zero $g_{0\alpha}$. We shall not discuss such cases in this book. If $g_{0\alpha} = 0$, then (13.1) reduces to

$$ds^2 = g_{00}(dx^0)^2 + g_{\alpha\beta} \, dx^\alpha \, dx^\beta. \qquad (13.2)$$

Suppose some observer is at position x^α at time x^0 and at the position $x^\alpha + dx^\alpha$ at time $x^0 + dx^0$. In the special relativistic situation, it is easy to show that the physical time interval $d\tau$ measured by the observer's clock is related to ds by

$$ds^2 = -c^2\,d\tau^2, \tag{13.3}$$

where ds^2 is given by the special relativistic metric (12.59) and (12.60). Even in a general relativistic situation, we can always introduce an inertial frame in a local region of spacetime where special relativity holds. We thus expect that the physical time $d\tau$ measured by the observer's clock should satisfy the same relation (13.3) in general relativity also. We now consider an observer at rest such that the position x^α does not change. It then follows from (13.2) and (13.3) that the physical time measured by the observer's clock is given by

$$d\tau = \frac{1}{c}\sqrt{-g_{00}}\,dx^0. \tag{13.4}$$

This is our first important relation. If an observer is at rest in a coordinate frame and an interval is dx^0 in the time-like coordinate x^0, then the actual physical time interval measured by a clock is given by multiplying dx^0 by $\sqrt{-g_{00}}/c$.

We now discuss how we can find the physical distance between the neighbouring points x^α and $x^\alpha + dx^\alpha$. Suppose a light signal is sent from the first point to the second point, where it is reflected back towards the first point immediately. The physical time interval between the moment when the light signal leaves the first point and the moment when the light signal comes back there should be equal to $2\,dl/c$, where dl is the physical distance between the two neighbouring points, if we assume that the light signal propagates at speed c. If two events are connected by a light signal, we know that in special relativity we have

$$ds^2 = 0. \tag{13.5}$$

The same consideration should hold in general relativity as well, since we can always introduce inertial frames in local regions. It follows from (13.2) and (13.5) that

$$dx^0 = \sqrt{\frac{-g_{\alpha\beta}\,dx^\alpha\,dx^\beta}{g_{00}}}. \tag{13.6}$$

Let us clarify the significance of this. A light signal starting from the first point x^α at $x^0 - dx^0$ should reach the second point $x^\alpha + dx^\alpha$ at x^0. If the light signal is immediately reflected back from the second point, it will again reach back to the first point at $x^0 + dx^0$. Thus the moment when the light signal leaves the first point and the moment when the light signal comes back there differ by $2\,dx^0$, with dx^0 given by (13.6). To get the physical time interval, we have to multiply this by $\sqrt{-g_{00}}/c$, as suggested in (13.4). On equating this physical time interval to $2\,dl/c$, we find

$$dl = \sqrt{g_{\alpha\beta}\,dx^\alpha\,dx^\beta}. \tag{13.7}$$

This second important relation gives the physical length between the neighbour-ing points x^α and $x^\alpha + dx^\alpha$. The length of a curve between two distant points can be obtained by integrating the length element given by (13.7), provided all the components of $g_{\alpha\beta}$ are independent of time. If $g_{\alpha\beta}$ changes with time during the propagation of the light signal from one point to another distant point, then we can meaningfully talk only about length elements dl along the path and not the length of the whole path. For the Robertson–Walker metric introduced in (10.19), the metric tensor evolves with time due to the time dependence of $a(t)$. Hence one has to take special care to treat the propagation of light in the expanding Universe or to talk about distances to faraway galaxies, as we shall see in §14.3 and §14.4.

The discussion of the previous paragraph also leads to the concept of simultaneity. An observer at the first point x^α sees the light signal leaving at $x^0 - dx^0$ and returning back at $x^0 + dx^0$. Since the median value of the time-like coordinate between these two moments is x^0, this observer would expect the light signal to reach the second point at time x^0 in his clock. We have already pointed out that the signal reaches the second point $x^\alpha + dx^\alpha$ when the time-like coordinate at the second point has the value x^0. This means that x^0 at the first point is simultaneous with x^0 at the second point. Extending this argument, events taking place at different spatial points are simultaneous if the time-like coordinate has the same value x^0 for these events. The coordinate x^0 is often called the *world time*. We need to consider the world time to figure out whether different events are simultaneous, whereas the physical time can be obtained from the world time by using (13.4). Since g_{00} will in general have different values at different spatial points, it is clear that clocks will run at different rates at different spatial points.

13.2 Gravitational redshift

We consider a constant gravitational field where the coordinates can be chosen in such a way that the metric tensor components are independent of time. Suppose a periodic signal is sent from point A to point B. Let a pulse be emitted at A at world time x_e^0 and reach B at world time x_r^0, the propagation time being $x_r^0 - x_e^0$. Suppose the next pulse is emitted at A at world time $x_e^0 + T^0$. For a constant gravitational field, the propagation time of this pulse to B will be the same as the propagation time of the first pulse. Hence the second pulse will reach B at the world time $(x_e^0 + T^0) + (x_r^0 - x_e^0) = x_r^0 + T^0$. This means that observers at A and B would both record the same world time difference T^0 between the two pulses. Using (13.4) to relate the world time to the physical time, we conclude that the physical period T_A inferred by observer A and the physical period T_B inferred by observer B should be related by

$$\frac{T_A}{T_B} = \sqrt{\frac{-(g_{00})_A}{-(g_{00})_B}} = \frac{\omega_B}{\omega_A}, \tag{13.8}$$

where ω_A and ω_B are the frequencies of some periodic signal measured at A and B respectively. This is the general expression showing how the frequency of a signal changes on propagating from one point in a gravitational field to another. We shall now consider some simplifications for weak gravitational fields.

For a weak gravitational field, we can use (12.70) so that

$$\sqrt{-g_{00}} = \sqrt{1 + \frac{2\Phi}{c^2}} \approx 1 + \frac{\Phi}{c^2}.$$

On using this, (13.8) gives

$$\frac{\omega_B}{\omega_A} = \frac{1 + \frac{\Phi_A}{c^2}}{1 + \frac{\Phi_B}{c^2}}, \tag{13.9}$$

where Φ_A, Φ_B are Newtonian gravitational potentials at A, B. For a weak gravitational field, (13.9) can further be written as

$$\omega_B = \omega_A \left(1 + \frac{\Phi_A - \Phi_B}{c^2}\right). \tag{13.10}$$

Suppose the point B is further away than A from the central region of the gravitational field. It is easy to check that the potential difference $\Phi_A - \Phi_B$ should be negative, making $\omega_B < \omega_A$. As a periodic signal makes its way out of a gravitational field, the frequency will decrease. For light coming out of a gravitational field, the spectrum should be shifted towards the red. This is the famous *gravitational redshift* predicted by general relativity. Pound and Rebka (1960) were able to verify the gravitational shift of wavelength by a brilliant terrestrial experiment, in which γ-rays from a source kept at the top of a tower were allowed to travel to the bottom where they were absorbed and analysed.

13.3 The Schwarzschild metric

We pointed out in §12.4.2 that Einstein's equation (12.96), which is the central equation of general relativity, is very difficult to solve and complete solutions are known only for a few cases of practical importance. The simplest gravitational problem one can think of is to find the gravitational field due to an isolated point mass M. Soon after Einstein's formulation of general relativity, Schwarzschild (1916) obtained the exact solution of this problem.

Let us choose the position of the mass M as the origin of our coordinate system and use spherical coordinates. According to the Newtonian theory of gravity, the gravitational potential at a distance r is given by

$$\Phi = -\frac{GM}{r}. \tag{13.11}$$

Far away from the mass point where the gravitational field is weak, the metric should be given by the expression (12.68) valid in the weak field limit. Substituting for Φ from (13.11) and using spherical coordinates, we write

$$ds^2(r \to \infty) = -\left(1 - \frac{2GM}{c^2 r}\right) c^2 dt^2 + dr^2 + r^2(d\theta^2 + \sin^2\theta \, d\phi^2). \tag{13.12}$$

Since there is no matter in space at points other than the point $r = 0$, the energy-momentum tensor given by (12.88) should be zero at all points except $r = 0$. Then, according to Einstein's equation (12.96), the Einstein tensor also must be zero at all points except $r = 0$. Our job now is to find a metric which tends to (13.12) as $r \to \infty$ and for which the Einstein tensor is zero at all points except $r = 0$. The metric satisfying these requirements is the famous *Schwarzschild metric*:

$$ds^2 = -\left(1 - \frac{2GM}{c^2 r}\right) c^2 dt^2 + \frac{dr^2}{\left(1 - 2GM/c^2 r\right)} + r^2(d\theta^2 + \sin^2\theta \, d\phi^2). \tag{13.13}$$

Calculating the Einstein tensor for this metric involves a huge amount of algebra. The first step in calculating the Einstein tensor is the calculation of the Christoffel symbols. As pointed out in §12.2.3, the Christoffel symbols have 40 components in four-dimensional spacetime. Some of these components turn out to be zero. Still one has to take care of many non-zero components. It is instructive to go through this tedious but straightforward algebra once in your lifetime and to show that all components of the Einstein tensor for the metric (13.13) are zero at points other than $r = 0$. When r is very large, $2GM/c^2 r$ becomes small compared to 1. If we neglect this in the coefficient of dr^2 in (13.13), we are led to (13.12). One may wonder whether we ought to keep $2GM/c^2 r$ in the coefficient of dt^2, while neglecting it in the coefficient of dr^2. It is not difficult to justify this. Suppose a particle is at the point r, θ, ϕ at time t and at the point $r + dr, \theta + d\theta, \phi + d\phi$ at time $t + dt$. If the particle is moving non-relativistically, then we must have $dr^2 \ll c^2 dt^2$. Hence the term involving dr^2 is itself small and a small term in its coefficient is of second order of smallness. When we neglect this term, we still have to keep the similar term in the coefficient of dt^2.

It may be noted that the coefficient of dr^2 in (13.13) diverges when r has the value

$$r_S = \frac{2GM}{c^2}. \tag{13.14}$$

This is called the *Schwarzschild radius*. We shall make some comments about the significance of the Schwarzschild radius in §13.3.3. We already pointed out in §1.5 that the effect of general relativity can be neglected if f defined in (1.11), which is equal to r_S/r, is small compared to 1. This means that at radial distances large compared to r_S general relativistic effects can be neglected. This also follows from the fact that the metric at such large distances can be approximated by (13.12), which leads to the same results as what we would get from the Newtonian theory of gravity (see §12.3).

We would expect the metric around a black hole to be given by (13.13). This is certainly true if the black hole is not rotating. In the Newtonian theory of gravity, the gravitational field due to a mass does not depend on whether the mass is rotating or not. One of the intriguing results of general relativity is that a rotating mass tries to drag bodies around it to rotate with it (Thirring and Lense, 1918). Kerr (1963) discovered the exact metric for rotating black holes. It has cross-terms between time and space coordinates, unlike the Schwarzschild metric which does not have such cross-terms. These cross-terms are responsible for the rotational dragging. We shall not discuss the *Kerr metric* in this elementary book.

13.3.1   Particle motion in Schwarzschild geometry. The perihelion precession

A particle will move along a geodesic in the Schwarzschild metric. One can use the geodesic equation (12.51) to study the motions of particles. We shall, however, present a discussion starting more from the basics.

Since the Schwarzschild metric is spherically symmetric, a particle moving in this metric should always lie in a plane passing through the origin. We leave it to the reader to find good arguments to justify this. We can choose the plane of motion to be the equatorial plane in which $\theta = \pi/2$ and $\sin\theta = 1$. A standard convention in general relativity is to choose units of length and time such that c and G turn out to be 1. Setting $c = 1$ and $G = 1$, it follows from (13.13) that the metric lying in the equatorial plane is given by

$$ds^2 = -d\tau^2 = -\left(1 - \frac{2M}{r}\right) dt^2 + \frac{dr^2}{\left(1 - \frac{2M}{r}\right)} + r^2 \, d\phi^2. \qquad (13.15)$$

If a particle moves from a spacetime point A to a spacetime point B, then the path length between them (which turns out to be the proper time measured in a clock carried with the particle) is given by

$$\int_A^B d\tau = \int_A^B L \, d\lambda, \qquad (13.16)$$

where λ is a parameter measured along the path of the particle and L is given by

$$L = \sqrt{\left(1 - \frac{2M}{r}\right)\left(\frac{dt}{d\lambda}\right)^2 - \frac{(dr/d\lambda)^2}{\left(1 - \frac{2M}{r}\right)} - r^2\left(\frac{d\phi}{d\lambda}\right)^2}. \tag{13.17}$$

The basic idea of general relativity is that the particle should follow a geodesic along which the integral given by (13.16) has to be an extremum. This requirement implies that L given by (13.17) should satisfy the Lagrange equation

$$\frac{d}{d\lambda}\left(\frac{\partial L}{\partial(dq^i/d\lambda)}\right) - \frac{\partial L}{\partial q^i} = 0,$$

where q^i can be t, r or ϕ (see, for example, Mathews and Walker, 1979, §12–1). It is seen from (13.17) that L is independent of t and ϕ. This suggests that we shall have the following two constants of motion

$$\frac{\partial L}{\partial(dt/d\lambda)} = \frac{\left(1 - \frac{2M}{r}\right)\frac{dt}{d\lambda}}{L} = \left(1 - \frac{2M}{r}\right)\frac{dt}{d\tau},$$

$$\frac{\partial L}{\partial(d\phi/d\lambda)} = -\frac{r^2\frac{d\phi}{d\lambda}}{L} = -r^2\frac{d\phi}{d\tau},$$

since $L = d\tau/d\lambda$. We denote these constants of motion by e and $-l$, i.e.

$$e = \left(1 - \frac{2M}{r}\right)\frac{dt}{d\tau}, \tag{13.18}$$

$$l = r^2\frac{d\phi}{d\tau}. \tag{13.19}$$

Dividing (13.15) by $d\tau^2$ and using these constants of motion, we get

$$\frac{e^2}{\left(1 - \frac{2M}{r}\right)} - \frac{(dr/d\tau)^2}{\left(1 - \frac{2M}{r}\right)} - \frac{l^2}{r^2} = 1.$$

On rearranging terms a little bit, this can be put in the form

$$\frac{e^2 - 1}{2} = \frac{1}{2}\left(\frac{dr}{d\tau}\right)^2 + V_{\text{eff}}(r), \tag{13.20}$$

where

$$V_{\text{eff}}(r) = -\frac{M}{r} + \frac{l^2}{2r^2} - \frac{Ml^2}{r^3}. \tag{13.21}$$

It is to be noted that the problem is now reduced to a one-dimensional problem of r as a function of τ, since the t and ϕ coordinates have been eliminated with the help of the two constants of motion.

To proceed further, it is now instructive to make some comparisons with the problem of particle motion in an inverse-square law force field in classical mechanics. This problem is often referred to as the *Kepler problem* and has been discussed by Goldstein (1980, §3–2, §3–3) and by Landau and Lifshitz (1976, §14, §15). Readers are urged to refresh their memories about this problem, since we are going to use many analogies with this problem. The classical Kepler problem also has two constants of motion – the angular momentum and the energy. Our constant of motion l given by (13.19) is clearly the general relativistic generalization of the classical angular momentum. To interpret e defined by (13.18), we consider the motion of a particle in the faraway regions where the gravitational field is weak. Then the relation between dt and $d\tau$ can be obtained from (12.74). On using (12.74) and (13.11), it readily follows from (13.18) that

$$e \approx 1 + \frac{\Phi}{c^2} + \frac{1}{2}\frac{v^2}{c^2}. \tag{13.22}$$

Here we have not set c equal to 1 to make the physics clearer. It is obvious that e multiplied by mc^2 would give the sum of the rest mass, potential and kinetic energies in the non-relativistic limit. It follows from (13.22) that

$$\frac{e^2 - 1}{2} \approx \frac{\Phi}{c^2} + \frac{1}{2}\frac{v^2}{c^2}.$$

The right-hand side is essentially the total energy (sum of potential and kinetic energies) used in classical mechanics calculations. We thus identify $(e^2 - 1)/2$ as the relativistic generalization of the classical energy. Now it is easy to interpret (13.20). The term $(1/2)(dr/d\tau)^2$ is like the kinetic energy. Then (13.20) implies that $(e^2 - 1)/2$, which is a constant and reduces to the classical energy in the non-relativistic limit, has to be equal to the sum of the kinetic energy and an effective potential $V_{\text{eff}}(r)$. The classical Kepler problem also gives rise to a one-dimensional equation exactly similar to (13.20), except that the effective potential does not have the last term $-Ml^2/r^3$ appearing in (13.21) (see, for example, Goldstein, 1980, §3–3; Landau and Lifshitz, 1976, §15). It is this last term $-Ml^2/r^3$ which makes results of general relativity different from the classical Kepler problem. When r is much larger than r_S (which is equal to $2M$ in our units), this last term in (13.21) becomes negligible compared to the previous term $l^2/2r^2$ and the general relativistic effects disappear. It is easy to check that, if we had used the metric (13.12) in our calculations rather than (13.13), then this last term would not be there in (13.21).

The values of r at which $V_{\text{eff}}(r)$ has extrema can be obtained from

$$\frac{dV_{\text{eff}}}{dr} = 0,$$

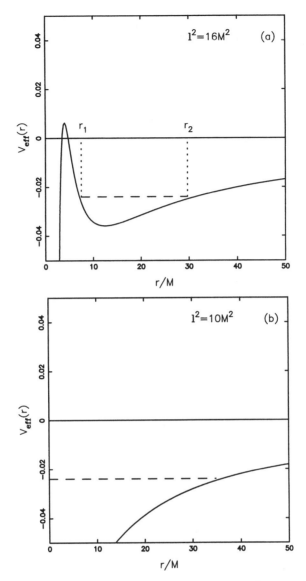

Fig. 13.1 Plots of $V_{\text{eff}}(r)$ given by (13.21) for the cases (i) $l^2 = 16M^2$ and (ii) $l^2 = 10M^2$. The dashed horizontal lines indicate possible values of $(e^2 - 1)/2$.

which gives

$$r = \frac{l^2}{2M} \left[1 \pm \sqrt{1 - 12 \left(\frac{M}{l} \right)^2} \right]. \tag{13.23}$$

If $l^2 > 12M^2$, then $V_{\text{eff}}(r)$ has extrema for two real values of r given by (13.23). On the other hand, if $l^2 < 12M^2$, then there is no extremum for any real value of r. Figures 13.1(a) and 13.1(b) respectively show plots of $V_{\text{eff}}(r)$ for one case with $l^2 > 12M^2$ and one case with $l^2 < 12M^2$. In Figure 13.1(a), a possible value of $(e^2 - 1)/2$ is indicated by a horizontal dashed line, which cuts the curve $V_{eff}(r)$ at two points r_1 and r_2. Since $(dr/d\tau)^2$ in (13.20) is always positive, we easily see that (13.20) can be satisfied only if r lies between r_1 and r_2. We must have $r = r_1$ and $r = r_2$ as the two turning points within which the orbit of the particle should be confined. In the case of Figure 13.1(b), it is not possible for the orbit to be confined from the lower side. This implies that a particle with $l^2 < 12M^2$ should keep falling inward till it hits the central mass. What does this signify? If we throw a particle with zero angular momentum towards a gravitating mass, the particle will fall into the gravitating mass. If the gravitating mass is a point, then in the Newtonian theory of gravity even a very small angular momentum can make sure that the particle does not fall into the gravitating mass. We see in general relativity that the particle has to have an angular momentum with amplitude larger than $2\sqrt{3}M$ in order not to fall into the central mass. There is one other important point to be made. When $l^2 > 12M^2$ and $V_{eff}(r)$ has two extremum points, it is possible for a particle to have a circular orbit if it is located at the minimum of $V_{eff}(r)$. The limiting circular orbit is obtained when $l^2 = 12M^2$. On substituting this in (13.23), we find

$$r = 6M = 3r_S \tag{13.24}$$

as the lowest value of r above which a circular orbit is possible. It is not possible for a particle to go around a black hole in a circular orbit of radius less than $3r_S$. The circular orbit of radius $3r_S$ is called the *last stable orbit*.

The determination of the orbit

Finally we now want to calculate the orbit of the particle, which should be in the form of a functional relation between r and ϕ. It follows from (13.20) that

$$\frac{dr}{d\tau} = \pm\sqrt{e^2 - 1 - 2V_{eff}(r)}. \tag{13.25}$$

From (13.19) we have

$$\frac{d\phi}{d\tau} = \frac{l}{r^2}.$$

On dividing (13.25) by this and then squaring, we get

$$\left(\frac{l}{r^2}\frac{dr}{d\phi}\right)^2 = e^2 - 1 - 2V_{eff}(r). \tag{13.26}$$

To proceed further, we make the substitution

$$r = \frac{1}{u}, \tag{13.27}$$

which is a standard procedure followed in the classical Kepler problem also. Then

$$\frac{dr}{d\phi} = -\frac{1}{u^2}\frac{du}{d\phi}$$

so that (13.26) becomes

$$l^2\left(\frac{du}{d\phi}\right)^2 = e^2 - 1 - 2V_{\text{eff}}(u).$$ (13.28)

Differentiating both sides with respect to ϕ, we get

$$2l^2\frac{du}{d\phi}\frac{d^2u}{d\phi^2} = -2\frac{dV_{\text{eff}}}{du}\frac{du}{d\phi}.$$ (13.29)

We cancel $2\,du/d\phi$ from both the sides and then write (13.21) in the form

$$V_{\text{eff}}(u) = -Mu + \frac{1}{2}l^2u^2 - Ml^2u^3$$

to calculate dV_{eff}/du. This gives

$$\frac{d^2u}{d\phi^2} + u = \frac{1}{p} + 3Mu^2,$$ (13.30)

where

$$p = \frac{l^2}{M}.$$ (13.31)

The orbit as a relation between u and ϕ can be obtained by solving the orbit equation (13.30).

The last term in (13.30) comes from the last term in (13.21). We have already identified this term as the contribution of general relativity. If this last term were not present in the orbit equation (13.30), then it would be the equation of an ellipse, as we find in the classical Kepler problem. Let us try to solve (13.30) for a situation where the general relativistic effect is small and the last term in (13.30) can be treated as a small perturbation compared to the other terms. The zeroth order solution of (13.30) in the absence of this last term would be

$$u_0 = \frac{1}{p}(1 + \epsilon\cos\phi).$$ (13.32)

This is the equation of an ellipse with eccentricity ϵ. Let us now try a solution of the form

$$u = u_0 + u_1,$$ (13.33)

where u_0 is given by (13.32). On substituting this in (13.30) and approximating the small perturbation term as $3Mu^2 \approx 3Mu_0^2$, we get

$$\frac{d^2u_1}{d\phi^2} + u_1 = 3Mu_0^2.$$

On substituting for u_0 from (13.32), we get

$$\frac{d^2u_1}{d\phi^2} + u_1 = \frac{3M}{p^2}(1 + 2\epsilon\cos\phi + \epsilon^2\cos^2\phi). \tag{13.34}$$

The term $2\epsilon\cos\phi$ in the right-hand side acts like a resonant forcing term, since it varies with ϕ the same way as u_0. As the effect of this term is going to be much more significant than that of the other two terms in the right-hand side of (13.34), we can neglect these other two terms. When we keep only the $2\epsilon\cos\phi$ term in the right-hand side of (13.34), its solution can be written down as

$$u_1 = \frac{3M\epsilon}{p^2}\phi\sin\phi, \tag{13.35}$$

which can be verified by substituting in (13.34). From (13.32), (13.33) and (13.35), we have

$$u = \frac{1}{p}\left[1 + \epsilon\cos\phi + \frac{3M\epsilon}{p}\phi\sin\phi\right].$$

When $3M\phi/p$ is small compared to 1, this can be written as

$$u = \frac{1}{p}\left[1 + \epsilon\cos\left\{\phi\left(1 - \frac{3M}{p}\right)\right\}\right]. \tag{13.36}$$

If the particle followed an exact elliptical path as given by (13.32), then the value of u would be repeated when ϕ changes by 2π. It follows from (13.36) that u would repeat when ϕ changes by $2\pi + \delta\phi$, where

$$\delta\phi = 2\pi\frac{3M}{p} = 6\pi\frac{M^2}{l^2}$$

on using (13.31). Clearly $\delta\phi$ is the angle by which the perihelion of the particle precesses during one revolution. If one puts back G and c which were set to 1 in our analysis, then the expression for the perihelion precession is given by

$$\delta\phi = 6\pi\left(\frac{GM}{cl}\right)^2. \tag{13.37}$$

For the planet Mercury, the perihelion precession rate turns out to be 43″ per century (Exercise 13.4). This provided one of the famous tests of general relativity.

13.3.2 Motion of massless particles. The bending of light

A photon or a massless particle moving with speed c follows a special geodesic for which

$$ds^2 = 0.$$

A geodesic with this property is called a *null geodesic*. For such a massless particle moving in the equatorial plane, (13.15) becomes

$$\left(1 - \frac{2M}{r}\right)dt^2 - \frac{dr^2}{\left(1 - \frac{2M}{r}\right)} - r^2 d\phi^2 = 0. \tag{13.38}$$

Even for a massless particle, we expect the energy and the angular momentum to be conserved because of the symmetry with respect to t and ϕ. However, e and l as defined in (13.18) and (13.19) tend to be infinite, since $d\tau \to 0$ along the trajectory of the particle as its mass goes to zero. But the ratio

$$\frac{e}{l} = \left(1 - \frac{2M}{r}\right)\frac{1}{r^2}\frac{dt}{d\phi} \tag{13.39}$$

would remain finite and constant even when the mass tends to zero. In the case of a particle with mass, we had used the proper time τ (which would be the time measured by a clock moving with the particle) as a label to mark the trajectory of the particle. For the massless particle, $d\tau = 0$ and τ can no longer be used to label the trajectory. So we introduced an *affine parameter* λ which increases along the trajectory of the massless particle in such a way that

$$e = \left(1 - \frac{2M}{r}\right)\frac{dt}{d\lambda} \tag{13.40}$$

remains constant. Then

$$l = r^2 \frac{d\phi}{d\lambda} \tag{13.41}$$

also has to be a constant to make the ratio given in (13.39) a constant. Thus, for a massless particle, we define e and l with the help of the affine parameter λ rather than the proper time τ as done in (13.18) and (13.19).

Dividing (13.38) by $d\lambda^2$, we get

$$\frac{e^2 - (dr/d\lambda)^2}{1 - 2M/r} = \frac{l^2}{r^2}$$

on using (13.40) and (13.41). From this

$$\frac{e^2}{l^2} - \frac{1}{l^2}\left(\frac{dr}{d\lambda}\right)^2 = \frac{1}{r^2}\left(1 - \frac{2M}{r}\right),$$

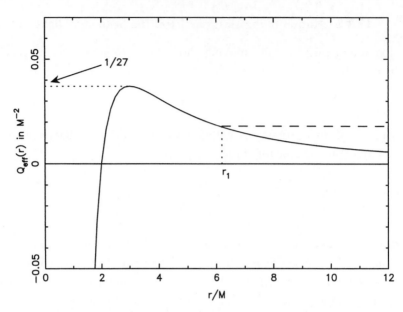

Fig. 13.2 A plot of the effective potential $Q_{\text{eff}}(r)$ for a massless particle given by (13.43). The dashed horizontal line indicates a possible value of $1/b^2$.

which can be written as

$$\frac{1}{b^2} = \frac{1}{l^2}\left(\frac{dr}{d\lambda}\right)^2 + Q_{\text{eff}}(r),\tag{13.42}$$

where $b = l/e$ is clearly a constant of motion and

$$Q_{\text{eff}}(r) = \frac{1}{r^2}\left(1 - \frac{2M}{r}\right),\tag{13.43}$$

which is plotted in Figure 13.2, has a maximum at $r = 3M$ with a value $(27\,M^2)^{-1}$. Figure 13.2 also shows a dashed horizontal line indicating a possible value of $1/b^2$ less than $(27\,M^2)^{-1}$. It easily follows from (13.42) that r has to be restricted to a lower limit r_1 if the trajectory is on the right side of the $Q_{\text{eff}}(r)$ curve. In other words, a massless particle coming from infinity would not approach any closer than r_1. However, if we have $b < 3\sqrt{3}M$, making $1/b^2$ larger than $(27\,M^2)^{-1}$, then the horizontal line corresponding to $1/b^2$ would be above the maximum of $Q_{\text{eff}}(r)$ and a massless particle coming from infinity would fall into the gravitating mass M.

To understand the significance of the important result that a massless particle coming from infinity with $b < 3\sqrt{3}M$ would be captured by the mass M, let us try to figure out the physical meaning of b. At r much larger than the Schwarzschild radius $2M$, it follows from (13.40) and (13.41) that

$$b \approx r^2\frac{d\phi}{dt}.\tag{13.44}$$

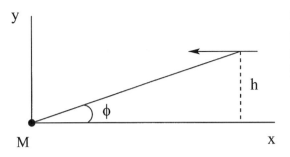

y

ϕ

M

X

h

Fig. 13.3 A massless particle approaching the central mass M from a large distance with an impact parameter h.

We now consider a massless particle approaching the central mass M from a large distance with the impact parameter h as shown in Figure 13.3. The x axis is chosen in such a way that the particle moves in the negative x direction. If the polar angle ϕ is measured with respect to the x axis, then we have

$$\tan \phi = \frac{h}{x},$$

which on differentiation with respect to t gives

$$\sec^2 \phi \frac{d\phi}{dt} = -\frac{h}{x^2} \frac{dx}{dt}.$$

Here $-dx/dt$ is the speed of the particle, which is $c = 1$ in our units. Hence

$$\frac{d\phi}{dt} = \frac{h}{(x \sec \phi)^2} = \frac{h}{r^2}.$$

Comparing it with (13.44), we at once see that

$$b = h,$$

which means that the parameter b is nothing but the impact parameter with which the massless particle approaches the mass M. When this impact parameter is less than $3\sqrt{3}M$, the massless particle or the photon gets captured by the central mass M.

The orbit of light

If the massless particle has an impact parameter much larger than $3\sqrt{3}M$, then its trajectory will be slightly bent. To calculate this bending, we first have to derive the orbit equation. From (13.42), we get

$$\frac{dr}{d\lambda} = \pm l \sqrt{\frac{1}{b^2} - Q_{\text{eff}}(r)}.$$

Substituting for l from (13.41), we have

$$\frac{dr}{d\phi} = \pm r^2 \sqrt{\frac{1}{b^2} - Q_{\text{eff}}(r)},$$

so that

$$\left(\frac{1}{r^2}\frac{dr}{d\phi}\right)^2 = \frac{1}{b^2} - Q_{\text{eff}}(r),\tag{13.45}$$

which can be compared with (13.26). We proceed to solve it in exactly the same way we solved (13.26). By introducing the variable $u = 1/r$ and using (13.43) to write

$$Q_{\text{eff}}(u) = u^2 - 2Mu^3,$$

we put (13.45) in the form

$$\left(\frac{du}{d\phi}\right)^2 = \frac{1}{b^2} - u^2 + 2Mu^3.$$

Differentiating with respect to ϕ and cancelling $2\,du/d\phi$ from both sides, we finally get the orbit equation

$$\frac{d^2u}{d\phi^2} + u = 3Mu^2,\tag{13.46}$$

which has to be solved to find u as a function of ϕ giving the orbit.

It is easy to check that the term $3Mu^2$ on the right-hand side of (13.46) is the general relativistic effect. When this term is small, we can solve (13.46) by following the same perturbative approach which we followed to solve the orbit equation (13.30) for particles with non-zero mass. When the $3Mu^2$ term is neglected, we can write the zeroth order solution as

$$u_0 = \frac{\cos\phi}{R},\tag{13.47}$$

where R is the distance of the closest approach and ϕ is defined in such a way that we have $\phi = 0$ at the point of closest approach. Again writing u in the form (13.33), we find that u_1 should satisfy the equation

$$\frac{d^2u_1}{d\phi^2} + u_1 = 3Mu_0^2 = \frac{3M}{R^2}\cos^2\phi,$$

of which a solution is

$$u_1 = \frac{M}{R^2}(1 + \sin^2\phi).\tag{13.48}$$

Then the full solution can be written down by adding (13.47) and (13.48), i.e.

$$u = \frac{\cos\phi}{R} + \frac{M}{R^2}(1 + \sin^2\phi).\tag{13.49}$$

Since we have chosen $\phi = 0$ at the point of closest approach, the incoming and the outgoing directions of the massless particle would have been $-\pi/2$ and $\pi/2$ respectively if the particle travelled in a straight line. This will be clear from

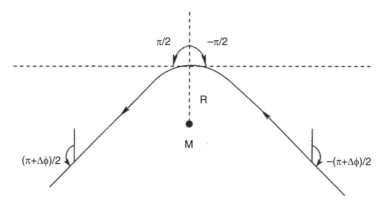

Fig. 13.4 The trajectory of light bent by an angle $\Delta\phi$ while passing by the side of the mass M. The angle ϕ is measured with respect to the vertical direction such that $\phi = 0$ at the point of closest approach.

Figure 13.4. If the particle undergoes a deflection $\Delta\phi$, then the incoming and outgoing directions are

$$\phi = \mp\left(\frac{\pi}{2} + \frac{\Delta\phi}{2}\right).$$

Then we have

$$\cos\phi = -\sin\frac{\Delta\phi}{2} \approx -\frac{\Delta\phi}{2}$$

on assuming $\Delta\phi$ to be small. When the massless particle initially starts from infinity or finally reaches infinity, we have $u \approx 0$ and $\sin\phi \approx 1$. Hence (13.49) gives

$$0 \approx -\frac{\Delta\phi}{2R} + \frac{M}{R^2}(1 + 1),$$

from which we finally have

$$\Delta\phi = \frac{4M}{R}.$$

On putting back G and c, this becomes

$$\Delta\phi = \frac{4GM}{c^2 R}. \tag{13.50}$$

If a light ray passes by the side of a mass M such that the closest distance of approach is R, then the light ray is bent by the amount $\Delta\phi$ given by the famous relation (13.50).

  If we substitute the mass and radius of the Sun for M and R in (13.50), then $\Delta\phi$ turns out to be $1.75''$. This means that light from a star at the edge of the solar disk will be bent in such a way that the star will appear to be shifted outward from the centre of the solar disk by $1.75''$. We of course cannot see stars

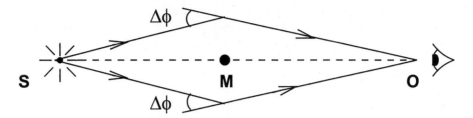

Fig. 13.5 An illustration of gravitational lensing by the mass M located symmetrically between the source S and the observer O.

at the edge of the solar disk under normal circumstances. Such stars, however, may become visible at the time of the total solar eclipse. Comparing their positions around the eclipsed Sun with their usual positions, one can determine whether a shift has taken place. Eddington and his colleagues carried out this exercise during an eclipse in 1919 and announced the discovery of the bending of light in conformity with general relativity (Dyson, Eddington and Davidson, 1920).

Gravitational lensing

There is one very important application of the gravitational bending of light in extragalactic astronomy. It is the phenomenon of *gravitational lensing.* Suppose there is a massive object M exactly between a source S and an observer O as shown in Figure 13.5. Light rays from the source S passing by M on different sides will be deflected by the angle $\Delta\phi$. The observer will then see the source S in the form of a ring. Such a ring is known as an *Einstein ring.* A few examples of nearly perfect Einstein rings are known. Figure 13.6 shows an almost complete Einstein ring. We expect to see a perfect ring only if the lensing-producing mass M is located symmetrically on the line of sight between S and O. In a less symmetric situation, we would see arcs of the ring rather than the whole ring. Many images of extragalactic sources in the forms of extended arcs are known, suggesting that gravitational lensing is a quite common phenomenon in the extragalactic world.

Let us also comment on another kind of gravitational lensing. As we have discussed in §9.2.2, the rotation curves of spiral galaxies suggest the presence of dark matter associated with these galaxies. One possibility is that the dark matter exists in the form of massive compact objects (from something like a large planet to something having a few solar masses) in the halo of the galaxy. Suppose one such object in the halo of our Galaxy comes between us and a star in a nearby galaxy. As gravitational lensing would amplify the light from the star, the star would appear brighter as long as the compact object remains between us and the star. Since such an event would be very rare and one cannot predict when a particular star is likely to be lensed, the best way of detecting such lensing events is to monitor a rich field of extragalactic stars for a long time to see if the brightness of any star is temporarily enhanced. Events of this kind in which stars in the Large Magellanic Cloud temporarily appeared brighter

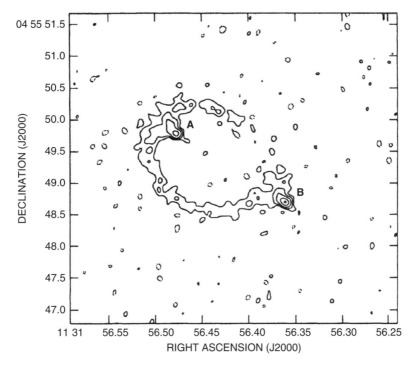

Fig. 13.6 A nearly complete Einstein ring MG 1131+0456 imaged at the radio frequency 15 GHz by the VLA (Very Large Array). From Chen and Hewitt (1993). (©American Astronomical Society. Reproduced with permission from *Astronomical Journal*.)

typically by 1 magnitude for a few days were first reported simultaneously by two groups (Alcock *et al.*, 1993; Aubourg *et al.*, 1993). From a study of such events, it seems that a part of the dark matter associated with our Galaxy, but probably not the whole of it, is in the form of compact objects in the galactic halo.

13.3.3 Singularity and horizon

It should be clear from the expression (13.13) of the Schwarzschild metric that g_{tt} becomes infinite at $r = 0$ and g_{rr} becomes infinite at the Schwarzschild radius r_S given by (13.14). This is the case when we use (t, r, θ, ϕ) coordinates. One can think of making transformations to other coordinate systems. It is found that some component of the metric tensor always has to become infinite at $r = 0$ in any coordinate system, whereas it is possible to find coordinate systems in which the metric tensor remains well defined at the Schwarzschild radius $r = r_S$. We believe that the singularity at $r = 0$ is a real essential singularity. On the other hand, the metric becomes singular at $r = r_S$ only in certain coordinate systems. This is, therefore, a coordinate singularity and not an essential singularity. One particularly convenient coordinate system in which there is no singularity at $r = r_S$ was discovered by Kruskal (1960).

Apart from a black hole, any other self-gravitating system (even a neutron star) would have a physical radius larger than the Schwarzschild radius r_S. In such a situation, the free space where the Schwarzschild metric holds will not include $r = r_S$. Only in the case of a black hole, do we need to worry about the physical significance of the radius $r = r_S$. An observer falling in a black hole will not feel anything unusual when crossing the Schwarzschild radius $r = r_S$. Let us consider an observer falling radially in a black hole starting from rest at infinity. It easily follows from (13.18) and (13.19) that $e = 1$ and $l = 0$ in this case. From (13.20) and (13.21), we then have

$$\frac{1}{2}\left(\frac{dr}{d\tau}\right)^2 - \frac{M}{r} = 0,$$

which leads to

$$\sqrt{r}\,dr = -\sqrt{2M}\,d\tau. \qquad (13.51)$$

It should be kept in mind that $d\tau$ is the time interval measured by the clock carried with the falling observer. Suppose we take $\tau = 0$ when the falling observer is at $r = r_0$. Then (13.51) can be integrated to give

$$\frac{2}{3}(r^{3/2} - r_0^{3/2}) = -\sqrt{2M}\tau. \qquad (13.52)$$

It is easy to see that the falling observer will reach the central singularity $r = 0$ in a finite time measured by his own clock.

Since the falling observer passes through the Schwarzschild radius $r = r_S$ without feeling anything special there, does it mean that the Schwarzschild radius has no particular physical significance? The physical significance of the Schwarzschild radius becomes clear as soon as we consider an observer who is trying to come out rather than an observer who is falling in. Not only an observer, but no signal including light can escape outside from $r < r_S$. Hence the surface $r = r_S$ is called the *horizon* of the black hole. A signal from the inside of the horizon can never come out, although things from the outside can fall through the horizon. The result that a light signal can escape outward only if it starts from $r > r_S$ can be shown by extending the analysis presented in §13.3.2. The reader is asked to do this calculation in Exercise 13.5.

Instead of considering the proper time τ measured by the clock of the observer falling in the black hole, if we consider the world time t, then we get a surprising result. Since we are considering $e = 1$, it follows from (13.18) that

$$\frac{dt}{d\tau} = \left(1 - \frac{2M}{r}\right)^{-1}. \qquad (13.53)$$

Then

$$\frac{dt}{dr} = \frac{dt}{d\tau}\frac{d\tau}{dr} = -\left(\frac{2M}{r}\right)^{-1/2}\left(1 - \frac{2M}{r}\right)^{-1} \tag{13.54}$$

on using (13.51) and (13.53). The solution of (13.54) is

$$t = t_0 + 2M\left[-\frac{2}{3}\left(\frac{r}{2M}\right)^{3/2} - 2\left(\frac{r}{2M}\right)^{1/2} + \log\left|\frac{(r/2M)^{1/2} + 1}{(r/2M)^{1/2} - 1}\right|\right], \tag{13.55}$$

which can be verified by substituting (13.55) into (13.54). It follows from (13.55) that t goes to infinity when $r = 2M$. In other words, an infinite amount of world time has to lapse before the falling observer reaches the horizon. To understand the significance of this result, let us consider another observer at rest far away from the black hole. The proper time and the world time of this faraway observer will be approximately the same. We have also argued in §13.1 that events taking place in different locations at the same world time are simultaneous. It then follows that, according to the faraway observer, the falling observer will take infinite time to reach the horizon at the Schwarzschild radius. It would appear to the faraway observer that the falling observer is forever hovering around the horizon, although the falling observer reaches $r = 0$ in a finite time according to his clock. To understand exactly how the falling observer appears to the faraway observer, it is necessary to analyse the propagation of light signals from the falling observer to the faraway observer. We shall not get into those details here.

13.4 Linearized theory of gravity

We pointed out in §10.2 that one of the implications of an action-at-a-distance theory is that the interaction has to propagate at an infinite speed. Our hope is that this problem would be rectified in a field theory. It is indeed a consequence of general relativity that gravitational interaction propagates at speed c. We now demonstrate this for a region where the gravitational field is weak so that the metric differs only slightly from the special relativistic metric η_{ik} given by (12.60). We write

$$g_{ik} = \eta_{ik} + h_{ik} \tag{13.56}$$

and assume that

$$|h_{ik}| \ll 1. \tag{13.57}$$

As a consequence of (13.57), we shall throw away terms quadratic or of higher order in h_{ij} compared to the first order terms. General relativity is basically a nonlinear field theory. On throwing away terms quadratic in h_{ij}, it reduces to a linear theory and will be seen to have many similarities with the electromagnetic theory.

We want to find out what happens to Einstein's equation (12.96) when we have a metric tensor of the form (13.56). For this purpose, we need to calculate the Einstein tensor G_{ij} arising out of the metric tensor (13.56). The first step is to calculate the Christoffel symbols by using (12.31). On substituting (13.56) into (12.31), we get

$$\Gamma^m_{ik} = \frac{1}{2}\eta^{ml}\left(\frac{\partial h_{li}}{\partial x^k} + \frac{\partial h_{lk}}{\partial x^i} - \frac{\partial h_{ik}}{\partial x^l}\right) \tag{13.58}$$

if we throw away the terms quadratic in h_{ij}. The next step is to calculate the Ricci curvature tensor R_{km} by using (12.41). Since it follows from (13.58) that the Christoffel symbols are linear in h_{ij}, we can throw away terms quadratic in the Christoffel symbols in (12.41) and then substitute from (13.58), which gives

$$R_{ik} = \frac{\partial\Gamma^l_{ik}}{\partial x^l} - \frac{\partial\Gamma^l_{il}}{\partial x^k} = \frac{1}{2}\eta^{lm}\left(\frac{\partial^2 h_{km}}{\partial x^i\partial x^l} - \frac{\partial^2 h_{ik}}{\partial x^l\partial x^m} - \frac{\partial^2 h_{lm}}{\partial x^i\partial x^k} + \frac{\partial^2 h_{il}}{\partial x^k\partial x^m}\right). \tag{13.59}$$

We now use the fact that $\eta^{lm}h_{km} = h^l_k$ and

$$\eta^{lm}\frac{\partial^2}{\partial x^l\partial x^m} = -\frac{1}{c^2}\frac{\partial^2}{\partial t^2} + \nabla^2 = \Box^2.$$

Then (13.59) becomes

$$R_{ik} = \frac{1}{2}\left(\frac{\partial^2 h^l_k}{\partial x^i\partial x^l} - \Box^2 h_{ik} - \frac{\partial^2 h}{\partial x^i\partial x^k} + \frac{\partial h^m_i}{\partial x^k\partial x^m}\right). \tag{13.60}$$

It now follows that the scalar curvature is given by

$$R = \eta^{ik}R_{ik} = \left(\frac{\partial^2 h^{km}}{\partial x^k\partial x^m} - \Box^2 h\right). \tag{13.61}$$

Substituting (13.60) and (13.61) in (12.44), the Einstein tensor is given by

$$G_{ik} = \frac{1}{2}\left(\frac{\partial^2 h^l_k}{\partial x^i\partial x^l} + \frac{\partial h^m_i}{\partial x^k\partial x^m} - \Box^2 h_{ik} - \frac{\partial^2 h}{\partial x^i\partial x^k} - \eta_{ik}\frac{\partial^2 h^{lm}}{\partial x^l\partial x^m} + \eta_{ik}\Box^2 h\right). \tag{13.62}$$

This is the expression of the Einstein tensor for the metric (13.56) if we throw away the quadratic and higher powers of h_{ik}. This looks like a complicated expression which becomes somewhat simplified if we introduce the variable

$$\overline{h}_{ik} = h_{ik} - \frac{1}{2}\eta_{ik}h, \tag{13.63}$$

from which it readily follows that

$$h_{ik} = \overline{h}_{ik} - \frac{1}{2}\eta_{ik}\overline{h} \tag{13.64}$$

on making use of the fact that $\eta_i^i = 4$. Substituting (13.64) in (13.62), we get

$$G_{ik} = \frac{1}{2} \left(\frac{\partial^2 \overline{h}_k^l}{\partial x^i \partial x^l} + \frac{\partial \overline{h}_i^m}{\partial x^k \partial x^m} - \eta_{ik} \frac{\partial^2 \overline{h}^{lm}}{\partial x^l \partial x^m} - \Box^2 \overline{h}_{ik} \right). \tag{13.65}$$

The expression (13.65) for the Einstein tensor is still rather complicated. However, if we could choose a coordinate system in which the divergence of $\overline{h}_{ik}$ is zero, then the first three terms on the right-hand side of (13.65) will become zero and the expression of the Einstein tensor will become very simple. We now show that any coordinate system can be slightly adjusted to make the divergence of $\overline{h}_{ik}$ zero. Let us consider introducing a new coordinate system in which

$$x'^i = x^i + \xi^i. \tag{13.66}$$

The introduction of such a new coordinate system is called a *gauge transformation* in general relativity. From (12.5), it follows that the metric tensor g_{ik} should get transformed in the new coordinate system to

$$g'_{ik} = g_{lm} \frac{\partial x^l}{\partial x'^i} \frac{\partial x^m}{\partial x'^k} = g_{lm} \left(\delta_i^l - \frac{\partial \xi^l}{\partial x'^i} \right) \left(\delta_k^m - \frac{\partial \xi^m}{\partial x'^k} \right) \tag{13.67}$$

on using (13.66). We shall see soon that ξ^i will be of order h_{ik} in our analysis. So we can throw away terms quadratic in them and replace $\partial \xi^l / \partial x'^i$ by $\partial \xi^l / \partial x^i$. Then (13.67) becomes

$$g'_{ik} = g_{ik} - \frac{\partial \xi_k}{\partial x^i} - \frac{\partial \xi_i}{\partial x^k}. \tag{13.68}$$

Using (13.56) and writing g'_{ik} in the same way as g_{ik}, we get

$$h'_{ik} = h_{ik} - \frac{\partial \xi_k}{\partial x^i} - \frac{\partial \xi_i}{\partial x^k}, \tag{13.69}$$

if we require η_{ik} to be the same in both coordinate systems. Note that h_{ik} itself does not transform like a tensor. If h_{ik} or $\overline{h}_{ik}$ transformed like tensors, then their divergences would transform like vectors, and it will not be possible to make this divergence zero in a frame when it is non-zero in other frames. Our aim now is to choose a coordinate system in which

$$\frac{\partial \overline{h}'_{ik}}{\partial x'_k} = 0, \tag{13.70}$$

which is equivalent to

$$\frac{\partial h'_{ik}}{\partial x_k} - \frac{1}{2} \frac{\partial h'}{\partial x^i} = 0$$

by virtue of (13.63). On substituting from (13.69) in this equation, we find after some easy algebra that

$$\Box^2 \xi_i = \frac{\partial \overline{h}_{ik}}{\partial x_k}. \tag{13.71}$$

Let us now try to understand the significance of this result. Suppose we have a coordinate system x^i in which h_{ik} or $\overline{h}_{ik}$ are known. By solving (13.71) we find ξ_i and then transform to a new coordinate system in accordance with (13.66). We shall have (13.70) satisfied in that coordinate system. Thus, by a suitable choice of gauge, we can find a coordinate system in which the divergence of $\overline{h}_{ik}$ vanishes. If we use such a coordinate system, then the Einstein tensor given by (13.65) should have a simple form

$$G_{ik} = -\frac{1}{2}\Box^2 \overline{h}_{ik}. \tag{13.72}$$

On substituting (13.72) in Einstein's equation (12.96), we find

$$-\Box^2 \overline{h}_{ik} = \frac{16\pi G}{c^4} T_{ik}, \tag{13.73}$$

which is the inhomogeneous wave equation. This is an equation which appears in electromagnetic theory and its solution is discussed in all standard textbooks of advanced electromagnetic theory (see, for example, Panofsky and Phillips, 1962, §14–2; Jackson, 2001, §6.4). We assume that readers know how to solve the inhomogeneous wave equation and merely quote the final result. The solution of (13.73) is

$$\overline{h}_{ik}(t, \mathbf{x}) = \frac{4G}{c^4} \int \frac{T_{ik}(t - |\mathbf{x} - \mathbf{x}'|/c, \mathbf{x}')}{|\mathbf{x} - \mathbf{x}'|} d^3x', \tag{13.74}$$

where $\mathbf{x}$ and $\mathbf{x}'$ are spatial coordinates of a field point and a source point. It is clear that T_{ik} acts as the source for $\overline{h}_{ik}$, which makes the metric different from a flat special relativistic metric. The solution (13.74) also implies that the information from the source to the field travels at speed c. Thus the gravitational interaction is restricted to propagate at speed c, which we enlisted in §10.2 as a requirement for a field theory of gravity.

We have seen that the metric tensor for a weak gravitational field is given by (12.70). We end our discussion of the linearized theory by showing that (13.74) is consistent with (12.70) for the case of a mass distribution at rest. It follows from (12.88) that

$$T_{00} = \rho c^2$$

for a static mass distribution, whereas the other components of T are zero. Then it follows from (13.74) that

$$\overline{h}_{00}(\mathbf{x}) = \frac{4G}{c^2} \int \frac{\rho(\mathbf{x}')}{|\mathbf{x} - \mathbf{x}'|} d^3x', \tag{13.75}$$

where we have not indicated the time dependence, since we are considering a static problem. All the other components of $\bar{h}_{ik}$ are zero. We know that the gravitational potential in the Newtonian theory of gravity is given by

$$\Phi(\mathbf{x}) = -G \int \frac{\rho(\mathbf{x}')}{|\mathbf{x} - \mathbf{x}'|} d^3 x'.$$

Comparing this with (13.75), we conclude that

$$\bar{h}_{00}(\mathbf{x}) = -\frac{4\Phi(\mathbf{x})}{c^2}. \tag{13.76}$$

It follows from (13.64) and (13.76) that

$$h_{00}(\mathbf{x}) = -\frac{2\Phi(\mathbf{x})}{c^2},$$

which implies the metric (12.70).

13.5 Gravitational waves

It is clear from (13.74) that a sudden change in the energy-momentum tensor T_{ik} would give rise to a signal propagating away at speed c. We also note that in a region of empty space (13.73) reduces to the wave equation

$$\Box^2 \bar{h}_{lm} = 0$$

suggesting the possibility of *gravitational waves*. We now work out some properties of such waves.

It follows from (13.64) that h_{lm} also satisfies the wave equation

$$\Box^2 h_{lm} = 0. \tag{13.77}$$

We also have to keep in mind that we are using a coordinate system in which $\bar{h}_{lm}$ should be divergence-free as in (13.70). Then (13.63) implies

$$\frac{\partial h_{lm}}{\partial x_m} = \frac{1}{2} \frac{\partial h}{\partial x^l}. \tag{13.78}$$

We can choose the propagation direction as the x_3 direction without any loss of generality. Since we are now considering a linearized theory, any arbitrary wave can be treated as a superposition of the various Fourier modes. A Fourier mode travelling in the x_3 direction can be written as

$$h_{lm} = A_{lm} e^{ik(ct - x_3)}. \tag{13.79}$$

The symmetric tensor A_{lm} has 10 components. Some of them have to be related, as we can see on substituting (13.79) in (13.78). For the four values of $l = 0, 1, 2, 3$, we get the following four conditions

$$A_{00} + A_{03} = -\frac{1}{2}A, \ A_{10} + A_{13} = 0, \ A_{20} + A_{23} = 0, \ A_{30} + A_{33} = \frac{1}{2}A,$$

$$(13.80)$$

where A is the trace

$$A = -A_{00} + A_{11} + A_{22} + A_{33}. \qquad (13.81)$$

Because of the four conditions (13.80), only six components of A_{lm} are independent. Let us take $A_{11}, A_{12}, A_{13}, A_{23}, A_{33}, A_{00}$ as the independent components.
The other components can be expressed in terms of them with the help of
(13.80) and (13.81). We need not concern ourselves with the exact expressions,
except to note that

$$A_{22} = -A_{11}, \qquad (13.82)$$

which follows from (13.80) and (13.81). We now think of carrying out another
gauge transformation such that the divergence of $\overline{h}_{lm}$ will be zero in the new
frame also. Since the divergence of $\overline{h}_{lm}$ is already zero in the frame we are
using, (13.71) reduces to

$$\Box^2 \xi_l = 0,$$

of which a solution is

$$\xi_l = i f_l e^{ik(ct - x_3)}. \qquad (13.83)$$

If we now make a coordinate transformation (13.66) with ξ_l given by (13.83),
then whatever we have been discussing should be valid in the new coordinate
system as well. It follows from (13.69) that

$$A'_{11} = A_{11}, \ A'_{12} = A_{12},$$

$$A'_{13} = A_{13} - kf_1, \ A'_{23} = A_{23} - kf_2,$$

$$A'_{33} = A_{33} - 2kf_3, \ A'_{00} = A_{00} + 2kf_0.$$

It is obvious that f_l can be chosen in such a way that $A_{13}, A_{23}, A_{33}, A_{00}$ turn out
to be zero in the new frame. It follows from (13.80) that A_{10}, A_{20} and A_{30} also
must be zero. The only non-zero components left are $A_{11} = -A_{22}$, which we
call a, and $A_{12} = A_{21}$, which we call b. This means that, in our chosen gauge,
(13.79) can be written as

$$h_{lm} = e^{ik(ct - x_3)} \begin{pmatrix} 0 & 0 & 0 & 0 \\ 0 & a & b & 0 \\ 0 & b & -a & 0 \\ 0 & 0 & 0 & 0 \end{pmatrix}. \qquad (13.84)$$

Since the amplitude of the wave involves only two independent variables a and
b, we can conclude that the gravitational wave has two possible polarizations.

We now want to figure out the physical characteristics of the two possible polarization modes of a gravitational wave. For this purpose, let us discuss what happens when the gravitational wave impinges on a set of particles. If a gravitational disturbance makes all the particles in a region, along with the observer, move in exactly the same way, then it will be difficult to ascertain the presence of the gravitational disturbance. On the other hand, if the gravitational disturbance causes a relative motion amongst different particles, then this relative motion can be used to detect the disturbance. We shall now study the relative motions amongst particles induced by a gravitational wave. Let x^i and $x^i + \delta x^i$ be the spacetime coordinates of two nearby particles. They would both satisfy the geodesic equation (12.51). By subtracting the one geodesic equation from the other, one can show that the relative separation δx^i would satisfy

$$\frac{D^2}{D\tau^2}\delta x^i = -R^i_{klm}\,\delta x^l\,\frac{dx^k}{d\tau}\frac{dx^m}{d\tau}. \tag{13.85}$$

This is called the *geodesic deviation equation*, which readers were asked to derive in Exercise 12.4. For slow motions in a region of weak gravity, we have (12.74), which implies that

$$\frac{dx^0}{d\tau} \approx c,$$

whereas the other spatial components are essentially the components of velocity which are much smaller. Hence the spatial components of (13.85) can be written as

$$\frac{d^2}{dt^2}\delta x^\alpha = -R^\alpha_{0\beta 0}\,\delta x^\beta\, c^2. \tag{13.86}$$

Remember that the Greek indices like α and β run over only values 1, 2, 3, corresponding to the spatial components. Here we are interested in finding the separation between the two particles by considering them at the same instant of time. So we take $\delta x^0 = 0$, which is used in obtaining (13.86). It now follows from (12.37) that

$$R^\alpha_{0\beta 0} = \frac{\partial \Gamma^\alpha_{00}}{\partial x^\beta} - \frac{\partial \Gamma^\alpha_{0\beta}}{\partial x^0}$$

on neglecting the terms quadratic in Christoffel symbols in our linearized analysis. Let us consider particles lying on the wavefront of the gravitational wave, which is assumed to propagate in the x^3 direction. Then δx^β in (13.86) has to lie in the (x^1, x^2) plane, which means that β can have values 1 or 2. It should be clear that differentiation with respect to x^β gives zero and we can write

$$R^\alpha_{0\beta 0} = -\frac{\partial \Gamma^\alpha_{0\beta}}{\partial x^0}. \tag{13.87}$$

The expression for the Christoffel symbol for the weak gravitational field is given by (13.58). Keeping in mind that $h_{0\beta} = 0$ according to (13.84), we find from (13.58) that

$$\Gamma^\alpha_{0\beta} = \frac{1}{2}\frac{\partial h^\alpha_\beta}{\partial x^0}.$$
(13.88)

On using (13.87) and (13.88), we obtain from (13.86) that

$$\frac{d^2}{dt^2}\delta x_\alpha = \frac{1}{2}\frac{\partial^2 h_{\alpha\beta}}{\partial t^2}\delta x^\beta.$$
(13.89)

This is finally the equation giving the relative motion amongst particles lying on the wavefront of a gravitational wave.

The obvious solution of (13.89) is

$$\delta x_\alpha = \delta x_{\alpha,0} + \frac{1}{2}h_{\alpha\beta}\,\delta x^\beta.$$
(13.90)

Let us consider the first polarization mode of the gravitational wave for which $b = 0$ in (13.84). On substituting from (13.84) into (13.90) with $b = 0$, we get

$$\delta x_1 = \delta x_{1,0}\left[1 + \frac{1}{2}ae^{ikct}\right],$$

$$\delta x_2 = \delta x_{2,0}\left[1 - \frac{1}{2}ae^{ikct}\right],$$
(13.91)

if we take $x_3 = 0$ on the wavefront where the particles are located. Let us consider a ring of particles in the (x_1, x_2) plane. It is clear from (11.90) that the particles on the x_1 axis move outward when the particles on the x_2 axis move inward. This means that the ring of particles would oscillate as shown in the upper row of Figure 13.7. Next we consider the other polarization mode for which we take $a = 0$ in (13.84). Again, on substituting from (13.84) into (13.90), we get

$$\delta x_1 = \delta x_{1,0} + \frac{1}{2}be^{ikct}\delta x_{2,0},$$

$$\delta x_2 = \delta x_{2,0} + \frac{1}{2}be^{ikct}\delta x_{1,0}.$$
(13.92)

The lower row of Figure 13.7 indicates how the ring of particles will oscillate in this case.

Apart from making the physical nature of the two polarization modes clear, Figure 13.7 suggests how gravitational waves may be detected. Suppose we have a gravitational wave with the first kind of polarization falling perpendicularly on a Michelson interferometer (see, for example, Born and Wolf, 1980, §7.5.4), of which the arms are along x_1 and x_2 directions. When one arm expands, the other arm contracts, causing a shift in the fringes. A periodically oscillating shift in the fringes will be the signal that a gravitational

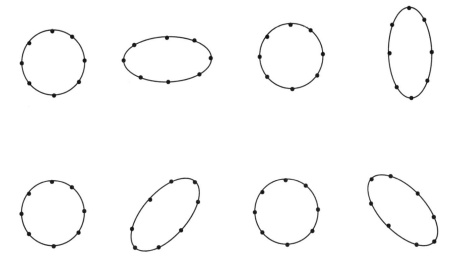

Fig. 13.7 A sketch indicating how a ring of particles will oscillate with time when gravitational waves of two kinds of polarization fall perpendicularly on the ring. The two rows correspond to the two polarizations.

wave is falling on the Michelson interferometer. It follows from (13.91) that the displacements of the arms will have the amplitude $\delta x_{\alpha,0} a$. Apart from being proportional to a, which is a measure of the strength of the wave, the amplitude is also proportional to the length of the arm. Other things being equal, a larger Michelson interferometer should have a larger displacement, causing a bigger fringe shift. If we want to detect a very faint signal, the Michelson interferometer has to be of a gigantic size.

There are several gravitational wave detection experiments under way. All of them essentially are huge Michelson interferometers with arm lengths of the order of kilometres. The most ambitious of these experiments is the Laser Interferometer Gravity-wave Observatory (LIGO) in the USA. At the time of writing this book, there is not yet any report of a positive detection. Only very violent motions involving large masses are expected to produce gravitational waves of sufficient intensity in faraway systems that would have a chance of being detected on the Earth by our present-day technologies. For example, a supernova in our Galaxy is likely to produce gravitational waves which should be detectable by the present generation of gravitational wave experiments.

Although we do not yet have an indisputable direct detection of gravitational waves, we pointed out in §5.5.1 that the binary pulsar provides indirect proof of the existence of gravitational waves. The continuous decrease in the orbital period implies a steady rate of energy loss, which agrees very well with the theoretical estimate of energy loss due to gravitational waves and provides another strong confirmation of general relativity. Anybody familiar with electromagnetic theory would know that an analysis of electromagnetic wave emission is much more complicated than an analysis of its properties. The same holds for the gravitational wave in general relativity. We have discussed the polarization characteristics and other properties of gravitational waves. But a calculation of gravitational wave emission by a system like the binary pulsar is a much more complicated problem and is beyond the scope of this elementary book.

Exercises

13.1 In a certain spacetime geometry, the metric is

$$ds^2 = -(1 - Ar^2)\, dt^2 + (1 - Ar^2)\, dr^2 + r^2(d\theta^2 + \sin^2\theta\, d\phi^2).$$

(a) Calculate the proper distance along a radial line from the centre $r = 0$ to a coordinate radius $r = R$. (b) Calculate the area of the sphere of coordinate radius $r = R$. (c) Calculate the 3-volume bounded inside the sphere of coordinate radius $r = R$. (d) Calculate the 4-volume of the four-dimensional tube bounded by a sphere of coordinate radius R and two $t = $ constant planes separated by T.

13.2 Show that the Schwarzschild metric (13.13) satisfies all the components of Einstein's equation in vacuum at all points other than $r = 0$.

13.3 If the general relativistic correction term in the analysis of particle motion in Schwarzschild geometry is neglected, show by substituting (13.32) in (13.28) that the eccentricity of the orbit is given by

$$\epsilon = \sqrt{1 + \frac{(e^2 - 1)l^2}{M^2}}$$

and figure out the condition for the orbit to be elliptical. If the orbit is elliptical, show that the semimajor axis is given by

$$a = \frac{l^2}{M(1 - \epsilon^2)}$$

and we have

$$\frac{e^2 - 1}{2} = -\frac{M}{2a}.$$

What is the significance of this last result?

13.4 (a) Calculate the perihelion precession in arc seconds per century for Mercury (semimajor axis $= 5.79 \times 10^7$ km, eccentricity $= 0.206$, period of revolution $= 88.0$ days). (b) Evaluate the maximum deflection of light by the Sun in seconds of arc.

13.5 Consider a light signal being emitted at a position $r = r_i$ satisfying $2M < r_i < 3M$ in Schwarzschild geometry, making an angle α with respect to the radial direction. Show that the light signal will be able to escape to infinity only if the following condition (in units $G = 1, c = 1$) is satisfied:

$$\sin \alpha < \frac{3\sqrt{3}M}{r_i} \left(1 - \frac{2M}{r_i}\right)^{1/2}.$$

Note that light can escape from the Schwarzschild radius $r = 2M$ only if it is emitted in the radially outward direction. [Hint: First argue that

$$\tan \alpha = \left(1 - \frac{2M}{r_i}\right)^{1/2} r_i \left(\frac{d\phi}{dr}\right)_i,$$

where $(d\phi/dr)_i$ is the initial value of $d\phi/dr$ along the light path when the light signal starts at r_i. Then you have to relate α to $b = l/e$ by making use of (13.38). Finally use the idea that a signal starting from the left side of the $Q_{\text{eff}}(r)$ curve in Figure 13.2 will be able to escape only if $1/b^2$ has a value higher than the maximum of the curve.]

13.6 Far away from a spherical star the gravitational field is weak and the linear theory should hold. (a) Find h_{ik} (i.e. the difference of the metric from a flat metric) at a far point in a suitable gauge. (b) Show that there is no gauge transformation which will cast this h_{ik} in a transverse traceless form (like the form in (13.84)).

13.7 A source of gravitational radiation is turned on for a finite time after which it no longer emits. A distant observer detects the radiation by watching the motion of two free particles initially at rest. Show that after the passage of the wave the observer finds the particles back in their original positions and at rest with respect to each other (to linear order in amplitude).

14

Relativistic cosmology

14.1 The basic equations

We have seen in Chapter 10 that certain aspects of spacetime dynamics of the Universe can be studied without a detailed technical knowledge of general relativity. However, some important topics in cosmology – especially those dealing with the analysis of high redshift observations – require general relativity for a proper understanding. After giving an introduction to general relativity in the previous two chapters, we are now ready to apply it to cosmology.

As pointed out in §10.3, it is convenient to use the co-moving coordinates in cosmology. If we neglect the motions which galaxies may have with respect to the expanding space, then a galaxy is at rest in this coordinate system and space is supposed to expand uniformly, carrying the galaxies with it. If matter is at rest in a coordinate system, then the energy-momentum tensor T_k^i in that system is given by (12.88). We clearly expect this to be the energy-momentum tensor of the Universe in the co-moving coordinates. Since we had tacitly assumed the coordinate system to be Cartesian when putting the classical hydrodynamic equations in the form (12.78), one may wonder if (12.88) is the expression of the energy-momentum tensor only in Cartesian coordinates. A little reflection should convince the reader that we can use (12.81) to introduce the generalized velocity in any coordinate system and (12.84) should give the general expression of the energy-momentum tensor in any coordinate system, leading to (12.88) in the special case when matter is at rest. It is easy to see that the energy-momentum tensor defined by (12.84) in different coordinate systems should transform according to the tensor transformation law (12.5).

We now want to apply Einstein's equation (12.96) to the Universe, which we write in the form

$$G_k^i = \frac{8\pi G}{c^4} T_k^i. \tag{14.1}$$

Since we already know that T_k^i is given by (12.88), we only need to obtain the Einstein tensor G_k^i of the Universe. For this we have to start from the expression of the metric. The metric of the Universe in the co-moving coordinates is the Robertson–Walker metric given by (10.19) or (10.20). Our job now is to calculate the Einstein tensor for this metric. This involves a large amount of straightforward algebra, since we first have to calculate the various Christoffel symbols by using (12.31) and then we have to calculate the Ricci tensor R_{ik} by using (12.41). It will be instructive for the reader to go through this algebra. We give the final result:

$$R_{tt} = -3\frac{\ddot{a}}{a}, \quad R_{\alpha\beta} = \left(\frac{\ddot{a}}{a} + 2\frac{\dot{a}^2}{a^2} + 2\frac{kc^2}{a^2}\right)\frac{g_{\alpha\beta}}{c^2}, \tag{14.2}$$

where the Greek indices α or β can have values 1, 2, 3, which we take to be r, θ, ϕ in the present case. It follows from (14.2) that the scalar curvature is given by

$$R = R_i^i = \frac{6}{c^2}\left(\frac{\ddot{a}}{a} + \frac{\dot{a}^2}{a^2} + \frac{kc^2}{a^2}\right). \tag{14.3}$$

By substituting (14.2) and (14.3) in (12.44), we finally get

$$G_t^t = -\frac{3}{c^2}\left(\frac{\dot{a}^2}{a^2} + \frac{kc^2}{a^2}\right), \quad G_r^r = G_\theta^\theta = G_\phi^\phi = -\frac{1}{c^2}\left(2\frac{\ddot{a}}{a} + \frac{\dot{a}^2}{a^2} + \frac{kc^2}{a^2}\right), \tag{14.4}$$

whereas all the off-diagonal elements of the Einstein tensor are zero. Note that we have to raise an index in accordance with (12.18) and use (12.16) in these calculations. Finally we have to substitute (12.88) and (14.4) in (14.1). The tt component gives

$$\frac{\dot{a}^2}{a^2} + \frac{kc^2}{a^2} = \frac{8\pi G}{3}\rho, \tag{14.5}$$

whereas the other three diagonal components give the identical equation

$$2\frac{\ddot{a}}{a} + \frac{\dot{a}^2}{a^2} + \frac{kc^2}{a^2} = -\frac{8\pi G}{c^2}P. \tag{14.6}$$

The above two equations are the basic equations giving the dynamics of spacetime.

From considerations of Newtonian cosmology in §10.4, we could write down the Friedmann equation (10.27), which is identical with (14.5). As we pointed out earlier, it is an astounding coincidence that we get exactly the same equation from a full general relativistic analysis and from very simple considerations of Newtonian cosmology (with some ad hoc assumptions). We now need to figure out the significance of the other equation (14.6). On differentiating (14.5) with respect to t, we get

$$\frac{2\dot{a}\ddot{a}}{a} - \frac{2\dot{a}^3}{a^2} - \frac{2kc^2\dot{a}}{a^2} = \frac{8\pi G}{3}\dot{\rho}a. \tag{14.7}$$

Assuming the expansion of the Universe to be adiabatic, the first law of thermodynamics $dQ = dU + P\,dV$ suggests

$$\frac{d}{dt}(\rho c^2 a^3) + P\frac{d}{dt}(a^3) = 0, \tag{14.8}$$

from which

$$c^2(\dot{\rho}a + 3\rho\dot{a}) = -3P\dot{a}.$$

Multiplying this by $8\pi G/3c^2$, we get

$$\frac{8\pi G}{3}\dot{\rho}a + 3\dot{a}\frac{8\pi G}{3}\rho = -\frac{8\pi G}{c^2}P\dot{a}.$$

We now substitute from (14.7) in the first term of this equation and substitute from (14.5) for $(8\pi G/3)\rho$ in the second term. One or two steps of algebra then lead to (14.6). This means that (14.6) can be obtained from (14.5) and (14.8). We can, therefore, regard (14.5) and (14.8) as our basic equations rather than regarding (14.5) and (14.6) as the basic equations.

We have already pointed out in §10.5 that (14.8) would lead to (10.37) if P is given by (10.36). For a Universe filled with matter and radiation, (10.37) becomes (10.50). In other words, the expression (10.50) for the density is equivalent to (14.8) for a Universe filled with matter and radiation. We are thus finally led to the conclusion that (14.5) and (10.50) constitute our basic equations. We have already discussed the solutions of these equations in §10.6 and §10.7 for the cases of the matter-dominated and the radiation-dominated Universe, corresponding respectively to the later and earlier epochs in the thermal history of the Universe. So we need not discuss again how the Universe expands with time. There was one important topic which could not be discussed satisfactorily within the framework of Newtonian cosmology. It is the propagation of light. We merely quoted (10.24) as an explanation of the redshift without proving it. We shall now study the propagation of light systematically and prove (10.24). Before taking up this subject in §14.3, we make a digression in §14.2 to a topic which suddenly seems to be taking centre stage in cosmology research in the last few years.

14.2   The cosmological constant and its significance

It is possible to put an extra term in Einstein's equation, which would make it

$$G_{ik} = \frac{8\pi G}{c^4}T_{ik} - \frac{\Lambda}{c^2}g_{ik}. \tag{14.9}$$

It follows from (12.27) that g_{ik} is also a divergenceless tensor like G_{ik} or T_{ik}. So, on taking the divergence of (14.9), each term will give zero if Λ is a constant. It is thus mathematically consistent to add the last term in (14.9). The constant Λ is called the *cosmological constant* because of the role it plays in cosmology as we shall see below.

If Einstein's equation is extended to (14.9) by including the cosmological constant term, then (14.5) and (14.6) also get modified to

$$\frac{\dot{a}^2}{a^2} + \frac{kc^2}{a^2} = \frac{8\pi G}{3}\rho + \frac{\Lambda}{3}, \tag{14.10}$$

$$2\frac{\ddot{a}}{a} + \frac{\dot{a}^2}{a^2} + \frac{kc^2}{a^2} = -\frac{8\pi G}{c^2}P + \Lambda. \tag{14.11}$$

On subtracting (14.10) from (14.11), we get

$$\frac{\ddot{a}}{a} = -\frac{4\pi G}{3}\left(\rho + \frac{3P}{c^2}\right) + \frac{\Lambda}{3}. \tag{14.12}$$

If $\Lambda = 0$, it is clear from (14.12) that a static solution in which a does not change with time is not possible.

Einstein (1917) first applied general relativity to cosmology at a time when the expansion of the Universe had not yet been discovered by Hubble (1929). Einstein (1917) wanted to construct a static solution of the Universe which is possible only with a non-zero Λ. Assuming $P \ll \rho c^2$, if the density has the value

$$\rho_0 = \frac{\Lambda}{4\pi G}, \tag{14.13}$$

it is easily seen from (14.12) that we get a static solution. One, however, finds that this static solution is unstable (Exercise 14.3). In other words, if this static Universe is disturbed from its initial static state, it will run away from this static state. After the expansion of the Universe was reported (Hubble, 1929), Einstein is said to have remarked that introducing the cosmological constant was the 'biggest blunder' of his life. Thereafter, the cosmological constant was almost banished from the literature of astrophysically motivated cosmology for several decades. Standard textbooks of cosmology would either ignore it or at most devote a small section to the cosmological constant as an odd curiosity. Some intriguing recent observations which will be discussed in §14.5, however, suggest that the cosmological constant may after all be non-zero. This is one of the most dramatic new developments in cosmology, leading to renewed interest in the cosmological constant. Here we present a brief discussion of how cosmological solutions get modified on including the cosmological constant.

Substituting for ρ from (10.50) in (14.10), we get

$$\frac{\dot{a}^2}{a^2} + \frac{kc^2}{a^2} = \frac{8\pi G}{3}\left[\rho_{M,0}\left(\frac{a_0}{a}\right)^3 + \rho_{R,0}\left(\frac{a_0}{a}\right)^4 + \rho_\Lambda\right], \tag{14.14}$$

where we have written

$$\rho_\Lambda = \frac{\Lambda}{8\pi G}. \tag{14.15}$$

It follows from (10.28) that we can write

$$\frac{8\pi G}{3} = \frac{H_0^2}{\rho_{c,0}}. \tag{14.16}$$

Then (14.14) can be put in the form

$$\frac{\dot{a}^2}{a^2} + \frac{kc^2}{a^2} = H_0^2 \left[\Omega_{M,0} \left(\frac{a_0}{a}\right)^3 + \Omega_{R,0} \left(\frac{a_0}{a}\right)^4 + \Omega_{\Lambda,0} \right], \tag{14.17}$$

where

$$\Omega_{M,0} = \frac{\rho_{M,0}}{\rho_{c,0}}, \quad \Omega_{R,0} = \frac{\rho_{R,0}}{\rho_{c,0}}, \quad \Omega_{\Lambda,0} = \frac{\rho_\Lambda}{\rho_{c,0}} \tag{14.18}$$

give the fractional contributions at the present time of matter, radiation and the cosmological constant to the critical density. It should be clear from (14.14) that the effect of the cosmological constant is like a fluid whose density ρ_Λ does not change with the expansion of the Universe. It follows from (10.37) that we should have $w = -1$ for such a fluid, suggesting a negative pressure $P = -\rho c^2$ on the basis of (10.36). This strange fluid-like entity with a negative pressure is often referred to as the *dark energy*.

We know that the contribution of radiation density has been negligible ever since the Universe became matter-dominated. We shall discuss some observations in §14.5 which additionally suggest that the Universe is nearly flat. If we neglect the radiation density and the curvature terms, then (14.17) can be written as

$$\frac{\dot{a}^2}{a^2} = H_0^2 \left[\Omega_{M,0} \left(\frac{a_0}{a}\right)^3 + \Omega_{\Lambda,0} \right]. \tag{14.19}$$

It is possible to write down an analytical solution of this equation, which is

$$\frac{a}{a_0} = \left(\frac{\Omega_{M,0}}{\Omega_{\Lambda,0}}\right)^{1/3} \sinh^{2/3} \left(\frac{3}{2}\sqrt{\Omega_{\Lambda,0}}H_0 t\right). \tag{14.20}$$

This can be verified by substituting this solution (14.20) into (14.19). It is instructive to consider the early time and the late time limits of the solution (14.20), which are

$$\frac{a}{a_0} \approx \left(\frac{3}{2}\Omega_{M,0}^{1/2} H_0 t\right)^{2/3} \quad \text{if} \quad \sqrt{\Omega_{\Lambda,0}}H_0 t \ll 1, \tag{14.21}$$

$$\frac{a}{a_0} \approx \left(\frac{\Omega_{M,0}}{4\Omega_{\Lambda,0}}\right)^{1/3} e^{\sqrt{\Omega_{\Lambda,0}}H_0 t} \quad \text{if} \quad \sqrt{\Omega_{\Lambda,0}}H_0 t \gg 1. \tag{14.22}$$

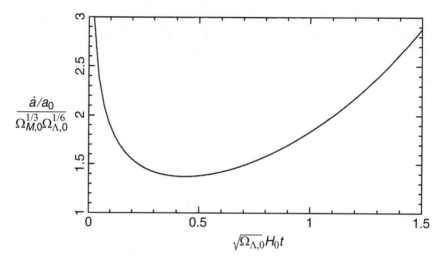

Fig. 14.1 A plot showing how the expansion rate $\dot{a}$ of a flat Universe ($k = 0$) with matter and a non-zero cosmological constant Λ changes with time.

To understand the significance of these limiting solutions, we look at (14.19). Since the matter density falls off as a^{-3}, it becomes less important with time as the Universe expands, whereas the term involving the cosmological constant Λ grows in relative importance. The solution (14.21) at early times is the matter-dominated Universe solution in the limit of early times, which was obtained earlier in (10.60). The fact that $\Omega_{\Lambda,0}$ cancels out of the equation justifies our assertion at the end of §10.6.2 that, even if Λ is non-zero, we do not make too much error in many calculations involving earlier times if we use the cosmological solution with $\Lambda = 0$. The values of $\Omega_{M,0}$ and $\Omega_{\Lambda,0}$ to be presented in §14.5 suggest that the present epoch of the Universe may actually be the epoch when the cosmological solution is making the transition from (14.21) to (14.22). The solution (14.22) for late times is essentially what we would get on neglecting the matter density term in (14.19) and keeping only the cosmological constant term. This exponential part of (14.22) follows directly from (14.10) if we neglect the curvature and the density terms, noting that

$$\sqrt{\Omega_{\Lambda,0}}\,H_0 = \sqrt{\frac{\Lambda}{3}} \tag{14.23}$$

by virtue of (14.15), (14.16) and (14.18). It is clear from (14.22) that the cosmological constant is of the nature of a cosmic repulsion which makes the Universe expand exponentially when it is the dominant term over density and curvature. The different behaviours at early and late times can be understood by considering how $\dot{a}$ changes with time. Figure 14.1 shows a plot of $\dot{a}$ obtained from the solution (14.20) plotted against the time t. At early times, the matter density is dominant and pulls back on the expanding Universe, making the expansion rate $\dot{a}$ decrease with time. On the other hand, when the Λ term dominates at late times, the Universe accelerates, making $\dot{a}$ increase with time.

The observations to be discussed in §14.5 suggest that at the present time the Universe may be making a transition from the matter-dominated era to the Λ-dominated era.

14.3   Propagation of light in the expanding Universe

We have already pointed out in §13.3.2 that a light signal travels along a null geodesic, i.e. a special geodesic along which $ds^2 = 0$. We now want to consider the propagation of a light signal from a distant galaxy to us. Let us take our position to be the origin of our coordinate system and let the position of the distant galaxy be at the radial co-moving coordinate $r = S(\chi)$, the metric of the Universe being the Robertson–Walker metric given by (10.19) or (10.20). Remember that $S(\chi)$ has to be χ, $\sin \chi$ or $\sinh \chi$ corresponding to the values $0, +1$ or -1 of k. From considerations of symmetry, we expect the light signal to propagate in the negative radial direction to reach us from the distant galaxy so that $d\theta = d\phi = 0$ along the path of light propagation. Putting further $ds^2 = 0$, we conclude from (10.20) that the light signal propagation is given by

$$-c^2 dt^2 + a(t)^2 d\chi^2 = 0.$$

If we now replace t by the coordinate η defined through (10.33), then we have

$$a(t)^2[-d\eta^2 + d\chi^2] = 0, \tag{14.24}$$

from which

$$d\chi = \pm d\eta.$$

As the light signal propagates towards us, the radial coordinate χ decreases with the increasing time. So we choose the minus sign and write the solution

$$\chi = \eta_0 - \eta, \tag{14.25}$$

where η_0 is the constant of integration. To understand the significance of this constant of integration, note that (14.25) gives the position χ of the light signal at time η. Since $\chi = 0$ at the time $\eta = \eta_0$, we easily see that η_0 is the time when the light signal reaches us.

We now consider a monochromatic light wave starting from the galaxy at χ. Let the two successive crests of the sinusoidal light wave leave the galaxy at times η and $\eta + \Delta\eta$. Suppose these successive crests reach us at η_0 and $\eta_0 + \Delta\eta_0$. For the second crest, it follows from (14.25) that

$$\chi = (\eta_0 + \Delta\eta_0) - (\eta - \Delta\eta).$$

Let us subtract from this (14.25) which is satisfied by the first crest. Then we get

$$\Delta\eta_0 = \Delta\eta. \tag{14.26}$$

In other words, if we were to use the time-like coordinate η to measure time, then the period of the wave would be the same when it was emitted and when it was received. But η does not give the physical time. For a stationary observer somewhere in the Universe, we have

$$ds^2 = -c^2 d\tau^2 = -c^2 dt^2$$

from (10.18). Thus the proper time τ at which the observer's clock runs coincides with t. Suppose Δt and Δt_0 are the periods measured by physical clocks when the light was emitted and when the light was received, the scale factors of the Universe being a and a_0 respectively at those times. It then follows from (10.33) that

$$c\Delta t = a\Delta\eta, \quad c\Delta t_0 = a_0 \Delta\eta_0,$$

from which

$$\frac{\Delta t_0}{\Delta t} = \frac{a_0}{a} \tag{14.27}$$

on making use of (14.26). Since an observer in any location in the Universe would think that light propagates at speed c, (14.27) obviously suggests (10.24). The significance of (10.24) is that light propagating in the expanding Universe gets stretched proportionately so that the wavelength of light expands the same way as the scale factor. The frequency of a photon should fall as a^{-1}.

We can now prove a very important result which we have been using throughout our discussion of cosmology. Blackbody radiation filling the expanding Universe continues to remain blackbody radiation even when there is no interaction with matter. We consider radiation within the frequency range v to $v + dv$ when the scale factor is a. If this radiation is initially blackbody radiation, then the energy $U_v dv$ in unit volume is given by the Planck distribution (2.1). After the Universe has expanded to scale factor a', the theory of light propagation suggests that the frequencies will change to

$$v \to v' = v\frac{a}{a'}, \quad v + dv \to v' + dv' = (v + dv)\frac{a}{a'}.$$

The radiation initially lying between frequencies v, $v + dv$ would now lie between frequencies v', $v' + dv'$ and the radiation initially occupying unit volume would now occupy a volume $(a'/a)^3$, having lost some energy in the work done for expansion. Exactly like what we have done at the beginning of §10.5, we can write down the $dU + P\,dV = 0$ relation for this radiation, which will lead to something like (10.37) with $w = 1/3$. In other words, we must have

$$U'_{\nu}d\nu' = U_{\nu}d\nu \left(\frac{a}{a'}\right)^4,$$

where $U'_{\nu}d\nu'$ is the energy in unit volume lying in the frequency range from ν' to $\nu' + d\nu'$ when the Universe has expanded to scale factor a'. If we substitute $\nu = \nu' (a'/a)$ along with $T = T' (a'/a)$ in the expression of U_{ν} as given by (2.1), it is straightforward to see that $U'_{\nu'}$ will have the same functional dependence on ν' that U_{ν} had on ν. This completes our proof that the radiation continues to remain blackbody radiation in the expanding Universe with the temperature falling as $T \propto a^{-1}$.

It is clear that light arriving from a galaxy at a certain location $r = S(\chi)$ would show a certain redshift z. We now try to find a functional relationship between r and z, which will be very useful in §14.4 where we shall consider various observational tests of cosmology. What we have discussed so far in this section holds whether the cosmological constant Λ is zero or non-zero. Now we shall consider the case of a matter-dominated Universe with $\Lambda = 0$, since it is possible to derive an elegant analytical expression relating r with z only in this case and it is useful to discuss this case before we get into a more general discussion with non-zero Λ.

Let us first consider a matter-dominated $\Lambda = 0$ Universe with positive curvature, i.e. $k = +1$. This case was discussed in §10.6.1 with the solution given by (10.56) which, in conjunction with (10.24), gives

$$\frac{1}{1+z} = \frac{\Omega_{M,0}}{2(\Omega_{M,0} - 1)}(1 - \cos\eta).$$

From this, we get

$$\cos\eta = \frac{\Omega_{M,0}(z - 1) + 2}{\Omega_{M,0}(1 + z)}. \tag{14.28}$$

Applying the rule $\sin^2\eta + \cos^2\eta = 1$, we then find

$$\sin\eta = \frac{2\sqrt{\Omega_{M,0} - 1}\sqrt{\Omega_{M,0}z + 1}}{\Omega_{M,0}(1 + z)}. \tag{14.29}$$

The light signal reaches us at time η_0, which means that $z = 0$ when $\eta = \eta_0$. Then (14.28) and (14.29) imply

$$\cos\eta_0 = \frac{2 - \Omega_{M,0}}{\Omega_{M,0}}, \quad \sin\eta_0 = \frac{2\sqrt{\Omega_{M,0} - 1}}{\Omega_{M,0}}. \tag{14.30}$$

In the $k = +1$ case we are considering, we have $r = \sin\chi$. Then (14.25) implies

$$r = \sin(\eta_0 - \eta) = \sin\eta_0\cos\eta - \sin\eta\cos\eta_0.$$

Substituting from (14.28), (14.29) and (14.30), we get

$$r = \frac{2\sqrt{\Omega_{M,0} - 1}[\Omega_{M,0}(z - 1) + 2 - \sqrt{\Omega_{M,0}z + 1}(2 - \Omega_{M,0})]}{\Omega_{M,0}^2(1 + z)}. \qquad (14.31)$$

From (10.31), we can write

$$\sqrt{\Omega_{M,0} - 1} = \frac{c}{a_0 H_0}.$$

Substituting this in (14.31), we finally get

$$r = \frac{2\Omega_{M,0}z + (2\Omega_{M,0} - 4)(\sqrt{\Omega_{M,0}z + 1} - 1)}{a_0 H_0 \Omega_{M,0}^2(1 + z)/c}. \qquad (14.32)$$

If one carries out a similar calculation for the $k = -1$ case starting from the solution (10.58) and taking $r = \sinh \chi$, then also one ends up with exactly the same relation (14.32) between r and z (Exercise 14.5). Hence we can take (14.32) as the general relation between r and z for a matter-dominated Universe with zero Λ. This relation (14.32) is known as *Mattig's formula* (Mattig, 1958).

When Λ is non-zero, it is not possible to derive such a nice analytical relation between r and z. The relation between r and z has to be expressed in the form of an integral in that case, as we shall see in §14.4.2.

14.4 Important cosmological tests

One of the most important observational laws in cosmology is Hubble's law (9.13). This law was established from the study of galaxies having $z \ll 1$ and we find that the linear relationship between distance and recession velocity holds at redshifts small compared to 1. An important question is whether we would theoretically expect any departures from this linear relationship at redshifts $z \approx 1$. The first hurdle before us is to pose the question properly. As pointed out in §9.3, only in the case of low redshifts can we interpret the redshifts in terms of recession velocities given by (9.12). Even the concept of distance involves complex issues when we consider faraway objects in the Universe. We pointed out in §13.1 that, when the metric tensor is a function of time as in the case of the Robertson–Walker metric, only length elements (given by (13.7)) and not finite lengths have proper meaning. So we first have to restate Hubble's law in terms of quantities which have precise meanings even when $z \approx 1$, before we can talk about departures from linearity. After restating Hubble's law in terms of suitable quantities, we shall see that the departures from linearity can give us clues about the values of important cosmological parameters like $\Omega_{M,0}$ and $\Omega_{\Lambda,0}$.

We shall first present our analysis for the case $\Lambda = 0$ where all the calculations can be done analytically. Then we shall discuss the case $\Lambda \neq 0$ which is more complicated.

14.4.1 Results for the case $\Lambda = 0$

To simplify our life, let us assume that all galaxies have the same intrinsic brightness and the same size. Then a galaxy at a certain radial co-moving coordinate r, which would correspond to a certain redshift z, will appear to have a certain definite apparent luminosity and apparent size. We thus expect the apparent luminosity and the apparent size to be functions of the redshift z. We can compare the theoretically derived functional relationships with observational data to put constraints on the parameters in the theoretical model. Since not all galaxies have the same intrinsic luminosity and intrinsic size, we expect the observational data points to show some scatter around the theoretically calculated functional relationships. However, an analysis involving a large number of galaxies should have a good statistical significance and should allow us to constrain the theoretical model.

The Hubble test

Consider a galaxy at position r. When light from this galaxy reaches us, the light passes through a spherical surface on which we lie and of which the galaxy is the centre. We first have to find the area of this spherical surface. By extending the discussion of length measurement presented in §13.1, we can easily conclude that an element of area on the spherical surface is $a(t)^2 r^2 \sin\theta \, d\theta \, d\phi$ if the metric of the Universe is given by (10.19). An integration of this over the entire spherical surface gives $4\pi a(t)^2 r^2$. Since we are considering light falling on this surface at the present time, we have to take $a(t)$ to be the present value a_0 of the scale factor. If $\mathcal{L}$ is the intrinsic luminosity of the galaxy (i.e. the rate of energy emission per unit time), then we may think that the flux received by us should be given by

$$\mathcal{F} = \frac{\mathcal{L}}{4\pi a_0^2 r^2}. \tag{14.33}$$

But this is not yet the final correct answer. A photon which originally had energy hc/λ at the time of emission undergoes redshift and has energy $hc/\lambda(1+z)$ when it reaches us. So we need to divide (14.33) by the factor $1 + z$ to get the flux corrected for photon redshifts. Another cause for concern is the time dilation given by (14.27). So the energy which was emitted by the galaxy in time Δt reaches us in time $\Delta t_0 = \Delta t(1+z)$. This would further reduce the flux by another factor of $1 + z$. The correct expression of flux is then given by

$$\mathcal{F} = \frac{\mathcal{L}}{4\pi a_0^2 r^2 (1+z)^2} \tag{14.34}$$

rather than (14.33).

We can write (14.34) in the form

$$\mathcal{F} = \frac{\mathcal{L}}{4\pi d_{\mathrm{L}}^2}, \tag{14.35}$$

where

$$d_{\mathrm{L}} = a_0 r (1+z) \tag{14.36}$$

is called the *luminosity distance*. This is an observationally measurable quantity. Once you measure the apparent luminosity, you can get d_{L} from (14.35) by assuming an average intrinsic luminosity $\mathcal{L}$ for all galaxies. When measuring the apparent luminosity, there is one other factor about which one has to be careful. Suppose you are measuring the apparent luminosity with the help of the energy flux reaching you in the optical band of the spectrum. For a galaxy at a significant redshift z, the original photons which were emitted in the optical band may now be redshifted to the infrared, whereas the photons which were originally emitted in the ultraviolet may now appear in the optical band and be detected by you. If the galaxy is intrinsically less luminous in the ultraviolet compared to the optical band, then this shifting of photon wavelengths may make it appear dimmer if you are measuring only the photons in the optical band. Assuming a standard shape for the typical galactic spectrum, one can correct for this. This is called the *K correction*. Whenever we talk about luminosity distance, it should be assumed that we are talking about K corrected luminosity distance.

Substituting for r from (14.32) in (14.36), we get

$$H_0 d_{\mathrm{L}} = \frac{c}{\Omega_{\mathrm{M},0}^2} [2\Omega_{\mathrm{M},0} z + (2\Omega_{\mathrm{M},0} - 4)(\sqrt{\Omega_{\mathrm{M},0} z + 1} - 1)]. \tag{14.37}$$

This is the functional relationship between d_{L} and z, telling us what would be the redshift z of a galaxy which is located at luminosity distance d_{L}. To see that Hubble's law follows from it at low redshifts, we consider $\Omega_{\mathrm{M},0} z < 1$. Then we can expand the square root by applying the binomial theorem. On keeping terms till z^2, it follows from (14.37) after a little algebra that

$$H_0 d_{\mathrm{L}} \approx c \left[z + \frac{1}{4}(2 - \Omega_{\mathrm{M},0}) z^2 \right]. \tag{14.38}$$

At low redshifts where we can neglect the z^2 term, we have a linear relation which is a restatement of Hubble's law in terms of the measurable quantities d_{L} and z. Theoretical considerations tell us that there should be a departure from the linear law at higher redshifts (unless $\Omega_{\mathrm{M},0} = 2$) and the amount of departure should depend on $\Omega_{\mathrm{M},0}$. Figure 14.2 shows theoretical plots showing

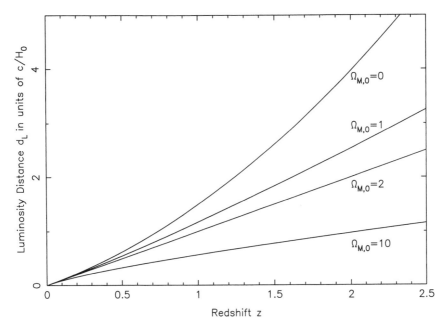

Fig. 14.2 The relation between the luminosity distance d_L and the redshift z of galaxies having the same intrinsic luminosity, for different values of $\Omega_{M,0}$, with $\Lambda = 0$.

the relation between d_L and z for various values of $\Omega_{M,0}$, as obtained from (14.37).

It may now seem that it would be straightforward to estimate $\Omega_{M,0}$. We have to determine the luminosity distances d_L of many galaxies having different redshifts z. Then we have to check if the observational data points lie close to one of the curves in Figure 14.2. We shall present some recent observational data in §14.5. Here let us just mention that the observational data seem not to fit any of the curves in Figure 14.2. In other words, a theoretical model with $\Lambda = 0$ does not provide a good fit to the observational data, urging astronomers to bring back the cosmological constant again to centre stage in cosmology.

The angular size test

Suppose a galaxy at redshift z has linear size $\mathcal{D}$. The galaxy will make up an arc of a circle passing through the galaxy with us at the centre. The angular size $\Delta\theta$ of the galaxy as seen by us can be obtained by equating $\Delta\theta/2\pi$ to the ratio of $\mathcal{D}$ to the circumference of this circle. From the metric (10.19), it is easy to argue that the circumference should be equal to $2\pi a(t)r$. Since we are considering the circle to pass through the galaxy, the appropriate value of $a(t)$ will be the scale factor a when light started from the galaxy. This is equal to $a_0/(1+z)$ so that the circumference is $2\pi a_0 r/(1+z)$. It is now easy to see that the angular size is given by

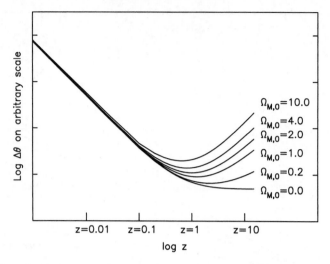

Fig. 14.3 The relation between the angular size $\Delta\theta$ and the redshift z of galaxies having the same intrinsic size, for different values of $\Omega_{M,0}$, with $\Lambda = 0$.

$$\Delta\theta = \frac{\mathcal{D}(1+z)}{a_0 r}. \tag{14.39}$$

We can write this as

$$\Delta\theta = \frac{\mathcal{D}}{d_A}, \tag{14.40}$$

where

$$d_A = \frac{a_0 r}{1+z} \tag{14.41}$$

is called the *angular size distance*.

    If we substitute for r from (14.32) in (14.39), it is easy to see that a_0 cancels out and we get an expression of $\Delta\theta$ as a function of z. Figure 14.3 shows $\Delta\theta$ as a function of z for different $\Omega_{M,0}$. This provides another possible test for the determination of $\Omega_{M,0}$. If we measure the angular sizes $\Delta\theta$ of many galaxies having different z, then we can try to fit the observational data with the theoretical curves in Figure 14.3, thereby allowing us to estimate $\Omega_{M,0}$.

The surface brightness test

Assuming that all galaxies have the same intrinsic surface brightness, we now ask the question as to what would be the apparent surface brightness of a galaxy at redshift z. The apparent surface brightness as seen by us is given by the quotient of the total flux received by us from the galaxy and the angular area of the galaxy as seen by us. Since the total flux received by us goes as $\propto d_L^{-2}$ and the angular area goes as $\propto d_A^{-2}$, we expect that the surface brightness should go as

$$S \propto \frac{d_A^2}{d_L^2}.$$

From (14.36) and (14.41), we then conclude

$$S \propto \frac{1}{(1+z)^4}. \tag{14.42}$$

This is a model-independent relation, which should hold if our basic ideas of the expanding Universe are correct and if galaxies are standard candles (i.e. if intrinsic brightnesses of galaxies systematically do not change with time and hence with redshift). In §2.2.2 we derived the constancy of specific intensity in a region free of matter, which implies that the surface brightness of an object should be independent of distance. This result certainly has to be modified when we consider general relativistic effects in an expanding Universe. In different portions of the book, we have mainly discussed the application of the radiative transfer equation (2.12) to interiors of stars or to interstellar medium within a galaxy – situations where the general relativistic effect given by (14.42) is utterly negligible. However, when we venture into the extragalactic world, (14.42) indicates that more distant galaxies should be dimmer. This also allows us to get around the Olbers paradox discussed in §6.1.1. Because of the inverse fourth law dependence seen in (14.42), the dimming of faraway galaxies is a rather drastic effect at high redshifts. Even a galaxy at a redshift of $z = 1$ would appear 16 times dimmer. If we want to study galaxies at redshift $z \approx 6$ as discussed in §11.8.1, we need very long exposure times due to the extremely low value of the apparent surface brightness.

14.4.2 Results for the case $\Lambda \neq 0$

Some of the results for the $\Lambda = 0$ case get carried over to the $\Lambda \neq 0$ case. For example, we define the luminosity distance d_L and the angular size distance d_A through (14.35) and (14.40) respectively even when the cosmological constant is non-zero (taking $\mathcal{F}$ as the observed flux and $\Delta\theta$ as the observed angular size). The expressions for d_L and d_A will also still be given by (14.36) and (14.41) respectively. However, r will no longer be related to the redshift z by (14.32) and hence a relation like (14.37), based on (14.32), will no longer hold. Before discussing how we relate r to z when $\Lambda \neq 0$, we point out that the surface brightness would still fall as $(1+z)^{-4}$ in accordance with (14.42).

When the cosmological constant Λ is non-zero, it is not possible to write down an analytical expression relating r with z. The relation between them has to be expressed in the form of an integral, which we now derive. Remember that

$$r = S(\chi), \tag{14.43}$$

where $S(\chi)$ has to be equal to $\sin \chi$, $\sinh \chi$ or χ, depending on whether k appearing in the Friedmann equation (10.27) is $+1$, -1 or 0. Since (14.24) and (14.25) describing the propagation of light are valid even when $\Lambda \neq 0$, we note from (14.25) that the position χ of a distant source of light is equal to the lapse in the time-like variable η between the emission of light by this source and its reception by us. From (10.33), it follows that

$$\eta_0 - \eta = c \int_{t_e}^{t_r} \frac{dt}{a(t)}, \tag{14.44}$$

where t_e and t_r are the values of time t when the light was emitted and when it was received. Using (14.25), we can put (14.44) in the form

$$\frac{\chi}{c} = \int_z^0 \frac{dz'}{a} \frac{dt}{da} \frac{da}{dz'}, \tag{14.45}$$

where the limits of the integration over the redshift denoted by z' are z and 0 corresponding to the emission and the reception of the light signal. From the relation (10.24) between the redshift and the scale factor a, it follows that

$$\frac{da}{dz'} = -\frac{a^2}{a_0}.$$

Substituting this in (14.45), we get

$$\frac{\chi}{c} = \frac{1}{a_0} \int_0^z \frac{dz'}{(\dot{a}/a)}. \tag{14.46}$$

For $(\dot{a}/a)$ in (14.46) we now have to substitute a general expression with non-zero Λ. We use (14.17), in which we neglect the term involving $\Omega_{R,0}$ which is very small compared to the other terms when the Universe is matter-dominated. By making use of (10.24), we write (14.17) in the form

$$\frac{\dot{a}^2}{a^2} = H_0^2 \left[\Omega_{M,0}(1+z)^3 + \Omega_{\Lambda,0} \right] + \kappa H_0^2 (1+z)^2, \tag{14.47}$$

where we have written κH_0^2 for $-kc^2/a_0^2$ so that

$$|\kappa| = \frac{c^2}{a_0^2 H_0^2}. \tag{14.48}$$

Since (14.47) is valid in our present epoch when $z = 0$ and $\dot{a}/a = H_0$, it easily follows from (14.47) that

$$\kappa = 1 - \Omega_{M,0} - \Omega_{\Lambda,0}. \tag{14.49}$$

We can use (14.49) to determine κ when $\Omega_{M,0}$ and $\Omega_{\Lambda,0}$ are given. From (14.46) and (14.47), we have

$$\chi = \frac{c}{a_0 H_0} \int_0^z [\Omega_{M,0}(1+z')^3 + \Omega_{\Lambda,0} + \kappa(1+z')^2]^{-1/2} dz'. \tag{14.50}$$

From (14.36), (14.43) and (14.50), we get

$$d_{\mathrm{L}} = a_0(1+z)S\left(\frac{c}{a_0 H_0}\int_0^z [\Omega_{\mathrm{M},0}(1+z')^3 + \Omega_{\Lambda,0} + \kappa(1+z')^2]^{-1/2}dz'\right).$$

If we use (14.48) to eliminate a_0 which is not directly observable, then we finally get

$$d_{\mathrm{L}} = \frac{(1+z)c}{H_0\sqrt{|\kappa|}}S\left(\sqrt{|\kappa|}\int_0^z [\Omega_{\mathrm{M},0}(1+z')^3 + \Omega_{\Lambda,0} + \kappa(1+z')^2]^{-1/2}dz'\right). \tag{14.51}$$

For given values of $\Omega_{\mathrm{M},0}$ and $\Omega_{\Lambda,0}$, we can evaluate (14.51) numerically to determine $H_0 d_{\mathrm{L}}$ as a function of redshift z. Observationally we can measure the redshifts z of a large number of galaxies and then determine their d_{L} from their observed apparent brightnesses by using (14.35). By comparing the observational data with the theoretical results, we can hope to determine the values of $\Omega_{\mathrm{M},0}$ and $\Omega_{\Lambda,0}$. We shall discuss the outcome of this exercise in the next section.

14.5 Cosmological parameters from observational data

Data of distant supernovae

As we pointed out in §4.7, Type Ia supernovae are believed to be caused by matter accreting onto a white dwarf having mass close to the Chandrasekhar mass. So we expect the maximum luminosity of a Type Ia supernova to have the same value everywhere and at all times. Hence such a supernova can be used as a standard candle. With the Hubble Space Telescope (HST), it has been possible to resolve and study Type Ia supernovae in distant galaxies. Once we measure the maximum apparent luminosity of the supernova when it is brightest, we can use (14.35) to calculate the luminosity distance d_{L} if we know the maximum absolute luminosity. From a knowledge of the redshift z of the galaxy in which the supernova took place, we get the z corresponding to this luminosity distance d_{L}. Hence, in a plot of d_{L} against z, each supernova will contribute one data point.

Sometimes, instead of d_{L}, one plots the equivalent quantity $m - M$ which is the difference between the apparent and absolute magnitudes of the Type Ia supernova (when it was brightest). By substituting d_{L} for d in (1.8), we can easily find how d_{L} is related to $m - M$. In a plot of $m - M$ against z in which the supernova data are represented by points, we can also put theoretical curves calculated from (14.51) for different combinations of $\Omega_{\mathrm{M},0}$ and $\Omega_{\Lambda,0}$, to find out which curve fits the observational data best. High-z supernova data from HST were analysed by two independent groups who carried out this exercise (Riess *et al.*, 1998; Perlmutter *et al.*, 1999). Figure 14.4 shows the result. The top panel

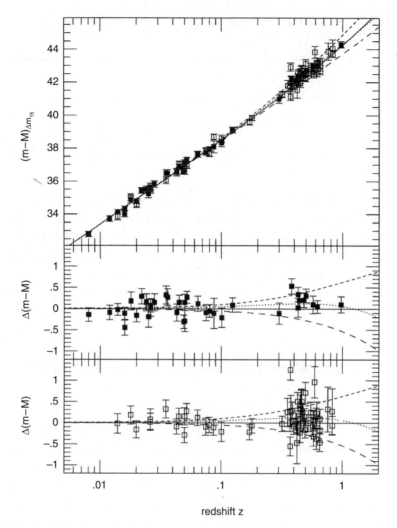

Fig. 14.4 The apparent luminosities of distant supernovae against redshifts z, along with theoretical curves for different combinations of the cosmological parameters. The cosmological parameters used for the different curves are: (i) solid line for $\Omega_{M,0} = 0$, $\Omega_{\Lambda,0} = 0$; (ii) long dashes for $\Omega_{M,0} = 1$, $\Omega_{\Lambda,0} = 0$; (iii) short dashes for $\Omega_{M,0} = 0$, $\Omega_{\Lambda,0} = 1$; (iv) dotted line for $\Omega_{M,0} = 0.3$, $\Omega_{\Lambda,0} = 0.7$. The lower two panels plot the deviation of $m - M$ from the solid line for $\Omega_{M,0} = 0$, $\Omega_{\Lambda,0} = 0$, showing the data of two groups separately: filled squares for the data of Riess *et al.* (1998) and open squares for the data of Perlmutter *et al.* (1999). From Leibundgut (2001). (©Annual Reviews Inc. Reproduced with permission from *Annual Reviews of Astronomy and Astrophysics*.)

shows data of both the groups along with some theoretical curves, the curve corresponding to the empty Universe ($\Omega_{M,0} = 0$, $\Omega_{\Lambda,0} = 0$) being indicated by the solid line. To make things clearer, the two bottom panels show the data of the two groups separately, with the vertical axis giving the differential $\Delta(m - M)$

with respect to the solid line for the empty Universe. It appears that the dotted line corresponding to $\Omega_{M,0} = 0.3$ and $\Omega_{\Lambda,0} = 0.7$ is the best theoretical fit to the observational data. This value of $\Omega_{M,0}$ agrees with what observers estimated from dynamical mass determinations of clusters of galaxies, as given by (10.42). However, the possibility that $\Omega_{\Lambda,0}$ may be non-zero sent a shock wave through the entire astrophysics community of the world, since it was generally believed for several decades that the cosmological constant Λ is zero.

Data of temperature anisotropies in CMBR

We now turn to a different kind of data. In §11.7.1 we discussed primary anisotropies in CMBR, as measured by the mission WMAP. The measured temperature variation as a function of the galactic coordinates is shown in Figure 11.3. Let us consider the temperature variation $\Delta T/T$ in a direction ψ. The temperature variation $\Delta T/T$ in a nearby direction $\psi + \theta$ is expected to be very similar if θ is sufficiently small, but will not be correlated with the temperature variation in direction ψ if θ is large. The angular correlation of the CMBR anisotropy is clearly given by

$$C(\theta) = \left\langle \frac{\Delta T}{T}(\psi) \frac{\Delta T}{T}(\psi + \theta) \right\rangle, \tag{14.52}$$

where the averaging is supposed to have been done for all possible values of ψ and possible values of θ around them. Since this correlation function $C(\theta)$ is a function of θ, we can expand it in Legendre polynomials (see, for example, Mathews and Walker, 1979, §7–1; Jackson, 1999, §3.2), i.e.

$$C(\theta) = \sum_l \frac{(2l + 1)}{4\pi} C_l P_l(\cos\theta), \tag{14.53}$$

where C_l is the coefficient of the l-th Legendre polynomial. Figure 14.5 plots C_l as a function of l. It is clear that there is a maximum around $l \approx 250$. To understand the significance of the maximum, note that $P_l(\cos\theta)$ has l nodes between 0 and π. When l is large, the first node is approximately located at $\Delta\theta \approx \pi/l$. A value of $l \approx 250$ would give a value of $\Delta\theta$ somewhat less than $1°$. This is the typical angular scale of the temperature anisotropies in CMBR. We now come to the question of what determines this angular scale.

Let us consider an object of linear size $\mathcal{D}$ on the last scattering surface from where the CMBR photons come and which, as we pointed out in §11.7, is at redshift $z_{dec} \approx 1100$. We now figure out the angular size $\Delta\theta$ which this object of size $\mathcal{D}$ will produce in the sky. The relation between $\mathcal{D}$ and $\Delta\theta$ is given by (14.39). Since we are considering the possibility of Λ not being zero, we should substitute for r from (14.43) and (14.50). To have a rough idea of how things go, let us substitute for r from (14.32) appropriate for the $\Lambda = 0$ case, since this will allow us to make some estimates analytically and a non-zero Λ does not

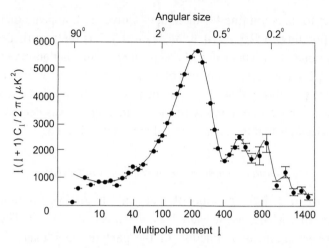

Fig. 14.5 The values of the coefficients C_l in the Legendre polynomial expansion of the angular correlation $C(\theta)$ of temperature anisotropies in the CMBR. Data for $l < 800$ come from WMAP, whereas data for higher l come from other experiments. The top of the figure indicates angular sizes corresponding to different values of l. Adapted from Bennett *et al.* (2003).

introduce too much error when calculating quantities relevant for early epochs. When $z \gg 1$, it follows from (14.32) that

$$r \rightarrow \frac{2c}{a_0 H_0 \Omega_{M,0}}.$$

Substituting this for r, we obtain from (14.39) that

$$\Delta\theta \approx \frac{\Omega_{M,0}}{2} \cdot \frac{\mathcal{D}z}{cH_0^{-1}}.$$

On substituting for H_0 from (9.17), this becomes

$$\Delta\theta \approx 34.4''(\Omega_{M,0}h)\left(\frac{\mathcal{D}z}{1 \text{ Mpc}}\right). \tag{14.54}$$

By putting $z_{\text{dec}} \approx 1100$ in (14.54), we can determine the linear size $\mathcal{D}$ of an object on the last scattering surface which would subtend an angle slightly less than $1°$ in the sky.

As we pointed out in §11.9, only perturbations larger than the Jeans length grow. So we may expect the Jeans length to give sizes of the typical perturbations on the last scattering surface. According to (11.40), the Jeans length was of the order of the horizon size ct till the decoupling of matter and radiation at $z_{\text{dec}} \approx 1100$. We now estimate the angle $\Delta\theta$ which the horizon on the last scattering surface would subtend to us today. We can use (10.60) to get the time when the decoupling took place, which turns out to be of order

$$t_{\text{dec}} \approx H_0^{-1}\Omega_{M,0}^{-1/2}z_{\text{dec}}^{-3/2}. \tag{14.55}$$

We would get the horizon by multiplying this by c. On substituting ct_{dec} for $\mathcal{D}$ in (14.54), we get

$$\Delta\theta \approx 0.87° \; \Omega_{M,0}^{1/2} \left(\frac{z_{\text{dec}}}{1100}\right)^{-1/2}. \qquad (14.56)$$

If we take $\Omega_{M,0}$ of order 1, then (14.56) gives an angular size comparable to the angular scale of anisotropies in the WMAP data. This suggests that the irregularities that we see on the last scattering surface correspond to the Jeans length, which was of the same order as the horizon till that time. It follows from (14.56) that a larger value of $\Omega_{M,0}$ would make $\Delta\theta$ larger, causing the peak in Figure 14.5 to shift leftward. The position of the peak would thus give the value of $\Omega_{M,0}$.

When we assume $\Lambda \neq 0$, the analysis becomes much more complicated and has to be done numerically. We shall not present that analysis here. The more complicated analysis with $\Lambda \neq 0$ suggests that the position of the peak in Figure 14.5 depends on $\Omega_{\Lambda,0} + \Omega_{M,0}$. The observed position of the peak turns out to be consistent with

$$\Omega_{\Lambda,0} + \Omega_{M,0} = 1. \qquad (14.57)$$

If $\Omega_{\Lambda,0} + \Omega_{M,0}$ were larger, then the peak would have shifted more towards the left.

The combined constraints

Figure 14.6 indicates the likely values of $\Omega_{\Lambda,0}$ and $\Omega_{M,0}$. The straight line corresponds to $\Omega_{\Lambda,0} + \Omega_{M,0} = 1$ concluded from the WMAP data of temperature anisotropies of CMBR. On the other hand, the ellipses indicate the best possible combinations of $\Omega_{\Lambda,0}$ and $\Omega_{M,0}$ which fit the data of distant supernovae. Constraints arising out of these two different sets of observational data are simultaneously satisfied if the values of our basic cosmological parameters are around

$$\Omega_{\Lambda,0} \approx 0.7, \;\; \Omega_{M,0} \approx 0.3. \qquad (14.58)$$

At the present time, these seem to be the values of these parameters accepted by most cosmologists. As we already pointed out, another independent confirmation of the above value of $\Omega_{M,0}$ comes from the virial mass estimates of clusters of galaxies. Two important conclusions follow from the values quoted in (14.58). Firstly, it follows from (14.49) that $\kappa \approx 0$, which means that our Universe must be nearly flat with very little curvature. Secondly, since $\Omega_{\Lambda,0}$ and $\Omega_{M,0}$ are of comparable values at the present time, it follows from (14.17) that the matter density was more dominant in the past when a was smaller than a_0, whereas the cosmological constant term will be more dominant in the future. It thus seems that we live in a flat Universe which is at present in the process of making a transition from a matter-dominated epoch to a Λ-dominated epoch.

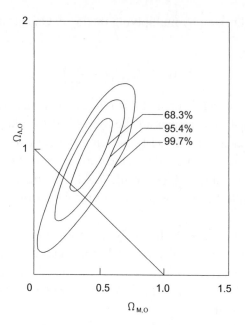

Fig. 14.6 Constraints on $\Omega_{\Lambda,0}$ and $\Omega_{M,0}$ jointly coming from the temperature anisotropies of the CMBR (the straight line corresponding to $\Omega_{\Lambda,0} + \Omega_{M,0} = 1$) and the data of distant Type Ia supernovae (the ellipses within which the values would lie at particular confidence levels). Adapted from Riess *et al.* (2004).

Exercises

14.1 Find the scalar curvature for the following three-dimensional metrics

$$ds^2 = a^2[d\chi^2 + \sin^2 \chi (d\theta^2 + \sin^2 \theta \, d\phi^2)],$$
$$ds^2 = a^2[d\chi^2 + \sinh^2 \chi (d\theta^2 + \sin^2 \theta \, d\phi^2)].$$

14.2 Calculate all the components of the Ricci tensor R_{ik} for the Robertson–Walker metric and verify (14.2).

14.3 Consider the static solution of the Universe when the cosmological constant Λ is assumed to be non-zero and the pressure is considered negligible (i.e. take $P = 0$). Let a_0 be the value of the scale factor a corresponding to this static solution. Assume that the scale factor is suddenly changed to $a_0 + a_1$. Show that a_1 will grow exponentially, implying that the static solution is unstable.

14.4 (a) Consider a free particle moving with respect to the co-moving coordinates in the Robertson–Walker metric without interacting with other particles. Without any loss of generality, we can choose the origin on the particle path such that χ changes along the path, but not θ or ϕ. By considering the extremum of $\int ds$, show that $a^2(d\chi/d\tau)$ will be a constant of motion. Argue from this that the physical velocity of the particle (with respect to the co-moving frame) will decrease with the expansion of the Universe as a^{-1}.

(b) Now consider a type of particles filling the Universe. If they are initially relativistic ($\kappa_B T \gg mc^2$) and are in thermodynamic equilibrium, we expect (11.7) to hold. If the particles then fall out of thermodynamic equilibrium and eventually become non-relativistic with the expansion of the Universe, show that (11.7) continues to hold if we assume T to fall as a^{-1}. This essentially implies that the distribution of the particles continues to look like Bose–Einstein or Fermi–Dirac distributions with energy given by pc (which is not true after the particles become non-relativistic). Discuss whether T appearing in the expressions of these distributions will have the usual significance of temperature. Should (11.8) also continue to hold after the particles become non-relativistic? If not, then how will you calculate the contribution of these particles to the density of the Universe?

14.5 For the case $k = -1$, derive the relation between the coordinate distance r and the redshift z of a galaxy (assuming $\Lambda = 0$). Show that the relation is the same as (14.32) derived for the $k = +1$ case.

14.6 If there are n galaxies per unit co-moving volume of the Robertson–Walker metric, find the number of galaxies which will be found within solid angle $d\Omega$ having redshifts in the range z to $z + dz$.

14.7 Using the integral relation (14.51), numerically evaluate and graphically plot the apparent luminosities of standard candles against their redshifts z for the following models: (i) $\Omega_M = 1.0$, $\Omega_\Lambda = 0.0$; (ii) $\Omega_M = 0.0$, $\Omega_\Lambda = 1.0$; (iii) $\Omega_M = 0.3$, $\Omega_\Lambda = 0.7$.

14.8 Consider an angular function

$$C(\theta) = \Gamma e^{-(\theta/\theta_0)^2}.$$

This function can be expanded in Legendre polynomials $P_l(\cos\theta)$ as in (14.53). Show that the coefficients C_l in the expansion are given by

$$C_l = 2\pi \int_{-1}^{+1} C(\theta) P_l(\cos\theta) \, d(\cos\theta).$$

Develop a numerical code to calculate these coefficients C_l. Run the code for a few selected values of θ_0 and find the corresponding values of l for which C_l is maximum (by plotting the values of C_l against l).

Appendix A

Values of various quantities

A.1 Physical constants

Speed of light	c	$=$	$3.00 \times 10^8 \, \mathrm{m \, s^{-1}}$
Gravitational constant	G	$=$	$6.67 \times 10^{-11} \, \mathrm{m^3 \, kg^{-1} \, s^{-2}}$
Planck constant	h	$=$	$6.63 \times 10^{-34} \, \mathrm{J \, s}$
Boltzmann constant	κ_B	$=$	$1.38 \times 10^{-23} \, \mathrm{J \, K^{-1}}$
Permeability of free space	μ_0	$=$	$1.26 \times 10^{-6} \, \mathrm{H \, m^{-1}}$
Permittivity of free space	ϵ_0	$=$	$8.85 \times 10^{-12} \, \mathrm{F \, m^{-1}}$
Charge of electron	e	$=$	$-1.60 \times 10^{-19} \, \mathrm{C}$
Mass of electron	m_e	$=$	$9.11 \times 10^{-31} \, \mathrm{kg}$
Mass of hydrogen atom	m_H	$=$	$1.67 \times 10^{-27} \, \mathrm{kg}$
Stefan–Boltzmann constant	σ	$=$	$5.67 \times 10^{-8} \, \mathrm{W \, m^{-2} \, K^{-4}}$
Constant in Wien's law	$\lambda_m T$	$=$	$2.90 \times 10^{-3} \, \mathrm{m \, K}$
Standard atmospheric pressure		$=$	$1.01 \times 10^5 \, \mathrm{N \, m^{-2}}$
1 electron volt	eV	$=$	$1.60 \times 10^{-19} \, \mathrm{J}$
1 angstrom	Å	$=$	$10^{-10} \, \mathrm{m}$
1 calorie		$=$	$4.19 \, \mathrm{J}$

A.2 Astronomical constants

1 astronomical unit	AU	$=$	$1.50 \times 10^{11} \, \mathrm{m}$
1 parsec	pc	$=$	$3.09 \times 10^{16} \, \mathrm{m}$
1 year	yr	$=$	$3.16 \times 10^7 \, \mathrm{s}$
Mass of Sun	$M_\odot$	$=$	$1.99 \times 10^{30} \, \mathrm{kg}$
Radius of Sun	$R_\odot$	$=$	$6.96 \times 10^8 \, \mathrm{m}$
Luminosity of Sun	$L_\odot$	$=$	$3.84 \times 10^{26} \, \mathrm{W}$
Mass of Earth	$M_\oplus$	$=$	$5.98 \times 10^{24} \, \mathrm{kg}$
Radius of Earth	$R_\oplus$	$=$	$6.37 \times 10^6 \, \mathrm{m}$

Appendix B

Astrophysics and the Nobel Prize

The Nobel Prize was not awarded to astrophysicists in the early decades of the twentieth century. The most impact-making astrophysicists of that period such as Eddington and Hubble did not win Nobel Prizes. From the middle of the twentieth century, several Nobel Prizes have been given for important discoveries in astrophysics. We list those Nobel Prizes below. It should, however, be emphasized that this list of Nobel laureates should *not* be considered to be some kind of list of greatest astrophysicists of our time. A few of the Nobel Prizes were given for serendipitous discoveries which had tremendous impact. On the other hand, some astrophysicists who are regarded amongst the greatest and most impact-making by common consent have not been honoured with the Nobel Prize. A reader of this book may be interested in making up a list of astrophysicists who, according to his/her opinion, should have been awarded the Nobel Prize.

1967	H. A. Bethe	Contributions to the theory of nuclear reactions, especially discoveries concerning energy production in stars
1970	H. Alfvén	Discoveries in magnetohydrodynamics
1974	M. Ryle	Observations and inventions of aperture-synthesis technique in radio astrophysics
	A. Hewish	Discovery of pulsars
1978	A. A. Penzias and R. W. Wilson	Discovery of cosmic microwave background radiation
1983	S. Chandrasekhar	Theoretical studies of the structure and evolution of stars
	W. A. Fowler	Theoretical and experimental studies of nuclear reactions of importance in the formation of chemical elements in the Universe
1993	R. A. Hulse and J. H. Taylor	Discovery of a new type of pulsar
2002	R. Davis and M. Koshiba	Detection of cosmic neutrinos
	R. Giacconi	Discovery of cosmic X-ray sources
2006	J. C. Mather and G. F. Smoot	Discovery of the blackbody form and anisotropy of cosmic microwave background radiation

The following is a list of Nobel laureates who have made important contributions in astrophysics, although they won Nobel Prizes for their works in other areas of physics.

1921   A. Einstein
1938   E. Fermi
1952   E. M. Purcell
1964   C. H. Townes
1979   S. Weinberg
1980   J. W. Cronin

Suggestions for further reading

The reader of this book is assumed to have a background of physics appropriate for the advanced undergraduate or the beginning graduate level. Knowledge is assumed of all the standard branches of physics which are usually covered at that level – classical mechanics, electromagnetic theory, special relativity, optics, thermal physics, statistical mechanics, quantum mechanics, atomic physics, nuclear physics and the standard mathematical tools often known collectively as the methods of mathematical physics. Since a reader of this book would be familiar with the standard textbooks on these different branches of physics and would have his/her own favourites, I do not make an attempt of listing textbooks of basic physics.

After giving some general references covering the whole of astrophysics, I provide references on the material covered in different chapters. The main aim of the references for different chapters is to help those readers of this book who want to go beyond what is covered here. No attempt at completeness is made here. It is not possible for an individual to be acquainted with everything written on all the topics covered in this book. I have mainly included those references which I myself have found useful and which are of the nature of pedagogical works at the immediate next level beyond this book, intentionally leaving out advanced monographs on specialized research topics. Some suitable references may not have been included merely due to the accident of my not knowing about them. I apologize to those authors who may feel that something written by them ought to have been referenced.

General references

There are several excellent elementary astronomy textbooks suitable for beginning undergraduate students where the authors assume very little knowledge of physics and not even a knowledge of calculus, since elementary astronomy courses at this level are popular in many undergraduate programmes. The pioneering classic astronomy text at this level was written by Abell (1964), which has later been updated by Morrison, Wolff and Fraknoi (1995). The other classic in this field is by Shu (1982). Although somewhat outdated by now, this amazing book manages to discuss many important physics aspects of basic astrophysics very clearly and at considerable depth with only high school mathematics. One of the more recent books is Gregory and Zeilik (1997). At a more advanced level (assuming more background of physics and mathematics), an outstanding textbook suitable for senior undergraduates is Carroll and Ostlie (2006). Although the descriptive and phenomenological topics are covered extremely well in this book, the treatment of conceptual topics is not always so satisfactory or adequate. The size of the book (more than 1300 pages) also makes it unsuitable for use in a one-semester astrophysics course.

After the elementary books mentioned above (there are many more elementary astronomy texts!), one can mention the many excellent graduate textbooks dealing with specific branches of astrophysics. However, there are not too many books which attempt to bridge the gap between these two kinds of books, by presenting the whole of astrophysics comprehensively in one volume assuming an advanced undergraduate or elementary graduate level of physics. The very few books of this kind include Unsöld and Baschek (2001), Harwit (2006), Shore (2002) and Maoz (2007). Since these authors more or less attempt the same thing which I attempt in this book, it is best that I do not comment on these books and leave it for readers to judge how successful which author has been. There is a two-volume introduction to astrophysics by Bowers and Deeming (1984a, 1984b) as well as a three-volume introduction to theoretical astrophysics by Padmanabhan (2000, 2001, 2002).

Apart from textbooks, one may mention other kinds of books which can be useful for students of astrophysics. A classic handbook compiling all kinds of astrophysical data was first brought out by Allen (1955) and is now updated by Cox (2000). Lang (1999) has collected many of the important astrophysical formulae in two volumes. The excellent glossary of astronomical terms compiled by Hopkins (1980) has unfortunately not been updated in many years. A collection of essays on several astrophysical topics by some of the world's leading astrophysicists has been edited by Bahcall and Ostriker (1997). One can question whether the title of the volume *Unsolved Problems in Astrophysics* is justified and whether it discusses unsolved problems from all areas of astrophysics which such a pretentious title would imply, but one cannot question the scholarship and authority of the essays included in this volume. At a more elementary level, articles on different aspects of astrophysics published in *Scientific American* till the mid-1970s were put together by Gingerich (1975). Several leading astrophysicists have written about their research fields in non-technical language in this volume.

Chapter 1

Since books on different astrophysical systems will be mentioned in the references for the various following chapters, here we only mention books which discuss how we obtain astronomical data with the help of the appropriate instruments. Perhaps Kitchin (2003), Roy and Clarke (2003) and Léna (1988) are amongst the most outstanding graduate level textbooks of this kind. There are also many books devoted specifically to optical, radio or X-ray telescopes. The present author, as a theoretician, is not very qualified to judge the relative merits and demerits of these books. So we do not attempt to provide a detailed bibliography of observational astronomy.

Chapter 2

The standard graduate textbook on radiative processes in astrophysics is the superbly written volume by Rybicki and Lightman (1979), which is justly regarded as a classic.

Since this book mainly deals with well-established principles, it has not dated with time. Other books covering this field are Tucker (1975) and Shu (1991). Serious students wishing to learn about radiative transfer in more detail should consult the famous monograph by Chandrasekhar (1950). The standard work on stellar atmospheres is Mihalas (1978).

Chapters 3–4

Since the study of stars has been the central theme in modern astrophysics for several decades, it is no wonder that there are many excellent books on stellar astrophysics. The classic volumes by Eddington (1930) and Chandrasekhar (1939), which played very important roles in the historical development of astrophysics, are not suitable for use as textbooks in modern courses on this subject. However, the other classic by Schwarzschild (1958) is still useful for pedagogical purposes, because the first two chapters of this book provide one of the clearest presentations of the basics of stellar astrophysics, although the later chapters dealing with details of stellar models are completely outdated now. One of the first books written after computers revolutionized the study of stellar structure is Clayton (1968), which has been used as the standard graduate textbook for many years. Amongst the more modern books on this subject, Kippenhahn and Weigert (1994) and Böhm-Vitense (1989a, 1989b, 1992) are very clearly written and are highly recommended. The beautifully written volume by Tayler (1994) introduces the subject at a more elementary level than the levels of the books cited earlier.

We now come to books dealing with specialized aspects of stellar astrophysics. Stix (2004) has written an excellent book on the Sun. Arnett (1996) discusses nuclear reactions during the advanced stages of stellar evolution and the physics of supernovae. The standard work on neutrino astrophysics is by Bahcall (1989).

Chapter 5

Shapiro and Teukolsky (1983) wrote the definitive graduate textbook on compact objects resulting from the end states of stellar collapse. In spite of the fact that many important developments have taken place in this field afterwards, this book is still highly recommended for the very clear discussion of the basics of both theory and observations. A more recent book on this subject is by Glendenning (2000). Although Longair (1994) covers many topics, a large part of his book is devoted to topics related to advanced stellar evolution and stellar collapse.

Chapter 6

The standard graduate textbook on galactic astronomy by Binney and Merrifield (1998) is supposed to be a replacement of the earlier volume by Mihalas and Binney (1981). In order to make room for the discussion of external galaxies, Binney and Merrifield (1998) condensed or deleted some topics pertaining to our Galaxy which were discussed by Mihalas and Binney (1981) at length. Mihalas and Binney (1981) may therefore be more useful than Binney and Merrifield (1998) for learning about our Galaxy. Another good book covering galaxies and interstellar matter is Scheffler and Elsässer (1988). The superb elementary book by Tayler (1993) would provide a good starting point to get into this subject.

There are several books dealing specifically with the interstellar matter. The famous classic by Spitzer (1978) is rather compactly written and beginners may find it a difficult book, although a perusal of this book can be a very rewarding experience. Osterbrock and Ferland (2005) cover many aspects of the subject, although one of the important components of interstellar matter – the HI component – is not discussed in this book. Other books dealing with the interstellar medium are Dyson and Williams (1997) and Dopita and Sutherland (2003).

Chapter 7

Very surprisingly, there seems to be only one really satisfactory modern textbook on the important subject of stellar dynamics. It is the monumental volume by Binney and Tremaine (1987). Readers may want to look at a classic – an old review article by Oort (1965). The dynamics of globular clusters is treated in a monograph by Spitzer (1987).

Chapter 8

Shu (1992) and Choudhuri (1998) are standard introductory texts on plasma astrophysics. Serious students may want to look up the classic in this field by Parker (1979). The topic of relativistic particles in astrophysics is covered thoroughly by Longair (1994). The applications of MHD to the Sun are discussed by Priest (1982). See Mestel (1999) for a masterly discussion of the role of magnetic fields in stellar astronomy.

Chapter 9

Several of the references for Chapter 6 cover external galaxies as well. We especially recommend Binney and Merrifield (1998). Another recent book on this rapidly developing subject is Schneider (2006). We recommend Peterson (1997) and Krolik (1999) as good introductions to the subject of active galaxies. Sarazin (1986) provides a comprehensive coverage of galaxy clusters.

Chapters 10–11

The brilliant small volume by Weinberg (1977) is a masterpiece of popular science and may be read profitably before delving into the more technical tomes. Somehow cosmology has been a popular subject for textbook writers and there is probably no other branch of astrophysics in which so many textbooks have been written and are still being written. We certainly cannot provide a complete bibliography of cosmology books here. So we selectively mention a few books in which authors stress the astrophysical aspects. After developing the necessary tools of general relativity, Narlikar (2002) gives a very clear introduction to the main themes of modern cosmology. In spite of the author's bias for the steady state theory, all the standard topics are covered satisfactorily. The volume by Peebles (1993) is a work of great scholarship and insight, but it is not particularly coherently written and beginners may get lost in this large book. Peacock (1999) covers various physics and astronomy topics which one needs to master in order to become a professional cosmologist. We recommend Kolb and Turner (1990) for topics pertaining to the early Universe.

Chapters 12–14

While general relativity also has been a favourite subject for textbook writers, the last few chapters of Landau and Lifshitz (1975) still provide one of the most beautiful and elegant introductions to this subject. Schutz (1985) and Hartle (2003) are two particularly user-friendly books in a subject generally regarded as difficult and abstruse. One famous (and bulky) introductory textbook is by Misner, Thorne and Wheeler (1973). But be cautioned that this book is not for you if you do not like very verbose and wordy presentations! Weinberg (1972) also wrote a famous book in which his aim was to develop general relativity without stressing the geometrical aspects of the theory. Readers desirous of learning the more formal aspects of general relativity may turn to Wald (1984).

References

Each reference is cited in the portions of the book indicated within square brackets. S. f. r. stands for *Suggestions for further reading*.

Aaronson, M. *et al.* 1982, *Astrophys. J.* **258**, 64. [§9.5]

Aaronson, M. *et al.* 1986, *Astrophys. J.* **302**, 536. [§9.2.2]

Abdurashitov, J. N. *et al.* 1996, *Phys. Rev. Lett.* **77**, 4708. [§4.4.2]

Abell, G. O. 1958, *Astrophys. J. Suppl.* **3**, 211. [§9.5]

Abell, G. O. 1964, *The Exploration of the Universe*. Holt, Rinehart and Winston. [S. f. r.]

Abramowitz, M. and Stegun, I. A. 1964, *Handbook of Mathematical Functions*. National Bureau of Standards. Reprinted by Dover. [§11.2]

Ahmad, Q. R. *et al.* 2002, *Phys. Rev. Lett.* **89**, 11301. [§4.4.2, §11.5]

Alcock, C. *et al.* 1993, *Nature*, **365**, 621. [§13.3.2]

Alfvén, H. 1942a, *Ark. f. Mat. Astr. o. Fysik* **29B**, No. 2. [§8.5]

Alfvén, H. 1942b, *Nature* **150**, 405. [Ex. 8.4]

Allen, C. W. 1955, *Astrophysical Quantities*. The Athlone Press. [S. f. r.]

Alpher, R. A. and Herman, R. C. 1948, *Nature* **162**, 774. [§10.5]

Anselmann, P. *et al.* 1995, *Phys. Lett.* **B342**, 440. [§4.4.2]

Arnett, D. 1996, *Supernovae and Nucleosynthesis*. Princeton University Press. [S. f. r.]

Atkinson, R. d'E. and Houtermans, F. G. 1929, *Zs. f. Phys.* **54**, 656. [§4.1]

Aubourg, E. *et al.* 1993, *Nature* **365**, 623. [§13.3.2]

Axford, W. I., Leer, E. and Skadron, G. 1977, *Proc. 15th International Cosmic Ray Conf.* **11**, 132. [§8.10]

Baade, W. 1944, *Astrophys. J.* **100**, 137. [§6.4]

Baade, W. 1954, *Trans. I. A. U.* **8**, 397. [§6.1.2]

Baade, W. and Zwicky, F. 1934, *Phys. Rev.* **45**, 138. [§5.4, §5.5]

Backer, D. C., Kulkarni, S. R., Heiles, C., Davis, M. M. and Goss, W. M. 1982, *Nature* **300**, 615. [§5.5.2]

Bahcall, J. N. 1989, *Neutrino Astrophysics*. Cambridge University Press. [S. f. r.]

Bahcall, J. N. 1999, *Current Science* **77**, 1487. [§4.4.2, Fig. 4.8]

Bahcall, J. N. and Ostriker, J. P. (Ed.) 1997, *Unsolved Problems in Astrophysics*. Princeton University Press. [S. f. r.]

Bahcall, J. N. and Ulrich, R. K. 1988, *Rev. Mod. Phys.* **60**, 297. [§4.4]

Becker, R. H. *et al.* 2001, *Astron. J.* **122**, 2850. [§11.8.2]

Bell, A. R. 1978, *Mon. Not. Roy. Astron. Soc.* **182**, 147 and 443. [§8.10]

Bennett, C. L. *et al.* 2003, *Astrophys. J. Suppl.* **148**, 1. [§11.7.1, Fig. 11.3, §14.5, Fig. 14.5]

Bernoulli, D. 1738, *Hydrodynamica*. [Ex. 8.1]

Bethe, H. 1939, *Phys. Rev.* **55**, 434. [§4.3]

Bethe, H. and Critchfield, C. H. 1938, *Phys. Rev.* **54**, 248. [§4.3]

Biermann, L. 1948, *Zs. f. Astrophys.* **25**, 135. [§3.2.4]

Binney, J. and Merrifield, M. 1998, *Galactic Astronomy*. Princeton University Press. [§6.1.2, §6.3.2, Fig. 6.8, Fig. 9.6, S. f. r.]

Binney, J. and Tremaine, S. 1987, *Galactic Dynamics*. Princeton University Press. [§7.3, §7.4, S. f. r.]

Blandford, R. D. and Ostriker, J. P. 1978, *Astrophys. J. Lett.* **221**, L29. [§8.10]

Blandford, R. D. and Rees, M. J. 1974, *Mon. Not. Roy. Astron. Soc.* **169**, 395. [§9.4.1]

Bless, R. C. and Savage, B. D. 1972, *Astrophys. J.* **171**, 293. [Fig. 6.3]

Böhm-Vitense, E. 1989a, *Stellar Astrophysics, Vol. 1: Basic Stellar Observations and Data*. Cambridge University Press. [§3.5.1, Fig. 3.4, S. f. r.]

Böhm-Vitense, E. 1989b, *Stellar Astrophysics, Vol. 2: Stellar Atmospheres*. Cambridge University Press. [S. f. r.]

Böhm-Vitense, E. 1992, *Stellar Astrophysics, Vol. 3: Stellar Structure and Evolution*. Cambridge University Press. [S. f. r.]

Boltzmann, L. 1872, *Sitzungsber. Kaiserl. Akad. Wiss. Wien* **66**, 275. [§7.5]

Boltzmann, L. 1884, *Wied. Annalen*, **22**, 291. [§2.4.1]

Bondi, H. 1952, *Mon. Not. Roy. Astron. Soc.* **112**, 195. [§4.6]

Bondi, H. and Gold, T. 1948, *Mon. Not. Roy. Astron. Soc.* **108**, 252. [Ex. 10.5]

Born, M. and Wolf, E. 1980, *Principles of Optics*, 6th edn. Pergamon. [§1.7.1, §13.5]

Bose, S. N. 1924, *Zs. f. Phys.* **26**, 178. [§11.2]

Bosma, A. 1978, *PhD Thesis*. Kapteyn Institute, Groningen. [Fig. 9.6]

Bowers, R. and Deeming, T. 1984a, *Astrophysics I: Stars*. Jones and Bartlett. [S. f. r.]

Bowers, R. and Deeming, T. 1984b, *Astrophysics II: Interstellar Matter and Galaxies*. Jones and Bartlett. [S. f. r.]

Brunish, W. M. and Truran, J. W. 1982, *Astrophys. J. Suppl.* **49**, 447. [Fig. 3.2, Fig. 3.3]

Bunsen, R. and Kirchhoff, G. 1861, *Untersuchungen über das Sonnenspecktrum und die Spektren der Chemischen Elemente*. [§1.2]

Burbidge, E. M., Burbidge, G. R., Fowler, W. A. and Hoyle, F. 1957, *Rev. Mod. Phys.* **29**, 547. [§4.3, §4.7]

Butcher, H. and Oemler, A. 1978, *Astrophys. J.* **219**, 18. [§9.5]

Byram, E. T., Chubb, T. A. and Friedman, H. 1966, *Science* **152**, 66. [§9.5]

Carroll, B. W. and Ostlie, D. A. 2006, *An Introduction to Modern Astrophysics*, 2nd edn. Benjamin Cummings. [S. f. r.]

Chadwick, J. 1932, *Proc. Roy. Soc.* **A 136**, 692. [§5.4]

Chandrasekhar, S. 1931, *Astrophys. J.* **74**, 81. [§1.5, §5.3]

Chandrasekhar, S. 1935, *Mon. Not. Roy. Astron. Soc.* **95**, 207. [§5.3]

Chandrasekhar, S. 1939, *An Introduction to the Study of Stellar Structure*. University of Chicago Press. Reprinted by Dover. [§3.3, S. f. r.]

Chandrasekhar, S. 1943, *Astrophys. J.* **97**, 255. [§7.4]

Chandrasekhar, S. 1950, *Radiative Transfer*. Clarendon Press. Reprinted by Dover. [§2.4, S. f. r.]

Chandrasekhar, S. 1952, *Phil. Mag. (7)* **43**, 501. [§8.6]

Chandrasekhar, S. 1984, *Rev. Mod. Phys.* **56**, 137. [Fig. 5.2]

Chandrasekhar, S. and Breen, F. H. 1946, *Astrophys. J.* **104**, 430. [§2.6.2]

Chen, G. H. and Hewitt, J. N. 1993, *Astron. J.* **106**, 1719. [Fig. 13.6]

Chevalier, R. A. 1992, *Nature* **355**, 691. [Fig. 4.14]

Chitre, S. M. and Antia, H. M. 1999, *Current Science* **77**, 1454. [Fig. 4.7]

Choudhuri, A. R. 1989, *Solar Phys.* **123**, 217. [§8.6]

Choudhuri, A. R. 1998, *The Physics of Fluids and Plasmas: An Introduction for Astrophysicists.* Cambridge University Press. [§4.4.1, §4.6, §6.7, §7.5, §8.5, §8.7, §8.10, §8.13, §9.5, S. f. r.]

Christoffel, E. B. 1868, *J. reine Angew. Math.* **70**, 46. [§12.2.3]

Clayton, D. D. 1968, *Principles of Stellar Evolution and Nucleosynthesis.* McGraw-Hill. [§2.6, S. f. r.]

Colless, M. *et al.* 2001, *Mon. Not. Roy. Astron. Soc.* **328**, 1039. [§9.6]

Compton, A. H. 1923, *Phys. Rev.* **21**, 715. [§11.7.2]

Copernicus, N. 1543, *De Revolutionibus Orbium Celestium.* [§1.2]

Cowie, L. L. and Binney, J. 1977, *Astrophys. J.* **215**, 723. [§9.5]

Cowsik, R. and McClelland, J. 1972, *Phys. Rev. Lett.* **29**, 669. [§11.5]

Cox, A. N. 2000, *Allen's Astrophysical Quantities*, 4th edn. Springer. [S. f. r.]

Cox, A. N. and Stewart, J. N. 1970, *Astrophys. J. Suppl.* **19**, 243. [§2.6]

Curtis, H. D. 1921, *Bull. Nat. Res. Council* **2**, 194. [§9.1]

Davis, L. and Greenstein, J. L. 1951, *Astrophys. J.* **114**, 206. [§6.7]

Davis, M., Efstathiou, G., Frenk, C. S. and White, S. D. M. 1985, *Astrophys. J.* **292**, 371. [§11.9]

Davis, R. Jr., Harmer, D. S. and Hoffman, K. C. 1968, *Phys. Rev. Lett.* **20**, 1205. [§4.4.2]

de Lapparent, V., Geller, M. J. and Huchra, J. P. 1986, *Astrophys. J. Lett.* **302**, L1. [§9.6, Fig. 9.19]

de Vaucouleurs, G. 1948, *Ann. Astrophys.* **11**, 247. [§9.2.1]

Dehnen, W. and Binney, J. J. 1998, *Mon. Not. Roy. Astron. Soc.* **298**, 387. [Fig. 6.7, Fig. 7.4]

Deshpande, A. A., Ramachandran, R. and Srinivasan, G. 1995, *J. Astrophys. Astron.* **16**, 69. [Fig. 5.5]

Dirac, P. A. M. 1926, *Proc. Roy. Soc.* **A 112**, 661. [§11.2]

Dopita, M. and Sutherland, R. S. 2003, *Astrophysics of the Diffuse Universe.* Springer. [S. f. r.]

Downs, G. S. 1981, *Astrophys. J.* **249**, 687. [Fig. 5.4]

Dreher, J. W. and Feigelson, E. D. 1984, *Nature* **308**, 43. [Fig. 9.10]

Dressler, A. 1980, *Astrophys. J.* **236**, 351. [§9.2.1]

Dreyer, J. L. E. 1888, *Mem. Roy. Astron. Soc.* **49**, 1. [§1.8]

Dyson, F. W., Eddington, A. S. and Davidson, C. 1920, *Phil. Trans. Roy. Soc.* **A 220**, 291. [§13.3.2]

Dyson, J. E. and Williams, D. A. 1997, *The Physics of the Interstellar Medium*, 2nd edn. Taylor & Francis. [S. f. r.]

Eddington, A. S. 1916, *Mon. Not. Roy. Astron. Soc.* **77**, 16. [§3.2.3]

Eddington, A. S. 1920, *Observatory* **43**, 353. [§4.1]

Eddington, A. S. 1924, *Mon. Not. Roy. Astron. Soc.* **84**, 104. [§3.6.1]

Eddington, A. S. 1926, *The Internal Constitution of the Stars.* Cambridge University Press. [§2.4.2, §3.1, §3.2.3, §5.3, Ex. 5.5, S. f. r.]

Edge, D. O., Shakeshaft, J. R., McAdam, W. B., Baldwin, J. E. and Archer, S. 1959, *Mem. Roy. Astron. Soc.* **68**, 37. [§1.8]

Einstein, A. 1916, *Ann. d. Phys.* **49**, 769. [§10.1, §10.2, §12.4.2]

Einstein, A. 1917, *Sitzungber. Preus. Akad. Wissen.*, 142. [§10.4, §14.2]

Einstein, A. 1924, *Sitzungber. Preus. Akad. Wissen.*, 261. [§11.2]

Einstein, A. and de Sitter, W. 1932, *Proc. Nat. Acad. Sci.* **18**, 213. [§10.6]

Emden, R. 1907, *Gaskugeln*. Teubner, Leipzig. [§5.3]

Euler, L. 1755, *Hist. de l'Acad. de Berlin*. [§8.2]

Euler, L. 1759, *Novi Comm. Acad. Petrop.* **14**, 1. [§8.2]

Ewen, H. I. and Purcell, E. M. 1951, *Nature* **168**, 356. [§6.5]

Faber, S. M. and Jackson, R. E. 1976, *Astrophys. J.* **204**, 668. [§9.2.2]

Feast, M. and Whitelock, P. 1997, *Mon. Not. Roy. Astron. Soc.* **291**, 683. [§6.2]

Felten, J. E., Gould, R. J., Stein, W. A. and Woolf, N. J. 1966, *Astrophys. J.* **146**, 955. [§9.5]

Fermi, E. 1926, *Rend. Lincei* **3**, 145. [§11.2]

Fermi, E. 1949, *Phys. Rev.* **75**, 1169. [§8.10]

Field, G. B. 1965, *Astrophys. J.* **142**, 531. [§6.8]

Field, G. B., Goldsmith, D. W. and Habing, H. J. 1969, *Astrophys. J.* **155**, 149. [§6.8]

Fowler, R. H. 1926, *Mon. Not. Roy. Astron. Soc.* **87**, 114. [§5.2]

Fowler, W. A. 1984, *Rev. Mod. Phys.* **56**, 149. [§4.3]

Francis, P. J. *et al.* 1991, *Astrophys. J.* **373**, 465. [Fig. 9.11]

Fraunhofer, J. 1817, *Gilberts Ann.* **56**, 264. [§1.2]

Freedman, W. L. *et al.* 2001, *Astrophys. J.* **553**, 47. [§9.3, Fig. 9.9]

Friedmann, A. 1924, *Zs. f. Phys.* **21**, 326. [§10.4]

Gamow, G. 1928, *Zs. f. Phys.* **51**, 204. [§4.1, §4.2]

Gamow, G. 1946, *Phys. Rev.* **70**, 572. [§4.3, §11.3]

Giacconi, R., Gursky, H., Paolini, F. R. and Rossi, B. B. 1962, *Phys. Rev. Lett.* **9**, 439. [§1.7.3, §5.6]

Giacconi, R. *et al.* 1972, *Astrophys. J.* **178**, 281. [§9.5]

Gilmore, G. and Reid, N. 1983, *Mon. Not. Roy. Astron. Soc.* **202**, 1025. [§6.1.2]

Gingerich, O. (Ed.) 1975, *New Frontiers in Astronomy*. W. H. Freeman and Company. [S. f. r.]

Glendenning, N. K. 2000, *Compact Stars: Nuclear Physics, Particle Physics and General Relativity*, 2nd edn. Springer. [S. f. r.]

Gold, T. 1968, *Nature* **218**, 731. [§5.5]

Goldreich, P. and Julian, W. H. 1969, *Astrophys. J.* **157**, 869. [§5.5]

Goldstein, H. 1980, *Classical Mechanics*, 2nd edn. Addison-Wesley. [§7.5, §12.3, §12.2.1, §13.3.1]

Gough, D. O. 1978, *Proc. Workshop on Solar Rotation*, Univ. of Catania, p. 255. [§4.8]

Gregory, S. A. and Zeilik, M. 1997, *Introductory Astronomy and Astrophysics*. Saunders. [S. f. r.]

Gunn, J. E. and Gott, J. R. 1972, *Astrophys. J.* **176**, 1. [§9.5]

Gunn, J. E. and Peterson, B. A. 1965, *Astrophys. J.* **142**, 1633. [§11.8.2]

Guth, A. H. 1981, *Phys. Rev.* **D23**, 347. [§11.6.1]

Hale, G. E. 1908, *Astrophys. J.* **28**, 315. [§4.8]

Hale, G. E., Ellerman, F., Nicholson, S. B. and Joy, A. H. 1919, *Astrophys. J.* **49**, 153. [§4.8]

Halliday, D., Resnick, R. and Walker, J. 2001, *Fundamentals of Physics*, 6th edn. John Wiley & Sons. [§9.3]

Hansen, C. J. and Kawaler, S. D. 1994, *Stellar Interiors*. Springer-Verlag. [Fig. 3.2, Fig. 3.3]

Hartle, J. B. 2003, *Gravity: An Introduction to Einstein's General Relativity*. Benjamin Cummings. [S. f. r.]

Hartmann, D. and Burton, W. B. 1997, *Atlas of Galactic Neutral Hydrogen*. Cambridge University Press. [Fig. 6.8]

Harwit, M. 2006, *Astrophysical Concepts*, 4th edn. Springer. [S. f. r.]

Hayashi, C. 1961, *Publ. Astron. Soc. Japan* **13**, 450. [§4.5]

Helmholtz, H. 1854, *Lecture at Kant Commemoration, Königsberg*. [§3.2.2]

Henriksen, M. J. and Mushotzky, R. F. 1986, *Astrophys. J.* **302**, 287. [Fig. 9.18]

Henyey, L. G., Vardya, M. S. and Bodenheimer, P. L. 1965, *Astrophys. J.* **142**, 841. [§3.3]

Herschel, W. 1785, *Phil. Trans.* **75**, 213. [§6.1]

Hertzsprung, E. 1911, *Potsdam Pub.* **63**. [§3.4]

Hess, V. F. 1912, *Sitzungsber. Kaiserl. Akad. Wiss. Wien* **121**, 2001. [§6.7, §8.10]

Hewish, A. S., Bell, J., Pilkington, J. D. H., Scott, P. F. and Collins, R. A. 1968, *Nature* **217**, 709. [§5.5]

Hewitt, A. and Burbidge, G. 1993, *Astrophys. J. Suppl.* **87**, 451. [Fig. 11.4]

Hiltner, W. A. 1954, *Astrophys. J.* **120**, 454. [§6.7]

Hirata, K. S. *et al.* 1990, *Phys. Rev. Lett.* **65**, 1297. [§4.4.2]

Homer Lane, J. 1869, *Amer. J. Sci.* **50**, 57. [§5.3]

Hopkins, J. 1980, *Glossary of Astronomy and Astrophysics*, 2nd edn. University of Chicago Press. [S. f. r.]

Hoyle, F. 1948, *Mon. Not. Roy. Astron. Soc.* **108**, 372. [Ex. 10.5]

Hoyle, F. 1954, *Astrophys. J. Suppl.* **1**, 121. [§4.3]

Hubble, E. P. 1922, *Astrophys. J.* **56**, 162. [§6.1.2, §9.1, §9.3]

Hubble, E. P. 1929, *Proc. Nat. Acad. Sci.* **15**, 168. [§9.3, §14.2]

Hubble, E. P. 1936, *The Realm of the Nebulae*. Yale University Press. [§9.2.1]

Huchtmeier, W. K. 1975, *Astron. Astrophys.* **45**, 259. [§9.2.2]

Hughes, D. H. *et al.* 1998, *Nature* **394**, 241. [§11.8.1]

Hulse, R. A. and Taylor, J. H. 1975, *Astrophys. J. Lett.* **195**, L51. [§1.6, §5.5.1]

Iben, I. 1965, *Astrophys. J.* **141**, 993. [Fig. 3.2, Fig. 3.3]

Iben, I. 1967, *Ann. Rev. Astron. Astrophys.* **5**, 571. [§4.5]

Iben, I. 1974, *Ann. Rev. Astron. Astrophys.* **12**, 215. [§4.5]

Jackson, J. D. 1999, *Classical Electrodynamics*, 3rd edn. John Wiley & Sons. [§5.2, §5.5, §8.10, §8.11, §13.4, §14.5]

Jaffe, W., Ford, H. C., Ferrarese, L., van den Bosch, F. and O'Connell, R. W. 1993, *Nature* **364**, 213. [§9.4.3, Fig. 9.14]

Jansky, K. G. 1933, *Proc. IRE* **21**, 1387. [§1.7.2]

Jeans, J. H. 1902, *Phil. Trans. Roy. Soc.* **A 199**, 1. [§6.8, §8.3, §11.9]

Jeans, J. H. 1922, *Mon. Not. Roy. Astron. Soc.* **82**, 122. [§7.6]

Jennison, R. C. and Dasgupta, M. K. 1953, *Nature* **172**, 996. [§9.4.1]

Johnson, H. L. and Morgan, W. W. 1953, *Astrophys. J.* **117**, 313. [§1.4]

Johnson, H. L. and Sandage, A. R. 1956, *Astrophys. J.* **124**, 379. [Fig. 3.8]

Joy, A. H. 1939, *Astrophys. J.* **89**, 356. [§6.2, Fig. 6.5]

Kant, I. 1755, *Allgemeine Naturgeschichte und Theorie des Himmels*. [§9.1]

Kapteyn, J. C. 1922, *Astrophys. J.* **55**, 302. [§6.1, §6.1.2]

Kapteyn, J. C. and van Rhijn, P. J. 1920, *Astrophys. J.* **52**, 23. [§6.1, §6.1.2]

Kelvin, Lord 1861, *Brit. Assoc. Repts., Part II*, p. 27. [§3.2.2]

Kerr, R. P. 1963, *Phys. Rev. Lett.* **11**, 237. [§13.3]

King, I. R. 1966, *Astron. J.* **71**, 64. [Ex. 7.4]

Kippenhahn, R. and Weigert, A. 1990, *Stellar Structure and Evolution*. Springer-Verlag. [§3.2.4, §3.3, §3.4, §4.5, Fig. 5.1, S. f. r.]

Kirchhoff, G. 1860, *Ann. d. Phys. u. Chemie* **109**, 275. [§2.2.4]

Kitchin, C. R. 2003, *Astrophysical Techniques*, 4th edn. Taylor & Francis. [S. f. r.]

Klebesadel, R. W., Strong, I. B. and Olson, R. A. 1973, *Astrophys. J.* **182**, 85. [§9.7]

Kolb, E. W. and Turner, M. S. 1990, *The Early Universe*. Addison-Wesley Publishing Company. [Fig. 11.1, §11.9, S. f. r.]

Konar, S. and Choudhuri, A. R. 2004, *Mon. Not. Roy. Astron. Soc.* **348**, 661. [§5.5.2]

Kramers, H. A. 1923, *Phil. Mag.* **46**, 836. [§2.6]

Krolik, J. H. 1999, *Active Galactic Nuclei*. Princeton University Press. [S. f. r.]

Kruskal, M. D. 1960, *Phys. Rev.* **119**, 1743. [§13.3.3]

Krymsky, G. F. 1977, *Dokl. Acad. Nauk. U.S.S.R.* **234**, 1306. [§8.10]

Landau, L. D. and Lifshitz, E. M. 1975, *The Classical Theory of Fields*, 4th edn. Pergamon. [S. f. r.]

Landau, L. D. and Lifshitz, E. M. 1976, *Mechanics*, 3rd edn. Pergamon. [§12.2.1, §12.3, §13.3.1]

Landau, L. D. and Lifshitz, E. M. 1980, *Statistical Physics*, 3rd edn. Pergamon. [§7.5]

Lang, K. R. 1999, *Astrophysical Formulae, Vols. I and II*, 3rd edn. Springer. [S. f. r.]

Langmuir, I. 1928, *Proc. Nat. Acad. Sci.* **14**, 627. [§8.13.1]

Laplace, P. S. 1795, *Le Systeme du Monde, Vol. II*. [§1.5]

Leavitt, H. S. 1912, *Harvard Coll. Obs. Circ.* **173**, 1. [§6.1.2]

Lee, B. W. and Weinberg, S. 1977, *Phys. Rev. Lett.* **39**, 165. [§11.5]

Leibundgut, B. 2001, *Ann. Rev. Astron. Astrophys.* **39**, 67. [Fig. 14.4]

Leighton, R. B., Noyes, R. W. and Simon, G. W. 1962, *Astrophys. J.* **135**, 474. [§4.4.1]

Lemaitre, G. 1927, *Ann. Soc. Sci. Bruxelles* **47A**, 49. [§10.5]

Léna, P. 1988, *Observational Astrophysics*. Springer-Verlag. [S. f. r.]

Lin, C. C. and Shu, F. 1964, *Astrophys. J.* **140**, 646. [§7.1]

Lindblad, B. 1927, *Mon. Not. Roy. Astron. Soc.* **87**, 553. [§6.2]

Longair, M. S. 1992, *High Energy Astrophysics, Vol. 1*, 2nd edn. Cambridge University Press. [§8.12]

Longair, M. S. 1994, *High Energy Astrophysics, Vol. 2*, 2nd edn. Cambridge University Press. [Fig. 4.9, Fig. 4.10, Fig. 5.7, §8.10, §8.11, S. f. r.]

Lynden-Bell, D. 1967, *Mon. Not. Roy. Astron. Soc.* **136**, 101. [§7.3]

Malmquist, K. G. 1924, *Medd. Lund Astron. Obs., Ser. II* **32**, 64. [§6.1.1]

Maoz, D. 2007, *Astrophysics in a Nutshell*. Princeton University Press. [S. f. r.]

Mather, J. C. *et al.* 1990, *Astrophys. J. Lett.* **354**, L37. [§10.5, Fig. 10.4, §11.7.1]

Mathews, J. and Walker, R. L. 1979, *Mathematical Methods of Physics*, 2nd edn. W. A. Benjamin. [§12.2.5, §13.3.1, §14.5]

Mathewson, D. S. and Ford, V. L. 1970, *Mem. Roy. Astron. Soc.* **74**, 139. [Fig. 6.13]

Mattig, W. 1958, *Astron. Nachr.* **284**, 109. [§14.3]

Maunder, E. W. 1904, *Mon. Not. Roy. Astron. Soc.* **64**, 747. [§4.8]

Maxwell, J. C. 1860, *Phil. Mag. (4)* **19**, 19. [§2.3.1]

Maxwell, J. C. 1865, *Phil. Trans. Roy. Soc.* **155**, 459. [§8.4]

Mayor, M. and Queloz, D. 1995, *Nature* **378**, 355. [§4.9]

McKee, C. F. and Ostriker, J. P. 1977, *Astrophys. J.* **218**, 148. [§6.6.5]

Mestel, L. 1999, *Stellar Magnetism*. Oxford University Press. [S. f. r.]

Meyer, P. 1969, *Ann. Rev. Astron. Astrophys.* **7**, 1. [Fig. 8.11]

Mihalas, D. 1978, *Stellar Atmospheres*, 2nd edn. Freeman. [§2.3.1, §2.4, §2.6, S. f. r.]

Mihalas, D. and Binney, J. 1981, *Galactic Astronomy*. Freeman. [Fig. 4.9, Fig. 4.10, §6.1.1, §6.2, S. f. r.]

Milne, E. A. and McCrea, W. H. 1934, *Q. J. Maths.* **5**, 73. [§10.1]

Misner, C. W., Thorne, K. S. and Wheeler, J. A. 1973, *Gravitation*. Freeman. [S. f. r.]

Mitchell, R. J., Culhane, J. L., Davison, P. J. N. and Ives, J. C. 1976, *Mon. Not. Roy. Astron. Soc.* **175**, 29. [§9.5]

Morrison, D., Wolff, S. and Fraknoi, A. 1995, *Abell's Exploration of the Universe*, 7th edn. Saunders. [S. f. r.]

Mouschovias, T. Ch. 1974, *Astrophys. J.* **192**, 37. [§8.8]

Muller, C. A. and Oort, J. H. 1951, *Nature* **168**, 357. [§6.5]

Nakajima, T., Oppenheimer, B. R., Kulkarni, S. R., Golimowski, D. A., Matthews, K. and Durrance, S. T. 1995, *Nature* **378**, 463. [§3.6.1]

Narlikar, J. V. 2002, *Introduction to Cosmology*, 3rd edn. Cambridge University Press. [§11.9, S. f. r.]

Oegerle, W. R. and Hoessel, J. G. 1991, *Astrophys. J.* **375**, 15. [Fig. 9.5]

Olbers, H. W. M. 1826, *Bode Jahrbuch* **110**. [§6.1.1]

Oort, J. H. 1927, *Bull. Astron. Inst. Netherlands* **3**, 275. [§6.2]

Oort, J. H. 1928, *Bull. Astron. Inst. Netherlands* **4**, 269. [§7.7, Fig. 7.5]

Oort, J. H. 1932, *Bull. Astron. Inst. Netherlands* **6**, 349. [§6.5, §7.6.1]

Oort, J. H. 1965, in *Galactic Structure* (ed. A. Blaauw and M. Schmidt), p. 455. University of Chicago Press. [S. f. r.]

Oort, J. H., Kerr, F. T. and Westerhout, G. 1958, *Mon. Not. Roy. Astron. Soc.* **118**, 379. [§6.5, Fig. 6.10]

Oppenheimer, J. R. and Volkoff, G. M. 1939, *Phys. Rev.* **55**, 374. [§5.4]

Osterbrock, D. E. 1978, *Proc. Nat. Acad. Sci.* **75**, 540. [§9.4.4]

Osterbrock, D. E. and Ferland, J. G. 2005, *Astrophysics of Gaseous Nebulae and Active Galactic Nuclei*, 2nd edn. University Science Books. [S. f. r.]

Ostriker, J. P. and Tremaine, S. 1975, *Astrophys. J. Lett.* **202**, L113. [§9.5]

Padmanabhan, T. 2000, *Theoretical Astrophysics, Vol. I: Astrophysical Processes*. Cambridge University Press. [S. f. r.]

Padmanabhan, T. 2001, *Theoretical Astrophysics, Vol. II: Stars and Stellar Systems*. Cambridge University Press. [S. f. r.]

Padmanabhan, T. 2002, *Theoretical Astrophysics, Vol. III: Galaxies and Cosmology*. Cambridge University Press. [S. f. r.]

Panofsky, W. K. H. and Phillips, M. 1962, *Classical Electricity and Magnetism*, 2nd edn. Addison-Wesley. [§2.6.1, §8.4, §8.11, §13.4]

Parker, E. N. 1955a, *Astrophys. J.* **121**, 491. [§8.6]

Parker, E. N. 1955b, *Astrophys. J.* **122**, 293. [§8.7]

Parker, E. N. 1957, *J. Geophys. Res.* **62**, 509. [§8.9]

Parker, E. N. 1958, *Astrophys. J.* **128**, 664. [§4.6]

Parker, E. N. 1966, *Astrophys. J.* **145**, 811. [§8.8]

Parker, E. N. 1979, *Cosmical Magnetic Fields*. Oxford University Press. [S. f. r.]

Pathria, R. K. 1996, *Statistical Mechanics*. Butterworth-Heinemann. [§5.2, §7.5]

Peacock, J. A. 1999, *Cosmological Physics*. Cambridge University Press. [S. f. r.]

Pearson, T. J. *et al.* 1981, *Nature* **290**, 365. [Fig. 9.12]

Peebles, P. J. E. 1966, *Astrophys. J.* **146**, 542. [§11.3]

Peebles, P. J. E. 1993, *Principles of Physical Cosmology*. Princeton University Press. [§11.8.2, S. f. r.]

Penzias, A. A. and Wilson, R. W. 1965, *Astrophys. J.* **142**, 419. [§10.5]

Perlmutter, S. *et al.* 1999, *Astrophys. J.* **517**, 565. [§14.5, Fig. 14.4]

Perryman, M. A. C. *et al.* 1995, *Astron. Astrophys.* **304**, 69. [§3.5.1, §3.5.2, Fig. 3.5]

Peterson, B. M. 1997, *An Introduction to Active Galactic Nuclei*. Cambridge University Press. [Fig. 11.4, S. f. r.]

Petschek, A. 1964, *AAS-NASA Symp. on Solar Flares*, NASA SP-50, p. 425. [§8.9]

Pierce, A. K., McMath, R. R., Goldberg, L. and Mohler, O. C. 1950, *Astrophys. J.* **112**, 289. [Fig. 2.6]

Planck, M. 1900, *Verhand. Deutschen Physik. Gesell.* **2**, 237. [§2.2.1, §2.3.1]

Pogson, N. 1856, *Mon. Not. Roy. Astron. Soc.* **17**, 12. [§1.4]

Popper, D. M. 1980, *Ann. Rev. Astron. Astrophys.* **18**, 115. [Fig. 3.4]

Pound, R. V. and Rebka, G. A. 1960, *Phys. Rev. Lett.* **4**, 337. [§13.2]

Priest, E. R. 1982, *Solar Magnetohydrodynamics*. D. Reidel Publishing Company. [S. f. r.]

Radhakrishnan, V., Murray, J. D., Lockhart, P. and Whittle, R. P. J. 1972, *Astrophys. J. Suppl.* **24**, 15. [Fig. 6.11]

Reber, G. 1940, *Astrophys. J.* **91**, 621. [§1.7.2]

Rees, M. J. 1966, *Nature* **211**, 468. [§9.4.2]

Reif, F. 1965, *Fundamentals of Statistical and Thermal Physics*. McGraw-Hill. [§2.2.1, §8.10, §11.2]

Rhoades, C. E. and Ruffini, R. 1974, *Phys. Rev. Lett.* **32**, 324. [§5.4]

Richtmyer, F. K., Kennard, E. H. and Cooper, J. N. 1969, *Introduction to Modern Physics*, 6th edn. McGraw-Hill. [§6.6]

Riemann, B. 1968, *Nachr. Ges. Wiss. Gött.* **13**, 133. [§12.2.4]

Riess, A. G. *et al.* 1998, *Astron. J.* **116**, 1009. [§14.5, Fig. 14.4]

Riess, A. G. *et al.* 2004, *Astrophys. J.* **607**, 665. [Fig. 14.6]

Roberts, M. S. and Whitehurst, R. N. 1975, *Astrophys. J.* **201**, 327. [§9.2.2]

Robertson, H. P. 1935, *Astrophys. J.* **82**, 248. [§10.3]

Rosseland, S. 1924, *Mon. Not. Roy. Astron. Soc.* **84**, 525. [§2.5]

Rots, A. H. 1975, *Astron. Astrophys.* **45**, 43. [Fig. 8.7]

Roy, A. E. and Clarke, C. 2003, *Astronomy: Principles and Practice*, 4th edn. Taylor & Francis. [S. f. r.]

Rubin, V. C. and Ford, W. K. 1970, *Astrophys. J.* **159**, L379. [§9.2.2]

Rubin, V. C., Ford, W. K. and Thonnard, N. 1978, *Astrophys. J. Lett.* **225**, L107. [§9.2.2, Fig. 9.7]

Russell, H. N. 1913, *Observatory* **36**, 324. [§3.4]

Russell, H. N. 1929, *Astrophys. J.* **70**, 11. [§4.3]

Russell, H. N., Dugan, R. S. and Stewart, R. M. 1927, *Astronomy, Vol. 2*. [§3.3]

Rybicki, G. B. and Lightman, A. P. 1979, *Radiative Processes in Astrophysics*. John Wiley & Sons. [§2.3.1, §2.6.1, §8.11, §8.12, S. f. r.]

Saha, M. N. 1920, *Phil. Mag. (6)* **40**, 472. [§2.3.1, §3.5.1]

Saha, M. N. 1921, *Proc. Roy. Soc.* **A 99**, 697. [§3.5.1]

Saha, M. N. and Srivastava, B. N. 1965, *A Treatise on Heat*, 5th edn. The Indian Press, Allahabad. [§2.2.1, §2.4.1, §8.10, §10.5]

Salpeter, E. E. 1952, *Astrophys. J.* **115**, 326. [§4.3]

Salpeter, E. E. 1955, *Astrophys. J.* **121**, 161. [§6.8]

Salpeter, E. E. 1964, *Astrophys. J.* **140**, 796. [§9.4.3]

Sandage, A. R. 1957, *Astrophys. J.* **125**, 435. [Fig. 3.9]

Sarazin, C. L. 1986, *Rev. Mod. Phys.* **88**, 1. [Fig. 9.17, S. f. r.]

Schechter, P. 1976, *Astrophys. J.* **203**, 297. [§9.2.1]

Scheffler, H. and Elsässer, H. 1988, *Physics of the Galaxy and Interstellar Matter*. Springer-Verlag. [S. f. r.]

Schmidt, M. 1963, *Nature* **197**, 1040. [§9.4.1]

Schneider, P. 2006, *Extragalactic Astronomy and Cosmology*. Springer. [S. f. r.]

Schuster, A. 1905, *Astrophys. J.* **21**, 1. [§2.2.2]

Schutz, B. 1985, *A First Course in General Relativity*. Cambridge University Press. [S. f. r.]

Schwarzschild, K. 1906, *Nachr. Ges. Wiss. Gött.*, 41. [§3.2.4]

Schwarzschild, K. 1907, *Nachr. Ges. Wiss. Gött.*, 614. [§6.3.3]

Schwarzschild, K. 1914, *Sitzungsber. Preus. Akad. Wiss.*, 1183. [§2.2.2]

Schwarzschild, K. 1916, *Sitzungsber. Preus. Akad. Wiss.*, 189. [§13.3]

Schwarzschild, M. 1958, *Structure and Evolution of the Stars.* Princeton University Press. Reprinted by Dover. [S. f. r.]

Seyfert, C. 1943, *Astrophys. J.* **97**, 28. [§9.4.1]

Shakura, N. I. and Sunyaev, R. A. 1973, *Astron. Astrophys.* **24**, 337. [§5.6]

Shapiro, S. L. and Teukolsky, S. A. 1983, *Black Holes, White Dwarfs, and Neutron Stars.* John Wiley & Sons. [§5.1, §5.4, §5.5.1, S. f. r.]

Shapley, H. 1918, *Astrophys. J.* **48**, 154. [§6.1.2]

Shapley, H. 1919, *Astrophys. J.* **49**, 311. [§6.1.2]

Shapley, H. 1921, *Bull. Nat. Res. Council* **2**, 171. [§6.1.2, §9.1]

Shectman, S. A. *et al.* 1996, *Astrophys. J.* **470**, 172. [§9.6, Fig. 9.19]

Shore, S. N. 2002, *The Tapestry of Modern Astrophysics*, Wiley-Interscience. [S. f. r.]

Shu, F. H. 1982, *The Physical Universe.* University Science Books. [Fig. 1.3, §7.1, S. f. r.]

Shu, F. H. 1991, *The Physics of Astrophysics, Vol. I: Radiation.* University Science Books. [S. f. r.]

Shu, F. H. 1992, *The Physics of Astrophysics, Vol. II: Gas Dynamics.* University Science Books. [S. f. r.]

Slipher, V. M. 1914, *Lowell Obs. Bull.* **2**, 62. [§9.3]

Smoot, G. F. *et al.* 1992, *Astrophys. J. Lett.* **396**, L1. [§11.7.1, Fig. 11.3]

Spitzer, L. 1978, *Physical Processes in the Interstellar Medium.* John Wiley & Sons. [§6.7, §8.3, S. f. r.]

Spitzer, L. 1987, *Dynamical Evolution of Globular Clusters.* Princeton University Press. [S. f. r.]

Spitzer, L. and Baade, W. 1951, *Astrophys. J.* **113**, 413. [§9.5]

Spitzer, L. and Jenkins, E. B. 1975, *Ann. Rev. Astron. Astrophys.* **13**, 133. [Fig. 2.9]

Spitzer, L. and Schwarzschild, M. 1951, *Astrophys. J.* **114**, 385. [§7.6.2]

Stefan, J. 1879, *Wien. Berichte* **79**, 391. [§2.4.1]

Stix, M. 2004, *The Sun*, 2nd edn. Springer. [S. f. r.]

Strömberg, G. 1924, *Astrophys. J.* **59**, 228. [§7.6.2]

Strömgren, B. 1939, *Astrophys. J.* **89**, 526. [§6.6.4]

Sunyaev, R. A. and Zeldovich, Ya. B. 1972, *Comm. Astroph. Space Sci.* **4**, 173. [§11.7.2]

Sweet, P. A. 1958, *Proc. IAU Symp.* **6**, 123. [§8.9]

Tayler, R. J. 1993, *Galaxies: Structure and Evolution.* Cambridge University Press. [S. f. r.]

Tayler, R. J. 1994, *The Stars: Their Structure and Evolution.* Cambridge University Press. [Fig. 2.8, §3.4, §3.5.2, Fig. 3.6, Fig. 4.5, §4.5, S. f. r.]

Taylor, J. H., Manchester, R. N. and Lyne, A. G. 1993, *Astrophys. J. Suppl.* **88**, 529. [Fig. 5.5]

Thirring, H. and Lense, J. 1918, *Phys. Z.* **19**, 156. [§13.3]

Thomson, J. J. 1906, *Phil. Mag.* **11**, 769. [§2.6.1]

Tonks, L. and Langmuir, I. 1929, *Phys. Rev.* **33**, 195. [§8.13, §8.13.1]

Toomre, A. and Toomre, J. 1972, *Astrophys. J.* **178**, 623. [§9.5]

Trumpler, R. J. 1930, *Lick Obs. Bull.* **14**, 154. [§6.1.3]

Tucker, W. H. 1978, *Radiation Processes in Astrophysics.* The MIT Press. [S. f. r.]

Tully, R. B. and Fisher, J. R. 1977, *Astron. Astrophys.* **54**, 661. [§9.2.2]

Unsöld, A. and Baschek, B. 2001, *The New Cosmos*, 5th edn. Springer. [S. f. r.]

van de Hulst, H. C. 1945, *Nederl. Tij. Natuurkunde* **11**, 201. [§6.5] van Paradijs, J. *et al.* 1997, *Nature* **386**, 686. [§9.7]

Vitense, E. 1953, *Zs. f. Astrophys.* **32**, 135. [§3.2.4]

Vogt, H. 1926, *Astron. Nachr.* **226**, 301. [§3.3]

von Weizsäcker, C. F. 1937, *Phys. Zs.* **38**, 176. [§4.3]

Wagoner, R. V. 1973, *Astrophys. J.* **179**, 343. [§11.3, Fig. 11.2]

Wagoner, R. V., Fowler, W. A. and Hoyle, F. 1967, *Astrophys. J.* **148**, 3. [§11.3]

Wald, R. M. 1984, *General Relativity*. The University of Chicago Press. [S. f. r.]

Walker, A. G. 1936, *Proc. Lond. Math. Soc. (2)* **42**, 90. [§9.3]

Weinberg, S. 1972, *Gravitation and Cosmology*. John Wiley & Sons. [S. f. r.]

Weinberg, S. 1977, *The First Three Minutes*. Basic Books. [S. f. r.]

Weiss, N. O. 1981, *J. Fluid Mech.* **108**, 247. [§8.6]

Wildt, R. 1939, *Astrophys. J.* **89**, 295. [§2.6.2]

Williams, R. E. *et al.* 1996, *Astron. J.* **112**, 1335. [§11.8.1, Fig. 11.5]

Wolfe, A. M., Turnshek, D. A., Lanzetta, K. M. and Lu, L. 1993, *Astrophys. J.* **404**, 480. [Fig. 11.6]

Yarwood, J. 1958, *Atomic Physics*. Oxford University Press. [§4.2, §11.7.2]

York, D. G. *et al.* 2000, *Astron. J.* **120**, 1579. [§9.6]

Zeldovich, Ya. B. and Novikov, I. D. 1964, *Usp. Fiz. Nauk* **84**, 377. [§9.4.3]

Zwaan, C. 1985, *Solar Phys.* **100**, 397. [Fig. 4.16]

Zwicky, F. 1933, *Helv. Phys. Acta* **6**, 110. [§9.5]

Index